Green Biotechnology for Herbal and Medicinal Plant Metabolites

Green Biotechnology for Herbal and Medicinal Plant Metabolites presents an innovative approach that leverages plant derived molecules for agriculture, medicine, and wellness in pursuit of sustainable health solutions. This comprehensive guide delves into the latest advancements in green biotechnology, showcasing in vitro, genetic, and biochemical strategies for enhancing the yield and potency of essential metabolites in medicinal plants. Designed to appeal to researchers, practitioners, and students alike, the book provides a balanced view of bioengineering plant metabolites' medicinal, economic, and environmental benefits. It illustrates how the potential of green biotechnology can unlock new frontiers in natural products, paving the way for innovative treatments and sustainable healthcare solutions.

The book explains how green biotechnology facilitates the sustainable extraction of plant secondary metabolites. Each chapter emphasizes the need to leverage eco-friendly technologies in metabolite extraction to maximize the benefits and opportunities provided by an integrative, risk-based approach to innovation, aligned with the Sustainable Development Goals. Additionally, the book underscores the role of bioinformatics tools in streamlining real-world solutions, helping to navigate the complexities of wet-lab experimentation with more uncertainties, interdependencies, and unforeseen risks.

Salient Features:

- Thoroughly explores the potential benefits of green biotechnology.
- Gives deep insight into the extraction strategies of plant metabolites.
- Chapters are designed in a way that provides comprehensive details on the industrial application of bio-compounds.
- Provides a comprehensive summary of the most recent findings in plant metabolite and green biotechnology.

Researchers working on green biotechnology for medicinal and herbal plants will undoubtedly find this book useful for developing a thorough understanding of the most recent developments and instruments in this field. The book can also be used as a thorough textbook for graduate-level classes in biomedical, health, and food and nutrition sciences. The book is a collection of chapters contributed by experts in medicinal plants.

Exploring Medicinal Plants

Series Editor: Azamal Husen, *Wolaita Sodo University, Ethiopia*

Medicinal plants render a rich source of bioactive compounds used in drug formulation and development; they play a key role in traditional or indigenous health systems. As the demand for herbal medicines increases worldwide, supply is declining as most of the harvest is derived from naturally growing vegetation. Considering global interests and covering several important aspects associated with medicinal plants, the Exploring Medicinal Plants series comprises volumes valuable to academia, practitioners, and researchers interested in medicinal plants. Topics provide information on a range of subjects including diversity, conservation, propagation, cultivation, physiology, molecular biology, growth response under extreme environment, handling, storage, bioactive compounds, secondary metabolites, extraction, therapeutics, mode of action, and healthcare practices.

Led by Azamal Husen, PhD, this series is directed to a broad range of researchers and professionals consisting of topical books exploring information related to medicinal plants. It includes edited volumes, references, and textbooks available for individual print and electronic purchases.

Therapeutic Medicinal Plants in Traditional Persian Medicine
Edited by Roja Rahimi and Roodabeh Bahramsoltani

Medicinal Plants for the Management of Neurodegenerative Diseases
Edited by Jamal Akhtar, Maryam Sarwat and Fouzia Bashir

Nutraceuticals Inspiring the Contemporary Therapy for Lifestyle Diseases
Edited by Mala Trivedi, Sachidanand Singh, Parul Johri, and Pedro López-Sánchez

Plants as Medicine and Aromatics
Uses of BotanicalsEdited by Mohd Kafeel Ahmad Ansari, Mushtaq Ahmad, and Gary Owens

Exploring Traditional Wild Edible Plants
Edited by Vibhor Agarwal, Sachidanand Singh and Rhaul Datta

Propagation to Pharmacopeia
Modern Approaches in Medicinal PlantsEdited by Abha Manohar K, Gopal Shukla, Saikat Gantait, A P Das, and Sumit Chakravarty

Exploring Medicinal Orchids
Edited by Azamal Husen and Manu Pant

Symbiotic Association of Microorganisms with Medicinal and Herbal Plants
Edited by Allah Ditta, Sajid Mehmood, and Azamal Husen

Tissue Culture Techniques and Medicinal Plants
Enhancing Propagation and ProductionEdited by Azamal Husen and Manu Pant

Comprehensive Guide to Hallucinogenic Plants
Edited by Noureddine Chaachouay, Abdelhamid Azeroual, and Lahcen Zidane

For more information about this series, please visit: www.routledge.com/Exploring-Medicinal-Plants/book-series/CRCEMP

Green Biotechnology for Herbal and Medicinal Plant Metabolites

Edited by
Sachidanand Singh, Parul Johri, Mala Trivedi,
and Pedro López-Sánchez

CRC Press
Taylor & Francis Group
Boca Raton London New York

CRC Press is an imprint of the
Taylor & Francis Group, an **informa** business

First edition published 2026
by CRC Press
2385 NW Executive Center Drive, Suite 320, Boca Raton FL 33431

and by CRC Press
4 Park Square, Milton Park, Abingdon, Oxon, OX14 4RN

CRC Press is an imprint of Taylor & Francis Group, LLC

ISBN: 978-1-032-49266-7 (hbk)
ISBN: 978-1-032-49268-1 (pbk)
ISBN: 978-1-003-39291-0 (ebk)

DOI: 10.1201/9781003392910

Typeset in Times
by Deanta Global Publishing Services, Chennai, India

Contents

Preface

Climate change refers to long-term shifts in temperatures and weather patterns that have come to define Earth's local, regional, and global climates. Scientists use observations from the ground, air, and space along with modeling simulations, to monitor and study past, present, and future climate change. Climate data records provide evidence of climate change through key indicators such as global land and ocean temperature increases, rising sea levels, and ice loss at Earth's poles and in mountain glaciers. This climate change has a significant impact on agriculture: it lowers yields, causes more wildfires, increases soil salinity, and results in higher temperatures, more floods, more droughts, and more pests and diseases in plants and crops. As climate change progresses, green biotechnology is uniquely positioned to improve industrial practices and processes with more sustainable biological alternatives.

Green biotechnology, also known as agricultural biotechnology, is a technology used to improve crops and plants. It is an amalgamation of various methods such as plant tissue culture, genetics, molecular markers, and many more. One of the essential applications of green biotechnology is the development of genetically modified crops that are resistant to diseases, pests, and various environmental stresses. This technology can also be used to develop crops with improved nutritional value or reduced allergenicity. Green biotechnology has played a significant role in increasing the production of food, fibers, bioactive compounds from various medicinal plants, green nanoparticles, and primary and secondary metabolites. It has also contributed to the reduced use of pesticides and herbicides, and has potential applications in phytoremediation, removing heavy metals, petroleum, and other pollutants from the environment, as well as applications in medicine and the production of pharmaceutical products. In addition, green biotechnology has the potential to reduce the use of nonrenewable resources such as fossil fuels. Collectively, we could say that green biotechnology is the lone savior of mankind.

The various chapters in this book have been precisely curated to provide readers with a wide range of perspectives, ensuring a thorough examination of the application of green biotechnology in herbal and medicinal plants. From plant primary and secondary metabolites, biosurfactants, and biochar to seaweeds and in silico analysis of phytoremediation, this book delineates the various approaches and applications of green biotechnology in herbal and medicinal plants.

We have had the honor of working with a group of respected professionals in the disciplines of clinical research, pharmacology, nutrition, and medicine who freely contributed their knowledge and experience to our project. Their input has been vital in transforming this book into a trustworthy and knowledgeable resource for comprehending and utilizing the power of green biotechnology.

About the Editors

Dr. Sachidanand Singh is an Associate Professor and Head of the Biotechnology Department at Pandit Deendayal Energy University, Gandhinagar, Gujarat. He earned a BTech in Biotechnology, an MTech in Bioinformatics, and a PhD in Bioinformatics, with postdoctoral research at Ohio State University's College of Pharmacy in the USA. Specializing in plant biotechnology, systems biology, and drug design, Dr. Singh has over 12 years of experience in research and academia. His career includes a decade with the School of Biotechnology and Health Sciences at Karunya University, Tamil Nadu. In 2018, he also served as a Statistical Genetics Consultant for Ohio State University's College of Pharmacy. Dr. Singh has published over 38 peer-reviewed research articles and has led significant projects, including a SERB-funded project from the Government of India and two university-sponsored research grants. Dr. Singh has guided 36 BTech and 10 MTech projects, and currently supervises six PhD students, with five PhD scholars having completed their research under his guidance. His work focuses on herbal products, drug design, and the integration of network biology, highlighting his commitment to advancing biotechnology and bioinformatics.

Dr. Parul Johri is currently working as a faculty member in the Department of Biotechnology at Dr. Ambedkar Institute of Technology for Divyangjan, Kanpur, Uttar Pradesh. She has an engineering background, with a BTech in Biotechnology, an MTech in Bioinformatics, and a PhD in Biotechnology from Amity University, Uttar Pradesh. She is also a recipient of a Gold Medal from Padmashree Dr. D. Y. Patil University, Navi Mumbai, Maharashtra. Dr. Johri has more than 15 years of teaching experience in various universities. She has published more than 30 research and review papers in various national and international journals of repute. Dr. Johri has edited three books and written more than 30 book chapters with reputed publication houses like Springer, Elsevier, Taylor & Francis, Wiley, and many more. She has six Indian patents published, of which three have been granted to her credit. Dr. Johri has received various funds from organizations such as DBT CTEP, NCW, and many more for conducting sensitization programs, seminars, and workshops. Her current research areas include algorithm development for protein secondary structure prediction, metagenomic analysis, transcriptome analysis, microarray data analysis, and genomic and proteomics studies.

Dr. Mala Trivedi is a Professor at the Amity Institute of Biotechnology, Amity University Uttar Pradesh, Lucknow, and was recipient of double Gold Medals from the University of Allahabad during her MSc in 1993. She completed her PhD in Plant Genetics and Breeding from the Central Institute of Medicinal and Aromatic Plants (CSIR-CIMAP), Lucknow. She served as visiting faculty and Research Associate at the Institute of Cell Biology, School of Life Sciences, Lanzhou University, Lanzhou, P.R. China. Dr. Trivedi has good experience in plant transgenics and has published more than 50 research papers in leading journals of national and international repute. She has also published three books and several book chapters with international publishers. Additionally, she holds seven Indian and two US patents. Dr. Trivedi is a member of many prestigious academic societies. Currently, she is involved in mass multiplication of economically important plants (medicinal and aromatic plants). She has successfully completed two international projects and is the Principal Investigator (PI) of DBT- and MSDE-funded projects. She has supervised 12 PhD scholars so far.

Dr. Pedro López Sánchez MD, MSc, and PhD in Pharmacology. Dr. López took a three-year postdoctoral position at Vanderbilt University, USA. He is a specialist in chronic degenerative illnesses, having spent more than 20 years studying diabetes mellitus, hypertension, and pregnancy-related hypertensive disorders in clinical and animal research. The renin-angiotensin-aldosterone system has been his interest for the last 15 years. Dr. López has conducted several research projects sponsored by the National Council of Science and Technology. He has published more than 60 research papers. He has patented several drug delivery systems based on nanoparticles. Dr. López has more than 25 years as a senior lecturer in pharmacology at the National Polytechnic Institute School of Medicine. He is a founding professor of the pharmacology MSc and PhD programs, coordinating both programs from 2009 to 2012 and from 2012 to 2015, respectively. He is a senior lecturer in molecular pharmacology in both programs. He is a member of several societies: Mexican Pharmacology Association, Mexican Society of Physiological Sciences, National Sciences Academy, American Society of Pharmacological and Experimental Therapeutics, and American Physiological Society. Dr. López is a regular reviewer for several reputed journals published by Elsevier, Taylor & Francis, Springer Nature, Oxford University Press, PLOS, John Wiley & Sons, and Spandidos.

Contributors

A. Puratchikody
Department of Pharmaceutical Technology University College of Engineering (BIT Campus)Anna University Tiruchirappalli, 620024 Tamil Nadu, India

A. Umamaheswari
Department of Pharmaceutical Technology University College of Engineering (BIT Campus) Anna University Tiruchirappalli, 620024 Tamil Nadu, India

Aishwarya Dwivedi
Amity Institute of Biotechnology Amity University Uttar Pradesh Lucknow Campus Lucknow UP, India

Amit Lath
Centre for Biotechnology Maharshi Dayanand University Rohtak, 124001 Haryana, India

Amrita Gupta
ICAR-National Bureau of Agriculturally Important Microorganisms Kushmaur, Maunath Bhanjan UP, 275103, India

Anil Patani
Department of BiotechnologySmt. S. S. Patel Nootan Science and Commerce College Sankalchand Patel University Visnagar, 384315 Gujarat, India

Anita Rani Santal
Department of Microbiology Maharshi Dayanand University Rohtak, 124001 Haryana, India

Anurag Chourasia
Quantum School of Health Science, Quantum University, Roorkee, India

Arun K. Singh Parihar
Drugs Testing Laboratory Avam Anusandhan Kendra Raipur, Chhattisgarh, India

Ashwini Kumar Dixit
Department of Botany Guru Ghasidas Vishwavidyalaya Koni, 495009 Bilaspur Chhattisgarh, India

Azza.S.Naik
School of Food Science and Environmental Health TU Dublin Grangegorman Dublin 7, Ireland

Barkha Kumari
Department of Botany Guru Ghasidas Vishwavidyalaya Koni, 495009 Bilaspur Chhattisgarh, India

Deeksha Singh
Department of Botany Dayalbagh Educational Institute Dayalbagh Agra, 282005 India

Devanshi Katiyar
Amity Institute of Biotechnology Amity University Uttar Pradesh Lucknow Campus Lucknow UP, India

Dinesh Kumar Chellappan
Department of Life Sciences School of Pharmacy International Medical University Bukit Jalil, 57000 Kuala Lumpur, Malaysia

Dilfuza Jabborova
Institute of Genetics and Plant Experimental Biology Uzbekistan Academy of Sciences Kibray, 111208 Uzbekistan

Durgesh Nandini Chauhan
Columbia College of Pharmacy Raipur, India

Garima Gupta
Amity Institute of Biotechnology Amity University Uttar Pradesh, Lucknow Campus Lucknow India, 226028

Hiral S. Desai
Department of Biotechnology Smt. S. S. Patel Nootan Science and Commerce College Sankalchand Patel University Visnagar, 384315 Gujarat, India

Jitendra Kumar Sharma
Department of Pulmonary Medicine All India Institute of Medical Sciences Patna, Bihar

Jyoti Gulia
Department of Microbiology Maharshi Dayanand University Rohtak, 124001 Haryana, India

Lovely Sinha
Department of Pulmonary Medicine All India Institute of Medical Sciences Patna, Bihar

Krishna Nanda Dhal
ICAR-National Bureau of Agriculturally Important Microorganisms Kushmaur, Maunath Bhanjan UP, 275103, India

Kamal Shah
Institute of Pharmaceutical Research GLA University Mathura, India

Kumar Abhinandan
Seed Labs Inc.AB
and
University of Calgary Department of Biological Sciences Calgary, AB

Mala Trivedi
Amity Institute of Biotechnology Amity University Uttar Pradesh Lucknow Campus Lucknow UP, India

Manish Singh Rajput
Department of Biotechnology Dr. Ambedkar Institute of Technology for Divyangjan Kanpur UP, 208024, India

Manish Dwivedi
Amity Institute of Biotechnology Amity University Uttar Pradesh, Lucknow Campus Lucknow UP, India

Mrunal Gokhale
School of Biotechnology and Bioinformatics D.Y. Patil Deemed to be University Navi Mumbai, India

Nagendra Singh Chauhan
Drugs Testing Laboratory Avam Anusandhan Kendra Raipur, Chhattisgarh, India

Nater Pal Singh
Centre for Biotechnology Maharshi Dayanand University Rohtak, 124001 Haryana, India

Nazleen Jahan
Department of Biotechnology Dr. Ambedkar Institute of Technology for Divyangjan Kanpur UP, 208024, India

Nikhil Kirtipal
School of Life Sciences Gwangju Institute of Science and Technology Gwangju, Republic of Korea, 61005

Noopur Khare
Bhai Gurdas Institute of Engineering and Technology Sangrur Punjab, India
and
Shri Ramswaroop Memorial University Barabanki Uttar Pradesh, India

Parul Johri
Department of Biotechnology Ambedkar Institute of Technology for Handicapped (AITH), Kanpur, India

Pramod Kumar Sahu
ICAR-National Bureau of Agriculturally Important Microorganisms Kushmaur, Maunath Bhanjan UP, 275103, India

Pragati Khare
Department of Pharmacy Bhagwant University Ajmer Rajasthan, India

Priyanshi Chaudhary
Department of Biotechnology Dr. Ambedkar Institute of Technology for Divyangjan Kanpur UP, 208024, India

R. Thirumurugan
Department of Pharmaceutical Chemistry, Sree Abirami College of Pharmacy (Affiliated to The Tamil Nadu Dr.M.G.R. Medical University, Chennai), Coimbatore 641021, Tamil Nadu, India

Rahul Dutta
Department of Agrochemistry Soil Science, Microbiology and Plant Nutrition, Faculty of Agrisciences Mendel University in Brno Zemedelska 1, Brno 61300 Czech Republic

Rajesh K Tiwari
Amity Institute of Biotechnology Amity University Uttar Pradesh

Rajiv Ranjan
Department of Botany Dayalbagh Educational Institute Dayalbagh Agra, 282005 India

Ritesh Pandey
Amity Institute of Biotechnology Amity University Uttar Pradesh

Rishika Nair
Department of Biotechnology Dr. Ambedkar Institute of Technology for Divyangjan Kanpur, UP, 208024 India

Roji Waghmare
Department of Food Engineering and Technology Institute of Chemical Technology Mumbai, Maharashtra, 400019 India

Sachidanad Singh
Department of Biotechnology School of Energy Technology Pandit Deendayal Energy University Gandhinagar, 382007 Gujarat, India

Satish Dubey
Department of Botany Guru Ghasidas Vishwavidyalaya Koni 495009 Bilaspur Chhattisgarh, India

Shivangi Mathur
Department of Botany Dayalbagh Educational Institute Dayalbagh Agra, 282005 India

Shraddha Manish Gupta
Department of Pharmaceutical Sciences School of Health Sciences and Technology UPES, Dehradun Uttarakhand, India

Shuchi Midha
SIES School of Business Studies Sector V, Nerul, Navi Mumbai Maharashtra, 400706 India

S. Lakshmana Prabu
Department of Pharmaceutical Technology University College of Engineering (BIT Campus)Anna University Tiruchirappalli, 620 024 Tamil Nadu, India

Sekhar Verma
SLT Institute of Pharmaceutical Sciences Guru Ghasidas Vishwavidyalaya Bilaspur Chhattisgarh, India

Umakant Sahu
University Institute of Pharmacy Pt. Ravishankar Shukla University Raipur Chhattisgarh, India

Vibha Gajbe
School of Biotechnology & Bioinformatics D.Y. Patil Deemed to be University Sec-15, CBD Belapur Navi Mumbai, 400614

Vibha Gajbe
Assistant Professor, School of Biotechnology & Bioinformatics D.Y. Patil Deemed to be University Sec-15, CBD Belapur Navi Mumba Maharashtra, 400614, India

Vishal Jain
University Institute of Pharmacy Pt. Ravishankar Shukla University Raipur Chhattisgarh, India

Vaishnavi Srivastav
Department of Biotechnology Dr. Ambedkar Institute of Technology for Divyangjan Kanpur, UP, 208024 India

Zaryab Shafi
ICAR-National Bureau of Agriculturally Important Microorganisms Kushmaur Maunath Bhanjan, UP, 275103 India
and
Department of Biosciences Faculty of Sciences Integral University Lucknow, UP, 226026 India

1 Overview of Green Technologies and Their Applications in Medicinal Plants

*Aishwarya Dwivedi, Devanshi Katiyar, Parul Johri, Sachidanad Singh, and Mala Trivedi**

1.1 INTRODUCTION

The term "green technologies" refers to a broad range of inventions that support sustainable practices and lessen negative effects on the environment. These technologies, which cover waste management, eco-friendly materials, renewable energy sources, and more, have found use in a variety of sectors, such as manufacturing, transportation, healthcare, and agriculture. They have the power to completely transform a number of elements of the cultivation, use, and preservation of medicinal plants. Here, the focus will be on how these technologies are applied at different stages of the plant lifecycle, from harvesting and growing to product creation and processing. Researchers, farmers, and industry stakeholders can collaborate to safeguard the long-term survival of medicinal plant species and enhance their therapeutic potential by embracing sustainable practices and leveraging cutting-edge tools and technology. In addition to improving human health, this would also help to protect biodiversity and maintain the planet's general health. Green technology applications in the field of medicinal plants examine how integrating green technology may lower carbon footprints, support sustainable livelihoods, and grow the global herbal medicine market. Here, we present the transformative potential of green technology in the field of medicinal plants by examining some intersections and advancements, thereby opening the door to a more affluent and sustainable future for both humans and the natural world.

1.2 GREEN EXTRACTION TECHNOLOGIES

The amount of solvents required and waste produced during the extraction process is reduced when phytoconstituents are extracted from natural sources using green extraction processes. Many environmental and health problems are caused

* Corresponding author

DOI: 10.1201/9781003392910-1

FIGURE 1.1 Green extraction technology.

by the large amount of solvent waste produced by traditional extraction methods. Furthermore, exposure to vapour and solvents is decreased by using automated modern techniques. Green extraction is predicated on analytical techniques that consume less energy, permit the use of various solvents and sustainable natural resources, and yield an extract or product that is both superior and safe (Figure 1.1). A life-cycle analysis of waste created during the extraction process with plants that produce active pharmaceutical ingredients (APIs) shows that 80% of the waste is connected to solvents.

Some important techniques used for processing products from medicinal plants are described in the sections that follow.

1.2.1 Hydro-Distillation

Hydro-distillation is an ancient method used for the extraction of essential oils and bioactive compounds from different plant parts, such as flower petals, stems, and roots. This technique was used by various ancient civilisations, such as those of Persia, China, India, and Egypt.

This technique is an alternative in cases where essential oil cannot be extracted by conventional means.

There are three ways to perform hydro-distillation:

- Water distillation
- Water and steam distillation
- Direct stream distillation (Azmir et al., 2013)

Hydro-distillation can be performed through any of these three ways. There are three kinds of physicochemical processes involved: hydro-diffusion, hydrolysis, and decomposition by heat (Oreopoulou et al.,2019).

Hot water and steam are the main media through which bioactive compounds and essential oils are released from the plant matrix and carried along. Indirect cooling condenses the vapour mixture, and water helps with the separation of oil and bioactive compounds. These compounds are commonly dried over anhydrous sodium sulphate. However, since hydro-distillation is conducted at temperatures above the boiling point of water, some volatile components, natural pigments, and heat-labile bioactive compounds may be lost.

1.2.2 Microwave-Assisted Extraction (MAE)

Microwave-assisted extraction (MAE) is a method that uses microwave energy to heat solvents placed with a sample and then to extract analytes from the sample matrix into the solvent.

Through microwave-assisted extraction, the sample and the solvent are rapidly heated, leading to better yield in many cases. In the case of closed vessels, extraction can be performed at a high temperature, thus accelerating the mass transfer of essential oil and bioactive compounds from the sample matrix. Usually, the extraction process takes 15–30 minutes and uses small solvent volumes in the range of 10–30 ml, which is about ten times less than the traditional method used (Eskilsson CS, Björklund E. et al.,2000).

Additionally, sample throughput is increased, as several samples can be extracted simultaneously. Microwave irradiation uses a specific frequency of the electromagnetic field in a way similar to a photochemical-activated reaction, approximated to be between 300 MHz and 300 GHz. Dielectric heating from MAE is suitable for heat-sensitive compounds. MAE is one of the contemporary methods employed to reduce extraction time and consume less solvent, and is a green technology used for extraction.

1.2.3 Supercritical CO_2 Extraction

Carbon dioxide supercritical extraction is one of the applications of the supercritical fluid extraction (SFE) process. In this technique, a supercritical fluid solvent is used for extraction, which provides numerous advantages. It eliminates the use of organic solvents, and limits the problem of storage and environmental risks compared to traditional Soxhlet extraction.

The diffusion coefficients of lipids and waxes in supercritical fluids are comparatively higher than in liquids; therefore, extraction can occur more quickly. As there is no surface tension and lower viscosity, the supercritical CO_2 is able to penetrate small pores in the plant and reach the inaccessible parts (Yang & Hu , 2014).

At 1071 psi and 31.1°C, carbon dioxide reaches its supercritical state. Any molecule in its supercritical state behaves both as a gas and a liquid. This helps CO_2 reach into small spaces as a gas does, but it can also dissolve and behave like a liquid solvent. The physical properties of supercritical CO_2 are easily manipulated; this allows us to use CO_2 and bind it to any molecule of our choice.

1.2.4 Natural Deep Eutectic Solvents (NDESs)

Natural deep eutectic solvents (NDESs) are a new alternative to toxic organic solvents. NDESs are made by mixtures of natural compounds such as sugars, organic acids, amino acids, and organic bases in certain ratios, which are abundant in organisms.

These solvents can dissolve natural or synthetic chemicals that have low water solubility and hence are a good replacement for organic solvents. However, unlike organic solvents, NDESs primarily depend on intermolecular interaction among their components (Yang Lui. et al.,2018).

In the past few years, NDESs have been used to extract bioactive compounds (e.g., flavonoids, saponins, polysaccharides, essential oils, and alkaloids) to enhance their yield, solubility, and even stability.

Additionally, there are many unique properties of NDESs, such as their biocompatibility and biodegradability, which have led to the popularization of these solvents for extraction processes.

1.2.5 Pressurized Liquid Extraction

This approach transfers solutes from a solid matrix during a separation process. The use of liquid solvents at high pressures and temperatures results in a decrease in the solvent's surface tension, which in turn makes it easier for the solvent to enter the pores of the matrix. This process disturbs the matrix, which enhances the analyte's mass transfer from the solvent sample (Garcia-Mendoza et al., 2015). Based on the required solute's solubility properties, the solvents are selected. This method of extraction has been used to successfully extract anthocyanins from a variety of plants. It has extracted gooseberry polysaccharides and examined antioxidant activity using the pressurised liquid method (Xu et al., 2016). An extract with antioxidant compounds from cranberry waste was obtained using this method. The only drawback of this method is that it needs advanced instrumentation because it uses higher pressure and higher extraction temperatures.

1.2.6 Pulse Electric Field-Assisted Extraction

The aim of pulsed electric field extraction is to increase extraction yield by electroporating the cell membrane. Across the cell membrane, an electric potential divides molecules based on their charge. This repulsion generates pores, in turn increasing their permeability. The success of the treatment will depend on the operating parameters, including the field intensity, input energy, pulse number, temperature, and material attributes (Azmir et al., 2013). It reduces energy usage and unit operations while using renewable plant resources and alternative solvents like water or agri-solvents (ethanol and methyl esters of fatty acids from vegetable oils), creating extracts of excellent quality and purity. It assists in anthocyanin release, increases the number of bioactive substances and vitamin C, and enhances the antioxidant activity of grape juice (Leong et al., 2016).

1.2.7 Ultrasound-Assisted Extraction

The frequency range of ultrasound, a distinct type of sound wave, spans 20 kHz to 100 MHz. Cavitation, a phenomenon caused by ultrasound-assisted extraction, involves the formation, expansion, and collapse of bubbles (Azmir et al., 2013). It is an efficient extraction method for a variety of analytes from various sample types. With the breakdown of plant cell walls, ultrasounds have the ability to increase heat and mass transfer, improving the release of the target compounds from a variety of natural sources. The effects of employing distilled water as the solvent for the ultrasound-assisted extraction of natural antioxidants from the *Eucommia ulmoides* Oliver plant have been documented (Xu et al., 2015). By supplying better yields and selectivity of natural antioxidants, ultrasonography has increased the efficacy of conventional treatments. This approach has been particularly successful for minimising the entrapment of phenolic chemicals in pine seeds (Corbin et al., 2015).

1.2.8 High-Voltage Electric Discharges

In this approach, energy is supplied directly into an aqueous solution through a plasma channel created by a current of high-voltage electrical discharge between two submerged electrodes (Barba et al., 2015). The propagation of the positive streamer from the negative electrode is initiated by an avalanche of electrons brought on by the strength of the electric field. Secondary events, such as bubble cavitation, turbulence, and pressure shock waves, contribute to the improvement of cell damage, promoting the release of chemicals and the extraction of biomolecules from the cytoplasm of the cells (Rajha et al., 2015). It is vital to optimise the extraction parameters for each product. The nature of the raw materials considerably impacts the efficiency of the treatment (Parniakov et al., 2014).

A ray of hope has appeared with the invention and widespread use of green technologies in a world confronted with hitherto unseen environmental concerns. Clean technology, sometimes referred to as eco-friendly technology or green technology, is a broad category of creative solutions intended to reduce environmental impact while advancing sustainable development.

1.3 ROLE OF GREEN TECHNOLOGY

Numerous environmental challenges confronting humanity include biodiversity loss, air and water pollution, resource depletion, and climate change. Unsustainable practices in the transportation, energy generation, agricultural, and industrial sectors are the main cause of these problems. The demand for green technology has increased in order to address these significant issues.

1.3.1 Mitigation of Climate Change

Green technology is essential to reducing the effects of climate change, which is mostly caused by greenhouse gas emissions. There are viable alternatives to fossil fuels in the form of solar panels, wind turbines, and sophisticated battery storage. These developments contribute to the transition of the world's energy system to a low-carbon economy by lowering carbon emissions.

1.3.2 Resource Conservation

The two main goals of green technology are the conservation and efficiency of resources. In order to limit resource consumption and waste generation, and save important resources for future generations, strategies such as waste reduction, sustainable agriculture, and circular economy models are used.

1.3.3 Pollutant Reduction

Using cutting-edge filtration and treatment techniques, green technology combats pollutants in the air and water.

1.3.4 Biodiversity Preservation

Sustainable agriculture practices, supported by green technology, help reduce habitat destruction and chemical pollution. This, in turn, supports biodiversity conservation and the long-term health of ecosystems.

1.4 CHALLENGES

Green technology has a lot of potential, but it also has certain drawbacks. These include opposition from sectors with entrenched interests in conventional methods, large starting expenses, and restricted scalability. It will take continual technical breakthroughs, public awareness campaigns, and government cooperation to overcome these obstacles.

Green technology is expected to become progressively more important in determining a sustainable future in the years to come. Innovations in fields like sustainable urban planning, energy storage, and carbon capture and utilisation will undoubtedly become more widely used. Governments, corporations, and individuals must use green technology in order to build an economically and environmentally sustainable world.

1.5 SUMMARY

An optimistic light in a society beset by environmental challenges is green technology. It provides creative ways to lessen pollution, alleviate climate change, save resources, and protect biodiversity. In order to ensure future generations have a sustainable future, green technology adoption is imperative. The advancement of environmentally conscious solutions brings us one step closer to a future where technology and sustainability coexist harmoniously, benefiting the Earth and its inhabitants.

REFERENCES

Azmir, J., Zaidul, I. S., Rahman, M. M., Sharif, K. M., Mohamed, A. A., Sahena, F., Jahurul, M., Ghafoor, K., Norulaini, N. A., & Omar, A. K. (2013). Techniques for extraction of bioactive compounds from plant materials: A review. *Journal of Food Engineering, 117*, 426–436.

Barba, F. J., Boussetta, N., & Vorobiev, E. (2015). Emerging technologies for the recovery of isothiocyanates, protein and phenolic compounds from rapeseed and rapeseed press-cake: Effect of high voltage electrical discharges. *Innovative Food Science and Emerging Technologies, 31*, 67–72.

Corbin, C., Fidel, T., Leclerc, E. A., Barakzoy, E., Sagot, N., Falguiéres, A., Renouard, S., Blondeau, J. P., Ferroud, C., Doussot, J., Lainé, E., & Hano, C. (2015). Development and validation of an efficient ultrasound assisted extraction of phenolic compounds from flax (Linum usitatissimum L.) seeds. *Ultrasonics Sonochemistry, 26*, 176–185. https://doi.org/10.1016/j.ultsonch.2015.02.008

Eskilsson, C. S., & Björklund, E. (2000). Analytical-scale microwave-assisted extraction. *Journal of Chromatography A, 902*(1), 227–250. https://doi.org/10.1016/s0021-9673(00)00921-3

Garcia-Mendoza, M. D., Paula, J. T., Paviani, L. C., Cabral, F. A., & Martínez-Correa, H. A. (2015). Extracts from mango peel by-product obtained by supercritical CO2 and pressurized solvent processes. *Lwt - Food Science and Technology, 62*, 131–137.

Leong, S. Y., Burritt, D. J., & Oey, I. (2016). Evaluation of the anthocyanin release and health-promoting properties of Pinot Noir grape juices after pulsed electric fields. *Food Chemistry, 196*, 833–841.

Liu, Y., Friesen, J. B., McAlpine, J. B., Lankin, D. C., Chen, S. N., & Pauli, G. F. (2018). Natural deep eutectic solvents: Properties, applications, and perspectives. *Journal of Natural Products, 81*(3), 679–690. https://doi.org/10.1021/acs.jnatprod.7b00945

Parniakov, O., Barba, F. J., Grimi, N., Lebovka, N. I., & Vorobiev, E. (2014). Impact of pulsed electric fields and high voltage electrical discharges on extraction of high-added value compounds from papaya peels. *Food Research International, 65*, 337–343.

Oreopoulou, A., Tsimogiannis, D., & Oreopoulou, V. (2019). Extraction of polyphenols from aromatic and medicinal plants: An overview of the methods and the effect of extraction parameters. In Watson, Ronald (ed.) *Polyphenols in Plants*, 2nd edition, pp. 243–260.

Rajha, H. N., Boussetta, N., Louka, N., Maroun, R. G., & Vorobiev, E. (2015). Effect of alternative physical pretreatments (pulsed electric field, high voltage electrical discharges and ultrasound) on the dead-end ultrafiltration of vine-shoot extracts. *Separation and Purification Technology, 146*, 243–251. https://doi.org/10.1016/j.seppur.2015.03.058

Xu, J. K., Li, M. F., & Sun, R. C. (2015). Identifying the impact of ultrasound-assisted extraction on polysaccharides and natural antioxidants from Eucommia ulmoides Oliver. *Process Biochemistry, 50*(3), 473–481. https://doi.org/10.1016/j.procbio.2014.12.021

Xu, Y., Cai, F., Yu, Z., Zhang, L., Li, X., Yang, Y., & Liu, G. (2016). Optimisation of pressurised water extraction of polysaccharides from blackcurrant and its antioxidant activity. *Food Chemistry, 194*, 650–658. https://doi.org/10.1016/j.foodchem.2015.08.061

Yang, Y., & Hu, B. (2014). Bio-based chemicals from biorefining: Lipid and wax conversion and utilization. In *Advances in Biorefineries: Biomass and Waste Supply Chain Exploitation* (pp. 693–720). Elsevier Ltd. https://doi.org/10.1533/9780857097385.2.693

2 Green Technology, the Lone Savior of Mankind

Anurag Chourasia

2.1 INTRODUCTION TO GREEN TECHNOLOGY

In this section, we embark on a journey to explore the realm of green technology, delving into its definition, historical context, evolution, and the significant importance it holds in addressing contemporary environmental challenges.

2.1.1 Defining Green Technology

Green technology, often referred to as environmental technology or cleantech, encompasses a wide array of innovations, practices, and methodologies aimed at reducing environmental impact, conserving resources, and promoting sustainability across various sectors. At its core, green technology seeks to develop solutions that mitigate or prevent environmental degradation while fostering economic growth and societal well-being.

2.1.2 Historical Context and Evolution

The roots of green technology can be traced back to early environmental movements and initiatives aimed at addressing pollution, resource depletion, and other environmental concerns. The modern concept of green technology emerged in the late 20th century, spurred by growing awareness of environmental issues and the need for sustainable development practices. Over the years, advancements in science, engineering, and policy have fueled the evolution of green technology, leading to the emergence of innovative solutions in renewable energy, waste management, transportation, and beyond.

2.1.3 Importance and Significance

The adoption of green technology is paramount in addressing pressing global challenges such as climate change, pollution, and resource depletion. By harnessing renewable energy sources, improving energy efficiency, and implementing eco-friendly practices, green technology offers pathways to mitigate environmental impact while promoting economic growth and social equity. Moreover, the

DOI: 10.1201/9781003392910-2

integration of green technology into various industries holds the potential to drive innovation, create green jobs, and enhance resilience to environmental risks.

In conclusion, the introduction of green technology represents a paradigm shift in our approach to development, emphasizing the imperative of sustainability and environmental stewardship. As we navigate the complexities of the 21st century, embracing green technology offers a pathway toward a more resilient, equitable, and sustainable future for generations to come.

2.2 THE ENVIRONMENTAL CRISIS: URGENCY FOR ACTION

In this section, we delve into the urgent environmental challenges facing our planet, emphasizing the critical need for immediate action to address climate change, pollution, biodiversity loss, and habitat destruction.

2.2.1 Climate Change: A Global Challenge

Climate change stands as one of the most pressing environmental issues of our time, with far-reaching implications for ecosystems, economies, and human well-being. The rise in greenhouse gas emissions, primarily from human activities such as burning fossil fuels and deforestation, has led to unprecedented changes in global temperatures, sea levels, weather patterns, and extreme weather events (Yadav et al. 2023). These changes pose significant risks to vulnerable communities, ecosystems, agriculture, and infrastructure, underscoring the imperative of reducing emissions, transitioning to renewable energy sources, and adapting to climate impacts.

2.2.2 Pollution and Environmental Degradation

Pollution, in its various forms, poses significant threats to environmental and human health. Air pollution, primarily from industrial emissions, vehicle exhaust, and biomass burning, contributes to respiratory diseases, cardiovascular problems, and premature mortality. Water pollution, stemming from agricultural runoff, industrial discharges, and improper waste management, contaminates freshwater sources, degrades aquatic ecosystems, and jeopardizes access to safe drinking water. Soil contamination, caused by chemicals, heavy metals, and improper waste disposal, compromises agricultural productivity, biodiversity, and food security. Addressing pollution requires comprehensive strategies, including stricter regulations, pollution control technologies, and public awareness campaigns.

2.2.3 Biodiversity Loss and Habitat Destruction

Biodiversity loss and habitat destruction pose grave threats to ecosystems, species, and ecological processes essential for life on Earth. Human activities such as deforestation, habitat fragmentation, overexploitation of natural resources, and invasive species introductions have resulted in unprecedented rates of species extinction and ecosystem degradation. Biodiversity loss not only diminishes the resilience of ecosystems but also undermines essential ecosystem services such as pollination, water purification, and climate regulation. Protecting and restoring habitats, implementing conservation measures, and promoting sustainable land management practices are crucial for safeguarding biodiversity and maintaining ecosystem health.

In summary, the environmental crisis presents a formidable challenge that demands urgent and concerted action from individuals, communities, governments, and the international community. By addressing climate change, pollution, biodiversity loss, and habitat destruction, we can work toward building a more sustainable and resilient future for all.

2.3 PRINCIPLES OF GREEN TECHNOLOGY

In this section, we explore the foundational principles of green technology, which serve as guiding frameworks for promoting sustainability, reducing environmental impact, and fostering a transition to a more resilient and eco-friendly society.

2.3.1 Sustainable Development

Sustainable development lies at the heart of green technology, emphasizing the need to meet the present needs of society without compromising the ability of future generations to meet their own needs. This principle underscores the importance of integrating environmental, social, and economic considerations into decision-making processes, thereby ensuring the long-term viability of human societies and ecosystems. Sustainable development aims to strike a balance between economic prosperity, social equity, and environmental stewardship, fostering holistic approaches to development that prioritize resource efficiency, environmental conservation, and social inclusivity.

2.3.2 Resource Conservation and Efficiency

Resource conservation and efficiency are fundamental tenets of green technology, emphasizing the responsible and efficient use of natural resources to minimize waste, pollution, and environmental degradation. This principle encompasses practices such as reducing energy consumption, optimizing water usage, and minimizing material waste through recycling and reuse. By adopting resource-efficient technologies and practices, individuals, businesses, and communities can lower their environmental footprint while maximizing the value derived from finite resources.

2.3.3 Renewable Energy Sources

The utilization of renewable energy sources lies at the forefront of green technology, offering sustainable alternatives to fossil fuels and mitigating greenhouse gas emissions associated with energy production. Renewable energy sources, such as solar, wind, hydroelectric, geothermal, and biomass, harness natural processes to generate clean and abundant energy. By transitioning to renewable energy sources, societies can reduce their dependence on finite fossil fuel reserves, enhance energy security, and mitigate the adverse impacts of climate change.

2.3.4 Waste Reduction and Recycling

Waste reduction and recycling play a crucial role in the circular economy model embraced by green technology, aiming to minimize waste generation, maximize resource recovery, and promote the sustainable use of materials. This principle advocates for reducing the generation of waste at the source, promoting product design for durability and recyclability, and implementing efficient recycling systems to recover valuable materials from waste streams. By adopting waste reduction and recycling practices, societies can conserve natural resources, reduce landfill emissions, and mitigate environmental pollution.

In summary, the principles of green technology provide a holistic framework for advancing sustainability, promoting resource efficiency, and mitigating environmental impacts across various sectors. By embracing these principles, individuals, businesses, and policymakers can contribute to building a more resilient, equitable, and sustainable future for generations to come.

2.4 GREEN TECHNOLOGY SOLUTIONS

In this section, we explore a range of green technology solutions that offer innovative approaches to address environmental challenges and promote sustainability across various sectors.

2.4.1 Renewable Energy Technologies

Renewable energy technologies harness natural resources such as sunlight, wind, water, and biomass to generate clean and sustainable energy. Solar photovoltaic (PV) systems convert sunlight into electricity, while wind turbines harness the kinetic energy of wind to generate power. Hydroelectric power plants utilize the energy of flowing water to produce electricity, and biomass energy systems convert organic materials into biofuels or biogas. These renewable energy technologies offer alternatives to fossil fuels, reducing greenhouse gas emissions and mitigating climate change.

2.4.2 Energy-Efficient Buildings and Infrastructure

Energy-efficient buildings and infrastructure aim to minimize energy consumption and reduce environmental impact throughout their lifecycle. Green building design principles incorporate features such as passive solar design, high-performance insulation, energy-efficient lighting, and advanced heating, ventilation, and air conditioning (HVAC) systems to optimize energy use and enhance indoor environmental quality. Additionally, sustainable construction practices, such as using recycled materials and minimizing construction waste, further contribute to resource conservation and environmental sustainability.

2.4.3 Sustainable Transportation Systems

Sustainable transportation systems prioritize modes of transportation that minimize environmental impact, reduce greenhouse gas emissions, and promote energy efficiency. Electric vehicles (EVs), powered by renewable energy sources, offer an eco-friendly alternative to conventional gasoline-powered vehicles, reducing air pollution and dependence on fossil fuels. Additionally, public transportation systems, cycling infrastructure, and pedestrian-friendly urban planning contribute to reducing traffic congestion, thereby improving air quality and enhancing overall urban livability.

2.4.4 Green Manufacturing and Production Processes

Green manufacturing and production processes aim to minimize environmental impact and resource consumption while maximizing efficiency and productivity. Sustainable manufacturing practices include optimizing resource use, minimizing waste generation, adopting clean production technologies, and promoting the recycling and reuse of materials. Additionally, eco-friendly certifications and standards, such as ISO 14001 and LEED certification, provide frameworks for organizations to implement environmentally responsible practices throughout their operations.

By embracing these green technology solutions, industries, communities, and policymakers can contribute to building a more sustainable and resilient future for generations to come.

2.5 CHALLENGES AND BARRIERS TO ADOPTION

In this section, we examine the challenges and barriers hindering the widespread adoption of green technology solutions, despite their potential to address environmental concerns and promote sustainability.

2.5.1 Economic and Financial Constraints

One of the primary barriers to the adoption of green technology is the economic and financial constraints associated with implementing these solutions. While green technologies offer long-term benefits such as energy savings and environmental protection, the initial investment costs can be prohibitive for many individuals, businesses, and governments. Additionally, limited access to financing, high upfront costs, and uncertainty about returns on investment can deter stakeholders from embracing green technology solutions.

2.5.2 Technological Limitations

Technological limitations and barriers pose significant challenges to the widespread adoption of green technology solutions. Some renewable energy technologies, such as solar and wind power, are intermittent and dependent on weather conditions, leading to concerns about reliability and grid integration. Additionally, advancements in energy storage, transmission, and grid infrastructure are needed to support the scalability and integration of renewable energy sources into existing energy systems.

2.5.3 Policy and Regulatory Issues

Policy and regulatory frameworks play a crucial role in shaping the adoption of green technology solutions by providing incentives, mandates, and support mechanisms for their deployment. However, inconsistent or inadequate policies, regulatory barriers, and political instability can impede progress in implementing green technology initiatives. A lack of policy coherence, regulatory uncertainty, and administrative burdens can create barriers to market entry and investment in green technology projects.

2.5.4 Societal Attitudes and Behavior

Societal attitudes and behavior also influence the adoption of green technology solutions, as public perceptions, cultural norms, and consumer preferences shape the demand for environmentally friendly products and services. Limited awareness, skepticism about the effectiveness of green technologies, and resistance to change can inhibit adoption rates. Additionally, social factors such as lifestyle choices, consumption patterns, and social norms may influence individual and collective behavior regarding the adoption of sustainable practices.

Addressing these challenges and barriers requires concerted efforts from policymakers, businesses, civil society organizations, and other stakeholders to overcome financial constraints, foster technological innovation, enact supportive policies, and promote sustainable behaviors. By addressing these barriers, we can accelerate the transition to a more sustainable and resilient future powered by green technology solutions.

2.6 SUCCESS STORIES AND CASE STUDIES

In this section, we delve into success stories and case studies that highlight the transition to renewable energy, sustainable urban planning, and corporate sustainability initiatives. By examining these examples, we can glean valuable insights into effective strategies and practices for creating a more sustainable future.

2.6.1 Transitioning to Renewable Energy: Lessons from Leading Nations

Transitioning to renewable energy is crucial for combating climate change and reducing dependence on fossil fuels. Several nations have emerged as leaders in this transition, implementing innovative policies and technologies to accelerate the shift toward renewable energy sources.

One notable success story is Germany's Energiewende, or energy transition, which aims to shift the country's energy supply toward renewable sources such as wind, solar, and biomass. Through a combination of policy incentives, investment in research and development, and public awareness campaigns, Germany has made significant progress in increasing the share of renewable energy in its energy mix. As of [2045], renewable energy sources accounted for [100%] of Germany's electricity generation, surpassing fossil fuels for the first time.

Similarly, Denmark has achieved remarkable success in harnessing wind power to meet its energy needs. Through strategic investments in wind infrastructure and offshore wind farms, Denmark now generates a substantial portion of its electricity from wind energy, making it a global leader in renewable energy adoption.

Lessons from these leading nations include the importance of long-term planning, collaboration between government, industry, and civil society, and sustained investment in renewable energy infrastructure. By learning from these examples, other countries can accelerate their own transition to a sustainable energy future.

2.6.2 Sustainable Cities and Communities: Models of Urban Planning

The rapid pace of urbanization presents both challenges and opportunities for creating sustainable cities and communities. Across the globe, innovative urban planning initiatives are transforming cities into more livable, resilient, and environmentally friendly spaces.

One inspiring example is the city of Curitiba, Brazil, renowned for its sustainable urban planning practices. Through initiatives such as integrated public transportation systems, green spaces, and waste management programs, Curitiba has successfully addressed urban challenges while promoting social equity and environmental sustainability.

In Europe, cities like Copenhagen and Amsterdam are leading the way in sustainable urban design. With a focus on pedestrian and cycling infrastructure, renewable energy, and green building standards, these cities have created vibrant urban environments with low carbon footprints and a high quality of life for residents.

Key lessons from these models of urban planning include the importance of holistic approaches that integrate social, economic, and environmental considerations, as well as community engagement and participatory decision-making processes. By adopting similar strategies, cities can become more resilient, inclusive, and sustainable in the face of urbanization pressures.

2.6.3 Corporate Sustainability Initiatives: Examples of Green Business Practices

As awareness of environmental issues grows, an increasing number of businesses are embracing sustainability as a core principle of their operations. From reducing carbon emissions to minimizing waste and promoting social responsibility, corporate sustainability initiatives are driving positive change across industries.

One standout example is Patagonia, a clothing company known for its commitment to environmental stewardship and social activism. Through initiatives such as the Worn Wear program, which encourages customers to repair and recycle their clothing, and the 1% for the Planet initiative, where the company donates a portion of its sales to environmental causes, Patagonia has demonstrated how businesses can align profit with purpose.

Another notable example is Unilever, a multinational consumer goods company that has set ambitious sustainability goals, including achieving net-zero emissions from its products by [2039]. Through initiatives such as the Sustainable Living Plan, which aims to improve the health and well-being of one billion people while reducing the company's environmental footprint, Unilever is leading the way in corporate sustainability.

These examples highlight the business case for sustainability, showing how companies can drive innovation, attract customers, and enhance their brand reputation by integrating environmental and social considerations into their business strategies. By following their lead, businesses of all sizes can contribute to a more sustainable and prosperous future for all.

2.7 THE ROLE OF INNOVATION AND COLLABORATION

Innovation and collaboration are essential drivers for addressing environmental challenges and achieving sustainability goals. In this section, we explore the role of research and development in green technologies, the importance of public-private partnerships for environmental innovation, and the benefits of international cooperation and knowledge sharing.

2.7.1 Research and Development in Green Technologies

Research and development (R&D) play a crucial role in advancing green technologies that can mitigate environmental impacts and promote sustainable development. From renewable energy to clean transportation and sustainable agriculture, R&D efforts are driving innovation across various sectors.

One notable area of R&D is in renewable energy technologies, such as solar photovoltaics, wind turbines, and energy storage systems. Investments in R&D have led to significant cost reductions and efficiency improvements, making renewable energy increasingly competitive with fossil fuels.

Similarly, R&D in sustainable transportation is yielding breakthroughs in electric vehicles, hydrogen fuel cells, and alternative fuels. These innovations are helping to decarbonize the transportation sector and reduce air pollution and greenhouse gas emissions.

Moreover, R&D in sustainable agriculture is enabling the development of technologies for precision farming, water conservation, and organic pest control. These innovations are essential for increasing food production while minimizing environmental impacts and preserving natural resources.

2.7.2 Public-Private Partnerships for Environmental Innovation

Public-private partnerships (PPPs) are powerful mechanisms for catalyzing environmental innovation by leveraging the strengths of both sectors and fostering collaboration toward common goals. By bringing together government, businesses,

academia, and civil society, PPPs can mobilize resources, share expertise, and scale up innovative solutions.

One example of a successful PPP is the Clean Energy Ministerial (CEM), a global initiative that brings together energy ministers from 25 countries and the European Union to collaborate on clean energy deployment. Through initiatives such as the CEM's Advanced Power Plant Flexibility Campaign and the Electric Vehicle Initiative, participating countries are working together to accelerate the transition to clean energy technologies.

Another example is the Global Alliance for Buildings and Construction (GlobalABC), a partnership between governments, businesses, and international organizations committed to promoting sustainable buildings and construction practices. Through initiatives such as the GlobalABC's Global Roadmap for Buildings and Construction, stakeholders are working together to improve energy efficiency, reduce carbon emissions, and promote green building standards worldwide.

PPPs can also facilitate technology transfer and capacity building in developing countries, where access to finance and expertise may be limited. By fostering collaboration between developed and developing countries, PPPs can accelerate the adoption of sustainable technologies and contribute to global efforts to address climate change and promote sustainable development.

2.7.3 International Cooperation and Knowledge Sharing

International cooperation and knowledge sharing are essential for tackling global environmental challenges that transcend national boundaries. By sharing best practices, expertise, and resources, countries can learn from each other's experiences and collaborate on solutions to common problems.

One example of international cooperation is the Paris Agreement, a landmark accord adopted by nearly 200 countries to combat climate change and accelerate the transition to low-carbon economies. Through mechanisms such as Nationally Determined Contributions (NDCs) and the Green Climate Fund, countries are working together to reduce greenhouse gas emissions, adapt to the impacts of climate change, and mobilize finance for sustainable development.

Another example is the United Nations Framework Convention on Climate Change (UNFCCC), which provides a platform for countries to negotiate and implement international climate agreements. Through processes such as the Conference of the Parties (COP), countries come together to discuss and negotiate climate policies, share experiences, and build consensus on collective action.

Additionally, initiatives such as the Global Environmental Facility (GEF) and the Intergovernmental Panel on Climate Change (IPCC) facilitate knowledge sharing and capacity building on environmental issues. By synthesizing scientific research, providing technical assistance, and supporting projects in developing countries, these initiatives contribute to global efforts to address environmental challenges and promote sustainable development.

In conclusion, innovation and collaboration are essential drivers for advancing environmental sustainability. By investing in R&D, fostering public-private partnerships, and promoting international cooperation and knowledge sharing, we can accelerate the transition to a more sustainable and resilient future for all.

2.8 THE FUTURE OF GREEN TECHNOLOGY

In this section, we explore the future of green technology, including emerging trends and innovations, potential impacts on socio-economic development, and opportunities for global leadership and stewardship in building a more sustainable world.

2.8.1 Emerging Trends and Innovations

The future of green technology is marked by rapid advancements and innovative solutions aimed at addressing pressing environmental challenges. Several emerging trends hold promise for transforming industries and shaping the way we interact with the environment.

One notable trend is the rise of artificial intelligence (AI) and machine learning in optimizing energy efficiency and resource management. AI-driven systems can analyze vast amounts of data to identify patterns, optimize energy consumption, and automate processes, leading to significant cost savings and environmental benefits.

Another trend is the development of advanced materials and nanotechnology for renewable energy generation and storage. Innovations such as perovskite solar cells, graphene-based batteries, and carbon capture materials are revolutionizing the energy landscape, making renewable energy more affordable, efficient, and scalable.

Furthermore, the Internet of Things (IoT) is enabling the creation of smart, interconnected systems that monitor and control environmental conditions in real time. From smart buildings and cities to precision agriculture and environmental monitoring, IoT technologies are enhancing sustainability and resilience across various sectors.

These emerging trends represent the forefront of green technology innovation, offering new opportunities to address environmental challenges and create a more sustainable future.

2.8.2 Potential Impacts on Socio-Economic Development

The adoption of green technology has the potential to generate significant socio-economic benefits, ranging from job creation and economic growth to improved public health and social equity. By promoting sustainable development and reducing environmental degradation, green technology can enhance the well-being of communities and societies.

One potential impact is the creation of green jobs across various sectors, including renewable energy, clean transportation, and environmental conservation. As investments in green technology increase, so do opportunities for employment in fields such as solar installation, electric vehicle manufacturing, and sustainable agriculture.

Moreover, the transition to green technology can stimulate economic growth by fostering innovation, attracting investment, and enhancing competitiveness in global markets. By investing in clean energy infrastructure, improving energy efficiency, and promoting sustainable practices, countries can unlock new sources of economic value and prosperity.

Additionally, green technology can contribute to improved public health outcomes by reducing air and water pollution, mitigating climate change impacts, and enhancing access to clean water and sanitation. By prioritizing sustainable development goals, governments and businesses can create healthier, more resilient communities for current and future generations.

These potential impacts underscore the transformative power of green technology in driving socio-economic development and advancing human well-being.

2.8.3 Opportunities for Global Leadership and Stewardship

As the world faces urgent environmental challenges, there is a growing need for global leadership and stewardship in promoting the adoption of green technology and advancing sustainability goals. By taking bold action and forging partnerships across borders, countries can seize opportunities to lead the transition to a more sustainable future.

One opportunity for global leadership is in setting ambitious targets and policies to accelerate the deployment of green technology. By committing to renewable energy targets, carbon neutrality goals, and sustainable development agendas, countries can signal their commitment to addressing climate change and promoting environmental stewardship.

Furthermore, international cooperation and knowledge sharing are essential for scaling up green technology solutions and addressing common challenges. By sharing best practices, facilitating technology transfer, and providing financial support to developing countries, global leaders can foster collaboration and solidarity in tackling global environmental issues.

Moreover, businesses have a vital role to play in driving innovation and sustainability through corporate leadership and responsible business practices. By integrating environmental considerations into business strategies, promoting supply chain transparency, and engaging stakeholders, companies can contribute to global efforts to build a more sustainable and equitable world.

Ultimately, global leadership and stewardship are essential for mobilizing collective action and achieving meaningful progress toward a more sustainable future for all.

2.9 ETHICAL AND SOCIAL CONSIDERATIONS

In this section, we delve into the ethical and social considerations surrounding the implementation of green technology and sustainable practices. We examine issues of environmental justice and equity, the ethical responsibilities of stakeholders, and strategies for addressing the needs of vulnerable communities.

2.9.1 Environmental Justice and Equity

Environmental justice refers to the fair treatment and meaningful involvement of all people, regardless of race, ethnicity, or socio-economic status, in the development, implementation, and enforcement of environmental policies and practices. It recognizes that marginalized communities often bear a disproportionate burden of environmental pollution and degradation, and advocates for policies and practices that promote equity and justice.

One key aspect of environmental justice is ensuring that the benefits and burdens of environmental policies and projects are distributed equitably across communities. This requires considering the social, economic, and health impacts of environmental decisions and prioritizing the needs of disadvantaged and marginalized groups.

Moreover, environmental justice entails empowering communities to participate in decision-making processes and advocating for their rights to a healthy environment. By amplifying the voices of affected communities and supporting grassroots movements, environmental justice advocates work to challenge inequitable environmental policies and practices, and promote greater inclusivity and representation.

2.9.2 Ethical Responsibilities of Stakeholders

Stakeholders involved in the development and implementation of green technology and sustainability initiatives have ethical responsibilities to consider the broader social and environmental implications of their actions. This includes governments, businesses, non-governmental organizations, and individual citizens.

Governments have a responsibility to enact policies and regulations that promote environmental protection, social equity, and sustainable development. This may involve setting emission targets, enforcing environmental laws, and investing in renewable energy infrastructure to mitigate climate change and protect natural resources.

Businesses have a responsibility to adopt ethical practices that prioritize environmental sustainability, social responsibility, and stakeholder engagement. This may include reducing carbon emissions, minimizing waste, respecting human rights, and contributing to community development through corporate social responsibility initiatives.

Non-governmental organizations (NGOs) play a crucial role in advocating for environmental and social justice, holding governments and businesses accountable, and mobilizing public support for sustainable solutions. By conducting research, raising awareness, and building coalitions, NGOs can amplify the voices of marginalized communities and drive positive change.

Individual citizens also have ethical responsibilities to minimize their environmental impact, advocate for sustainable practices, and support policies and initiatives that promote social and environmental justice. This may involve adopting sustainable lifestyles, participating in community activism, and voting for political leaders who prioritize sustainability and equity.

2.9.3 Addressing the Needs of Vulnerable Communities

Vulnerable communities, including low-income populations, indigenous peoples, and marginalized groups, often bear the brunt of environmental degradation and climate change impacts. It is essential to recognize and address their unique needs and vulnerabilities in the design and implementation of green technology and sustainability initiatives.

One approach to addressing the needs of vulnerable communities is through inclusive decision-making processes that prioritize community engagement, participation, and empowerment. By involving affected communities in the planning, implementation, and evaluation of environmental projects, stakeholders can ensure that their voices are heard and their concerns are addressed.

Moreover, targeted policies and programs can help mitigate the disproportionate impacts of environmental degradation on vulnerable communities. This may include providing access to clean water and sanitation, improving air quality, expanding access to renewable energy, and enhancing disaster preparedness and resilience.

Furthermore, efforts to address the needs of vulnerable communities should be guided by principles of equity, justice, and solidarity. This may involve addressing underlying social and economic inequalities, promoting cultural sensitivity and respect for indigenous knowledge, and fostering partnerships between governments, civil society organizations, and community leaders.

In conclusion, addressing ethical and social considerations is essential for promoting equity, justice, and sustainability in the development and implementation of green technology and sustainability initiatives. By recognizing the rights and needs of vulnerable communities, upholding ethical responsibilities, and fostering inclusive and participatory processes, stakeholders can work toward a more just and sustainable future for all.

2.10 CONCLUSION: EMBRACING THE GREEN REVOLUTION

In this concluding section, we reflect on the key insights gleaned from our exploration of green technology and sustainability initiatives. We issue a call to action for individuals, businesses, governments, and global institutions to embrace the green revolution and work collectively toward a sustainable future. Finally, we highlight the promise of green technology in preserving our planet for future generations.

2.10.1 Recapitulation of Key Insights

Throughout this book, we have examined the transformative potential of green technology and sustainability initiatives in addressing pressing environmental challenges. From renewable energy and sustainable urban planning to corporate sustainability initiatives and international cooperation, we have explored various strategies and practices for building a more sustainable world.

Key insights include:

- The importance of transitioning to renewable energy sources to mitigate climate change and reduce dependence on fossil fuels.
- The role of sustainable urban planning in creating livable, resilient, and environmentally friendly cities and communities.
- The business case for corporate sustainability initiatives, including the benefits of reducing carbon emissions, minimizing waste, and promoting social responsibility.
- The significance of innovation, collaboration, and international cooperation in driving progress toward sustainability goals.
- By synthesizing these insights, we gain a deeper understanding of the opportunities and challenges inherent in the green revolution and the pathways toward a more sustainable future.

2.10.2 Call to Action: Toward a Sustainable Future

As we look to the future, it is clear that urgent action is needed to address environmental degradation, climate change, and social inequality. We call upon individuals, businesses, governments, and global institutions to take bold and decisive steps toward sustainability.

Individuals can contribute to the green revolution by adopting sustainable lifestyles, reducing energy consumption, minimizing waste, and advocating for policies that promote environmental and social justice.

Businesses have a critical role to play in driving innovation, adopting sustainable practices, and integrating environmental considerations into their business strategies. By investing in renewable energy, reducing emissions, and promoting responsible supply chain management, businesses can contribute to a more sustainable and equitable economy.

Governments must enact policies and regulations that support renewable energy deployment, promote sustainable development, and protect natural resources. By setting ambitious targets for emissions reduction, investing in green infrastructure, and fostering collaboration with stakeholders, governments can create an enabling environment for sustainability initiatives to thrive.

Global institutions, including the United Nations, international organizations, and multilateral development banks, have a responsibility to facilitate cooperation, provide financial support, and coordinate efforts to address global environmental challenges. By mobilizing resources, sharing knowledge, and promoting dialogue and cooperation, global institutions can help accelerate the transition to a sustainable future.

2.10.3 The Promise of Green Technology in Preserving Our Planet

Despite the daunting challenges we face, the promise of green technology offers hope for a brighter future. From renewable energy and clean transportation to sustainable agriculture and circular economy solutions, green technology holds the key to preserving our planet for future generations.

By harnessing the power of innovation, collaboration, and collective action, we can overcome barriers and unlock the full potential of green technology to create a more sustainable, equitable, and prosperous world for all.

In conclusion, embracing the green revolution is not only an imperative but also an opportunity to chart a course toward a sustainable future. By acting decisively and collectively, we can build a world where people and planet thrive in harmony.

BIBLIOGRAPHY

Adger, W. N., Arnell, N. W., & Tompkins, E. L. (2005). Successful Adaptation to Climate Change Across Scales. *Global Environmental Change*, 15(2), 77–86.

Bullard, R. D. (1990). *Dumping in Dixie: Race, Class, and Environmental Quality.* Westview Press.

Clean Energy Ministerial. (2022). About CEM. Retrieved from https://www.cleanenergyministerial.org/about-cem

Danish Energy Agency. (2022). Danish Energy Policy: A Summary of the Energy Policy and Key Energy Statistics. Retrieved from https://ens.dk/en/our-services/facts-analysis/statistics-key-figures-and-graphs/danish-energy-statistics

Ellen MacArthur Foundation. (2019). The Circular Economy: A Transformational Journey. Retrieved from https://www.ellenmacarthurfoundation.org/assets/downloads/Ellen-MacArthur-Foundation-Circular-Economy.pdf

European Commission. (2019). Attitudes of Europeans Towards the Environment. Retrieved from https://ec.europa.eu/environment/efe/themes/climate-change-and-energy/en

European Environment Agency. (2020). Renewable Energy in Europe: Key Trends and Statistics. Retrieved from https://www.eea.europa.eu/data-and-maps/indicators/renewable-gross-inland-energy-consumption/renewable-gross-inland-energy-consumption-14

European Environment Agency. (2022). Sustainable Urban Development in Europe. Retrieved from https://www.eea.europa.eu/publications/sustainable-urban-development-in-europe

Germany Trade & Invest. (2022). Energy Transition. Retrieved from https://www.gtai.de/gtai-en/invest/industries/energy/energy-transition-230850

Global Alliance for Buildings and Construction. (2022). About Us. Retrieved from https://globalabc.org/about-us/

Global Environment Facility. (2022). About Us. Retrieved from https://www.thegef.org/about-us

Intergovernmental Panel on Climate Change. (2018). Global Warming of 1.5°C. Retrieved from https://www.ipcc.ch/sr15/

International Energy Agency. (2019). Renewable Energy Policies in a Time of Transition. Retrieved from https://www.iea.org/reports/renewable-energy-policies-in-a-time-of-transition

International Energy Agency. (2020). Global EV Outlook 2020: Entering the Decade of Electric Drive? Retrieved from https://www.iea.org/reports/global-ev-outlook-2020

International Energy Agency. (2022). Tracking Clean Energy Progress. Retrieved from https://www.iea.org/reports/tracking-clean-energy-progress

International Renewable Energy Agency. (2019). Renewable Energy Market Analysis: Southeast Asia. Retrieved from https://www.irena.org/-/media/Files/IRENA/Agency/Publication/2019/Apr/IRENA_RE_Market_Analysis_SEAsia_2019.pdf

International Renewable Energy Agency. (2020). Renewable Power Generation Costs in 2019. Retrieved from https://www.irena.org/-/media/Files/IRENA/Agency/Publication/2020/Jun/IRENA_Renewable-Power-Generation-Costs-in-2019.pdf

International Renewable Energy Agency. (2022). Innovation Outlook: Renewable Energy. Retrieved from https://www.irena.org/-/media/Files/IRENA/Agency/Publication/2022/Feb/IRENA_Renewable_Energy_Innovation_Outlook_2022.pdf

IPCC Special Report on Renewable Energy Sources and Climate Change Mitigation. (2011). Retrieved from https://www.ipcc.ch/report/srren/

Millennium Ecosystem Assessment. (2005). Ecosystems and Human Well-being: Biodiversity Synthesis. Retrieved from https://www.millenniumassessment.org/documents/document.354.aspx.pdf

Patagonia. (n.d.). Environmental and Social Initiatives. Retrieved from https://www.patagonia.com/environmental-social-initiatives.html

Schieder, R., Abdul-Rashid, S. H., & Bekhet, H. A. (2020). Theoretical Framework of Green Technology for Sustainable Development: A Systematic Literature Review. *Journal of Advanced Research in Dynamical and Control Systems*, 12(7), 36–47.

Schlosberg, D. (2013). Theorising Environmental Justice: The Expanding Sphere of a Discourse. *Environmental Politics*, 22(1), 37–55.

Schlosberg, D., & Collins, L. B. (2014). From Environmental to Climate Justice: Climate Change and the Discourse of Environmental Justice. *Wiley Interdisciplinary Reviews: Climate Change*, 5(3), 359–374.

Singer, P., & O'Neill, O. (Eds.). (2003). *The Ethical Responsibilities of Multinational Corporations*. Oxford University Press.

Smil, V. (2002). *Enriching the Earth: Fritz Haber, Carl Bosch, and the Transformation of World Food Production*. MIT Press.

Unilever. (2022). Our Planet & Society. Retrieved from https://www.unilever.com/sustainable-living/our-strategy/planet-and-society/

United Nations. (1987). Report of the World Commission on Environment and Development: Our Common Future. Retrieved from https://sustainabledevelopment.un.org/content/documents/5987our-common-future.pdf

United Nations. (2022a). Sustainable Development Goals. Retrieved from https://sdgs.un.org/goals

United Nations. (2022b). The Sustainable Development Goals Report. Retrieved from https://unstats.un.org/sdgs/report/2022/

United Nations Environment Programme. (2018a). Towards a Pollution-Free Planet. Retrieved from https://www.unenvironment.org/resources/report/towards-pollution-free-planet

United Nations Environment Programme. (2018b). Towards a Zero-Emission, Efficient, and Resilient Buildings and Construction Sector. Retrieved from https://wedocs.unep.org/bitstream/handle/20.500.11822/25596/1_ZEB%20Factsheet.pdf?sequence=1&isAllowed=y

United Nations Environment Programme. (2019). Resource Efficiency: Potential and Economic Implications. Retrieved from https://www.unenvironment.org/resources/report/resource-efficiency-potential-and-economic-implications

United Nations Environment Programme. (2022a). Global Trends in Renewable Energy Investment. Retrieved from https://www.unep.org/resources/global-trends-renewable-energy-investment

United Nations Environment Programme. (2022b). Green Jobs: Towards Decent Work in a Sustainable, Low-Carbon World. Retrieved from https://www.unenvironment.org/resources/report/green-jobs-towards-decent-work-sustainable-low-carbon-world

United Nations Framework Convention on Climate Change. (2022). About the Convention. Retrieved from https://unfccc.int/process-and-meetings/about-the-convention

United Nations Industrial Development Organization. (2019). Green Industry: Pathway to Sustainable Development. Retrieved from https://www.unido.org/sites/default/files/2019-12/Green%20Industry.pdf

Werhane, P. H. (2007). Ethical Issues in Environmental Management. In *Encyclopedia of Life Support Systems (EOLSS)*. Retrieved from https://www.eolss.net/Sample-Chapters/C10/E6-33-03-05.pdf

World Bank. (2019). Curitiba – A Sustainable Model of Urban Development. Retrieved from https://www.worldbank.org/en/news/feature/2019/06/20/curitiba-a-sustainable-model-of-urban-development

World Bank. (2022). Climate Change. Retrieved from https://www.worldbank.org/en/topic/climatechange

World Economic Forum. (2022a). Harnessing the Fourth Industrial Revolution for the Earth. Retrieved from https://www.weforum.org/reports/harnessing-fourth-industrial-revolution-earth

World Economic Forum. (2022b). The Global Competitiveness Report. Retrieved from https://www.weforum.org/reports/the-global-competitiveness-report-2022-2023

World Health Organization. (2016). Ambient Air Pollution: A Global Assessment of Exposure and Burden of Disease. Retrieved from https://www.who.int/airpollution/en/

World Health Organization. (2022). Health and Climate Change. Retrieved from https://www.who.int/health-topics/climate-change#tab=tab_1

Yadav P, Singh S. R., Trivedi, M. (2023). Characterization of Fungi Isolated from Cattle Dung with Potential Lignocellulolytic Activities. *Ecology Environment and Conservation*, 29(April Suppl. Issue), S144–S151. http://doi.org/10.53550/EEC.2023.v29i02s.027

3 Green Technologies in the Extraction of Medicinal Plant Bioactive Compounds

*Deeksha Singh, Shivangi Mathur, Kumar Abhinandan, and Rajiv Ranjan**

3.1 INTRODUCTION

Ever since the evolution of mankind, plants have been used as a great source for extracting pharmacologically active agents. Although potential bioactive compounds from various plants were discovered long ago, assuring the safety and quality of the bioactive compounds obtained by extraction has become a significant challenge for researchers. The inaccessibility of high-performing and reliable extraction technologies, as well as the processes required to extract biocompounds for use in herbal medicines of quality and purity is problematic (Pérez et al., 2021; Verma et al., 2021). Due to these drawbacks, herbal drugs are still struggling to be accepted worldwide. In research involving the discovery of new bioactive phytoconstituents, extraction plays a crucial role in initiating processes such as isolation and purification. A plant may comprise phytoconstituents in abundant or insufficient quantities, necessitating techniques with exceptional performance for efficient extraction and analysis (Alara et al., 2021). Several factors vary the quantity of bioactive compound extracted, such as unprocessed materials, extraction method, and the solvent used during the process (Tiwari, 2015). Conventional extraction techniques are time-consuming and expensive. Therefore, there is an urgent need for highly advanced extraction processes in pharmaceutical industries to reduce time consumption (Chuo et al., 2022).

The extraction process is one of the most critical steps in studying biologically active compounds in plants (Gupta et al., 2022). Extraction efficiency is influenced by several variables, including solvent type and concentration, time spent extracting, temperature, and pH. The solvent used during the process exerts a significant influence. Currently, various solvents are used based on the polarity and chemical properties of compounds that need to be extracted. Basically, there are conventional and non-conventional techniques for the extraction of bioactive compounds. The conventional processes include maceration, soxhlet, and hydro-distillation, which require

* Corresponding author

 DOI: 10.1201/9781003392910-3

temperature, organic solvents, and agitation. Non-conventional or green extraction processes are environmentally friendly. These modern extraction processes include using eco-friendly organic solvents and minimal energy consumption (Rodriguez-Perez et al., 2015; Jabborova et al., 2022). Numerous studies have discussed the methods and significance of modern techniques for extracting bioactive compounds from herbal plants (Barba et al., 2016; Boussetta and Vorobiev., 2014; Dahmoune et al., 2015; Chemat et al., 2017; Mustafa and Turner, 2011; Donga and Chanda, 2020; Abderrezag et al., 2021). However, it is impossible to find a review in the literature that covers the various techniques in the same scientific paper and that suggests different extraction techniques according to the target biocompounds.

The chapter discusses applications using non-conventional energy sources, such as supercritical extraction, pressurized liquid, ultrasound-assisted, microwave, pulsed electric field, and electric high-voltage discharges in the extraction of bioactive compounds from medicinal plants.

3.2 CONVENTIONAL TECHNIQUES

Hydro-distillation, soxhlet, and maceration are some of the main conventional extraction techniques (Singh et al., 2019). Soxhlation involves using a small quantity of the sample, through which the desired solvent is passed inside the equipment, and the process is repeated until complete extraction takes place. Several references in the literature describe how this extraction system is optimized and provide practical examples of how favorable conditions can be achieved (Cravotto et al., 2011; Xhaxhiu et al., 2013; Tyagi and Ranjan, 2022). This technique uses prolonged extraction with large quantities of solvent (Heleno et al., 2016). The maceration process involves making small particles of the sample so that they have a larger surface area to be mixed with the solvent efficiently. Through agitation, the maceration process enhances the extraction process by increasing the efficiency of diffusion and eliminating the concentrated solution from the sample surface. Maceration has been used to extract bioactive compounds and essential oils for decades (Azmir et al., 2013; Patani et al., 2023). Polyphenols were extracted using the conventional extraction method from the fruit of *Aronia melanocarpa* (Ćujić et al., 2016). The extraction process was affected by four factors including solid–solvent ratio, solvents, extracting time, and particle size. A two-level factorial design was used to investigate these parameters independently. The study concluded that the steeping technique was more efficient in extricating anthocyanins and total phenols from chokeberry fruit (Ćujić et al., 2016).

The process of extracting volatile compounds from foods using distilled water is called hydro-distillation. The process is usually completed in 6–8 hours, and no organic solvents are used (Gangwar et al., 2023). The process is classified into three physicochemical processes: hydro diffusion, hydrolysis, and heat decomposition. This technique is characterized by high temperatures, which are a significant disadvantage. During extraction, some compounds that are sensitive to heat (heat labile) can degrade and will not be extracted (Wu et al., 2015a; Sao et al., 2022).

The hydro-distillation method uses distilled water to extract volatile components from food; it typically takes 6–8 hours and does not use organic solvents. A primary physicochemical process in this technique is hydro diffusion, followed by hydrolysis and decomposition. This technique is restricted due to the possibility of compounds degrading during extraction at high temperatures (Wu et al., 2015a). As a complete process, hydro-distillation extracts and physically separates volatile and non-volatile organic compounds (Vishwakarma et al., 2024). An azeotropic distillation of the volatile organic compounds is followed by their condensing, collection, and separation in a Florentine flask. Thc non-volatile organic compounds are extracted from the matrix inside the alembic using boiling water. However, hydro-distillation consumes a great deal of energy and takes a long time (Petigny et al., 2014). In conventional extraction methods, the polarity of the compound and the solvent are used to determine effectiveness since different solvents are needed for identification and isolation. Because compound polarities differ, it is challenging to develop an efficient extraction method for all compounds. An effective solvent is non-toxic, has a low boiling point, allows rapid mass transfer, is preservative, and does not dissociate the complex extract. The yield and quantity of the extract can also be influenced by many additional factors, including extract type, temperature, and extraction time (Da Silva et al., 2016).

3.3 GREEN TECHNOLOGIES

The development of alternative extraction techniques, such as "green" or "clean" technologies that reduce or eliminate the use of organic solvents, has been prompted by the numerous risks associated with solvents, including their toxicity, the presence of residues in extracts, and low yields (Tiwari., 2015). In such techniques, the stability of the extracted compounds is not affected, and there is a reduction in the amount of energy expended during extraction, as shown in Table 3.1.

Various green extraction methods aim to increase the extraction rate, reduce equipment size, improve mass and heat transfer, and decrease the number of steps in the process (Jacotet-Navarro et al., 2016). It is also intended that these technologies will contribute to preserving the environment and the resources it provides (Da Silva et al., 2016; Maddala et al., 2024).

3.4 DEFINITION OF GREEN EXTRACTION

Green chemistry is defined as chemical processes and products intended to lessen or eliminate hazardous substances. Green extraction is defined as "a method for reducing energy consumption, using alternative solvents and renewable natural products, and ensuring that the extract/product is safe and of high quality," to make it more concise (Chemat et al., 2012; Chouhan et al., 2023). Three approaches have been identified for designing and demonstrating green extraction at the laboratory and industrial scale in order to achieve optimal raw material, solvent, and energy consumption: (1) improving and optimizing existing processes; (2) using non-dedicated equipment; and (3) implementing new processes and procedures as well as finding alternative solvents.

TABLE 3.1
Overview of Green Extraction Methods for Extraction of Bioactive Compounds from Medicinal Plants

Conditions	UAE	MAE	PLE	PEFE	SCFE	Enzymatic Hydrolysis
Extraction solvent	Acidified water, dichloromethane, acetone, ethanol, hexane	Hexane, water, dichloromethane, acetone, ethanol	H_2O/EtOH, ethanol	EtOAc/MeOH/H_2O, CO_2/MeOH, olive oil	SC-CO_2, SC-CO_2/MeOH SC-CO_2/EtOH	Proteolytic enzymes to digest the material
Extraction temperature	10–70°C	80–350°C	50–190°C	<100°C	30–100°C	40–60°C
Extraction time	10–90 min	3–30 min	<1 h	<1 h	<1 h	1–4 h
Power amount	48–600 W	180–1200 W	Moderate	30 kW	Moderate	Low
Cost-effective	High-cost	Low-cost	Low-cost	Low-cost	Relatively low	High-cost
Industrial sectors	• Food • Pharmaceutical • Chemical • Textile	• Refinery • Food • Pharmaceutical • Chemical • Textile	• Food • Pharmaceutical	• Refinery • Food • Pharmaceutical	• Refinery • Food • Pharmaceutical • Chemical	• Food • Pharmaceutical

(Continued)

TABLE 3.1 (CONTINUED)
Overview of Green Extraction Methods for Extraction of Bioactive Compounds from Medicinal Plants

Conditions	UAE	MAE	PLE	PEFE	SCFE	Enzymatic Hydrolysis
Advantages	Enhanced penetration of selected solvent into cellular material, decreased extraction time, and reduced solvent consumption.	Improved extraction efficiency, lower solvent consumption, and greater penetration into cell material enhance the release of cell content. Loses an insignificant amount of heat to the surrounding environment. Increases extraction rates and lowers temperatures.	Increased efficiency and precision of analyte recovery with reduced solvent consumption and extraction time.	Since it does not require dewatering or drying, its operational cost is reduced. Non-thermal equipment does not require heating; therefore, it consumes less energy. No effect on product quality. Less time-consuming and high scalability.	It is fast. Since organic solvents are not needed, extracts are very pure. Contains no heavy metals or inorganic salts. Yields are high.	Organic solvents are not required. Low-cost commercial proteases provide an attractive alternative.
Drawbacks	Consumption of power is high. Scalability is challenging.	Consumption of power is high. Material and solvents that are polar are affected only by heating. Unsaturated fatty acids can be oxidized due to heat generation. Volatile solvent efficiency is low.	There is still a need for sample clean-up after extraction, which is the main disadvantage.	Air bubbles in the treatment chamber can cause a dielectric breakdown and cause the PEF treatment to become less uniform. Electric field strength and electrode gap strongly influence the efficiency of the PEF.	Equipment that operates at elevated pressures is very expensive and complex. The extraction of CO_2 is highly selective—it does not extract polar substances. Clean CO_2 must be supplied. Intensive power consumption.	Scalability is difficult/ expensive.
References	Chaves et al. (2020); Kumar et al. (2021)	Chaturvedi (2018); Llompart et al. (2019); Bagade and Patil (2021)	Wianowska and Gil (2019)	Joannes et al. (2015)	Chaves et al. (2020)	Aitta et al. (2021)

3.5 THE SIX PRINCIPLES OF GREEN EXTRACTION

The "Principles of Green Extraction of Natural Products" point industry and scientists in the right direction for developing innovative and environmentally friendly labels, standards, and directive principles. The principles accurately represent how solid-liquid extraction can be innovated in all aspects, not just the process.

Principle 1: Utilization of renewable plant resources and selection of crop varieties to innovate.
Principle 2: Use of unconventional solvents, primarily agro-solvents or water.
Principle 3: Recovery of energy and use of innovative technologies to lower energy consumption.
Principle 4: Co-product production instead of waste, including agro- and bio-refining.
Principle 5: Making processes safer, more robust, and more controlled by reducing unit operations.
Principle 6: Striving for a biodegradable and nondenatured extract that is free from contaminants.

These principles are not established as rules but rather as models to follow, and have been discovered by scientists and successfully implemented by industry.

3.6 GREEN TECHNOLOGIES IN THE EXTRACTION OF MEDICINAL PLANT BIOACTIVE COMPOUNDS

3.6.1 Ultrasound-Assisted Extraction (UAE)

Unlike sound waves we can hear, ultrasound is a form of sound beyond our perception. It usually ranges between 20 kHz and 100 MHz in chemistry. It is also characterized by expansion and compression as it passes through a medium. As a result of this process, a phenomenon known as cavitation occurs only in liquids, including liquids containing solid matter. Cavitation occurs when bubbles are produced, grow, and collapse. As kinetic energy is converted into heat, a great amount of energy can be generated. According to Shingare and Shingate (2012), bubbles are usually at about 5000 K, have a pressure of 1000 atm, and have heating and cooling rates of around 1010 K/s. It is based on this principle that UAE was developed. In solid-liquid extraction at 20 kHz to 100 MHz, UAE is an efficient, suitable, kinetic, and cost-effective non-conventional extraction technique that produces minimal damage to heat-labile compounds (Roselló-Soto et al., 2015). There are several advantages to using the UAE technique over conventional technologies, including that it is eco-friendly, cost-effective, and can be applied to a variety of industries, such as food, pharmaceutical, chemical, and textile. A pictorial representation of the UAE system is shown in Figure 3.1.

In plant samples, UAE is most useful for leaching organic and inorganic compounds from the plant matrix (Pico, 2013). The principal mechanism appears to

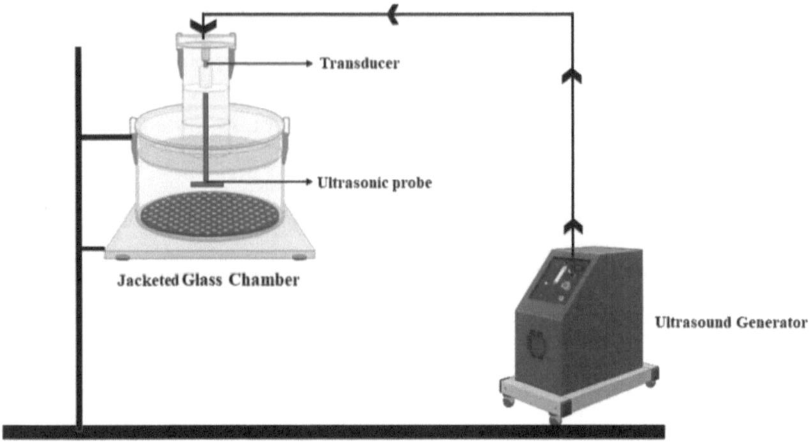

FIGURE 3.1 Diagrammatic representation of UAE system.

be the intensity of mass transfer and the accelerated access to solvent by the cell walls of plant parts caused by ultrasound. During ultrasound extraction, two types of physical phenomena occur: (a) diffusion of the extract across the cell wall and (b) rinsing of the extract after breaking the cell walls. Obtaining efficient and effective extraction requires a greater understanding of the milling degree, the moisture content in the sample, the particle size, and the solvent. A number of factors govern the ultrasound's actions, including pressure, temperature, time, and frequency. UAE has also incorporated various classical techniques to enhance the efficiency of conventional systems. An ultrasound device is placed in a suitable position in the extraction unit to improve solvent extraction efficiency (Azmir et al., 2013). Among UAE's advantages are reduced energy use, extraction time, and solvent consumption. In addition to facilitating better mixing, ultrasound energy facilitates faster energy transfer, reduced thermal gradients and extraction temperatures, reduced equipment size, quicker response to extraction control, selective extraction, quicker start-up, enhanced production, and termination process steps (Cravotto and Binello, 2016).

Herbal extracts can be extracted effectively using UAE. In a 2015 study, Vuong et al. showed that euphol was efficiently extracted from *Euphorbia tirucalli*. The most effective solvent for the extraction of euphol was determined by evaluating several solvents. Subsequently, the response surface methodology (RSM) was used to optimize ultrasound-aided extraction conditions, including temperature, time, and power, to extract maximum amounts of euphol. A solvent-to-fresh sample ratio of 100:32 mL/g, a temperature of 60°C for 75 min, and an ultrasonic power of 60% were the ideal ultrasonic extraction conditions for euphol.

Ultrasound-assisted extraction of antioxidants from the *Piper betle* was investigated by Ali et al. (2018) using three variables: temperature, solute-to-solvent ratio, and ethanol concentration. A combination of temperature at 50°C, a 1.2 g/mL solute-to-solvent ratio, and 80% ethanol concentration produced the best results in terms of yield, total phenolic content, and antioxidant activity. According to Yang and Zhang (2008), the ultrasonic sonication method has superior extraction efficiency compared to conventional methods for extracting quercetin and rutin from *Euonymus alatus* (Thund.) Sieb. Several alkaloids (vindoline, vinblastine, and catharanthine) have been extracted from *Catharanthus roseus* using UAE based on ionic liquids (Yang et al., 2011). UAE was used to extract phenolic compounds and anthocyanins from grape peels, which was optimized by considering temperature, solvent, and duration (Ghafoor et al., 2011). The ionic liquid-based UAE technique proved to be highly efficient and faster than conventional extraction methods in obtaining carnosic acid, phenol carboxylic acid, and rosmarinic acid from *Rosmarinus officinalis* (Zu et al., 2012).

3.6.2 Pulsed Electric Field Extraction (PEF)

During the last decade, pulsed electric fields (PEFs) have been recognized as useful techniques to improve drying, pressing, diffusion, and extraction processes (Vorobiev and Lebovka, 2006; Lakka et al., 2021; Athanasiadis et al., 2022). The underlying mechanism of this technique to enhance the extraction process is by rupturing the cell membrane. When a living cell is suspended in an electric field, an electric potential passes through the membrane. Electric potentials separate molecules within a cell membrane according to their charge, based on the dipole nature of membrane molecules. When the transmembrane potential reaches a critical point of approximately 1 V, the molecules that carry charges begin to repel each other, causing pores to form in weak areas of the membrane and dramatically increasing permeability (Figure 3.2). For PEF treatment of plant materials, a simple circuit with exponential decay pulses is typically used. Plant materials are placed in a chamber

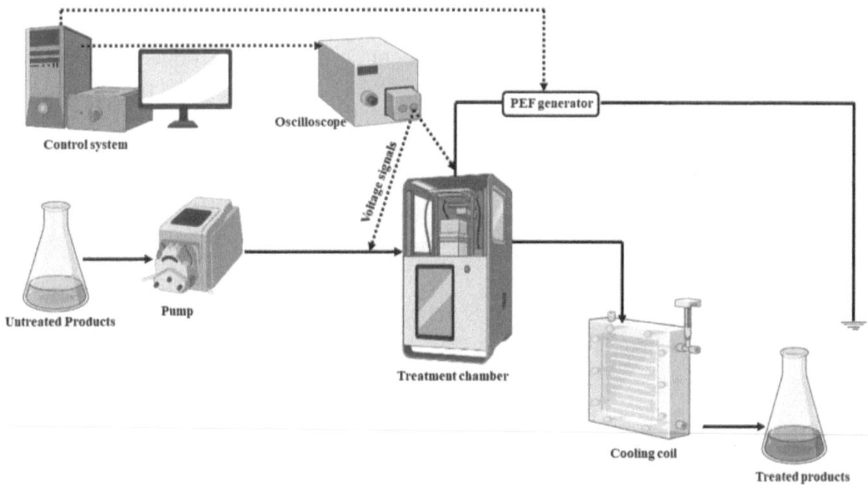

FIGURE 3.2 Schematic diagram of dynamic PEF system.

with two electrodes. PEF processes can be operated continuously or batch-wise, depending on their design (Puertolas et al., 2010).

Process parameters such as pulse number, field strength, treatment temperature, specific energy input, and material properties are crucial to determining the effectiveness of PEF treatment (Timmermans et al., 2019). PEF technology aids in mass transfer by breaking down the membrane structure of plant materials during extraction, thus enhancing extraction and reducing the time it takes. The increased permeability of cell membranes by PEF results in an improved release of compounds (intracellular) from plant tissues (Andreou et al., 2021). With a moderate electric field (500 or 1000 V/cm; for treatment time within 10^{-4}–10^{-2} s), PEF treatment damages the cell membrane of plant tissue with a slight increase in temperature (Vorobiev et al., 2009). Due to this, PEF can be effectively used to extract heat-labile compounds (Kate et al., 2016).

To reduce the extraction load in conventional techniques, the pretreatment of PEF is also practiced on plant tissues (Chemat et al., 2020). An increase in treatment time could result in an increase in the product temperature. The results showed that longer pulses at 0.9 kV/cm for 0.5 s and 1.1 kV/cm for 3 s maximized the extraction rate of polyphenols from fresh tea leaves (Zderic et al., 2016). In the study by Leong et al. (2016), anthocyanins were extracted from grape juice using PEF technology with a pulse rate of 20 s, a frequency of 50 Hz, and a voltage of 1.5 kV/cm. Compared to untreated samples, extraction assisted with PEF improved the release of vitamin C, anthocyanins, and antioxidant activity of grape juice. Using a pulsed electric field (PEF) treatment to permeabilize Merlot skin led to increased extraction of anthocyanin and polyphenol (Delsart et al., 2012).

3.6.3 Microwave-Assisted Extraction (MAE)

A novel form of extraction using microwave energy is microwave-assisted extraction (MAE), which can extract soluble compounds from a wide range of materials into a fluid (Vinatoru et al., 2017). Electromagnetic fields with frequencies between 300 MHz and 300 GHz are called microwaves (Figure 3.3). Usually, they consist of two perpendicular oscillating fields, magnetic and electric. Using microwaves as a heating method relies on polar materials being directly affected (Li et al., 2013b). The dipole rotation mechanism and ionic conductance are mainly responsible for forming heat energy from electromagnetic energy. The flow of ions during the ionic conductance creates resistance, which produces heat.

Ions follow rapidly changing field signs. As a result of this frequent change of direction, molecules collide, and heat is generated. According to Yahya et al. (2018), MAE involves three sequential steps:

1. Solutes are separated from active sites in a sample matrix under increased pressure and temperature.
2. The solvent diffuses across the matrix.
3. Solutes are released into the solvent from the sample matrix.

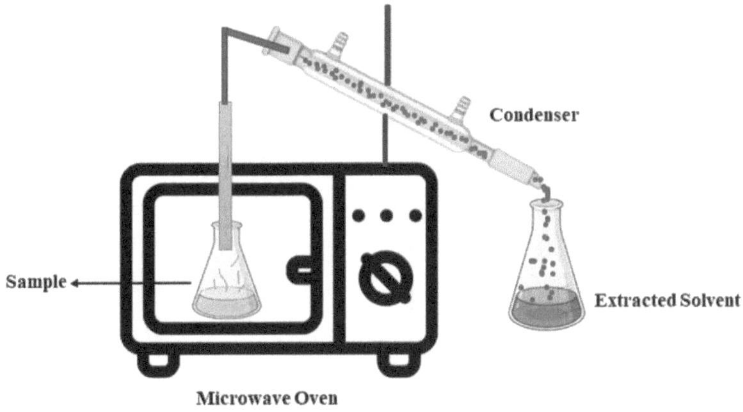

FIGURE 3.3 Pictorial representation of MAE system.

Mashuni et al. (2020) state that MAE has several advantages, including faster heating, a reduction in thermal gradients, a reduction in equipment size, and an increase in extract yield. Compared to conventional extraction processes, MAE can extract bioactive compounds more rapidly and with greater recovery.

This technique is also efficient in isolating organometallic and organic compounds. The use of organic solvents is reduced by MAE, making it a green technology (Alupului, 2012). A landmark study by Asghari et al. (2011) demonstrated that, compared to conventional extraction methods, MAE is faster and easier for extracting bioactive compounds (e.g., tannin and Z- and E-guggulsterone, cinnamaldehyde) from plants under optimal conditions. The intact phenolic compounds of maize and sorghum were extracted using MAE (Chiremba et al., 2012). There are numerous medicinal properties associated with *Berberis*, which is one of the largest sources of berberine and polyphenols. Under MAE conditions, Belwal et al. (2020) extracted these compounds from *Berberis jaeschkeana* roots. Using optimized MAE conditions of a 100% methanol solution with a microwave power of 598 W and a microwave exposure time of 2 min, 46.38 mg of berberine and 20.54 mg of palmatine were determined. Furthermore, by optimizing extraction time, solvent concentration, and microwave power, higher polyphenol yields from leaves of *Pistacia lentiscus* L. and fruit extract could be improved (Elez Garofulic et al., 2020).

3.6.4 Pressurized Liquid Extraction (PLE)

Richter first explained the pressurized liquid extraction (PLE) process in 1996. The technique is also known as ASE (accelerated fluid extraction), HPSE (high-pressure solvent extraction), PFE (pressurized fluid extraction), and ESE (enhanced solvent extraction) (Nieto et al., 2010). In PLE, high pressure is applied to keep the solvent liquid well past its standard boiling point. The process of extraction is expedited by using high pressure (Figure 3.4). In addition to reduced extraction time and solvent amount, automation techniques are the main reasons for the success of PLE-based

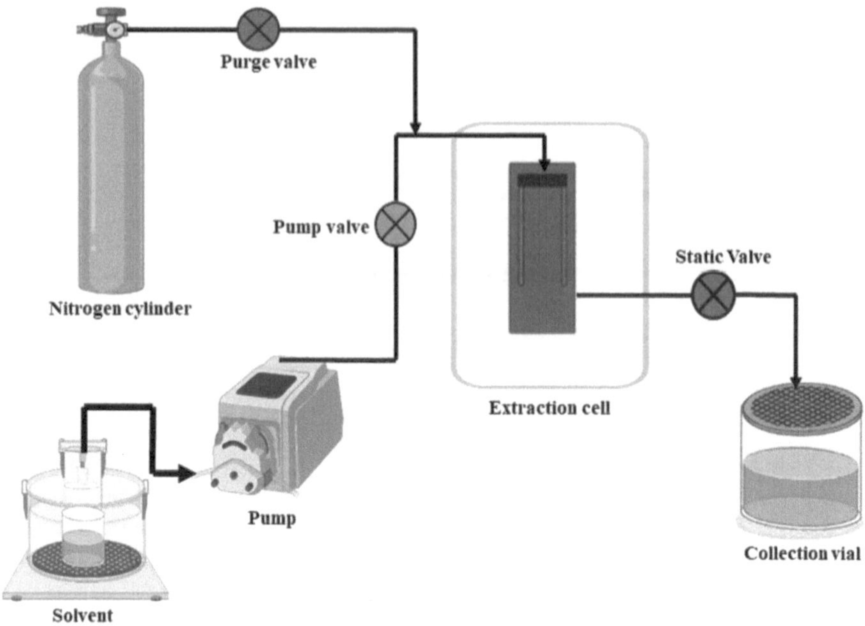

FIGURE 3.4 Systematic representation of PLE system.

techniques. High temperature and pressure decrease the amount of solvent used during the process, which helps to accelerate extraction. An increased extraction temperature promotes mass transfer rates and solubility, resulting in higher analyte solubility while lowering the surface tension and viscosity of the solvents used during the extraction process (Ibanez et al., 2012). Compared to the conventional soxhlet method, PLE is considered efficient because it drastically decreases the quantity of solvent and the extraction time. It is becoming increasingly common to consider PLE as an alternative to supercritical fluid extraction when attempting to extract polar compounds (Andreu and Pico, 2019). Furthermore, PLE can obtain organic pollutants from environmental matrices when they are stable at high temperatures (Wianowska and Gil, 2019). The bioactive components of marine sponges can also be extracted using PLE (Ibanez et al., 2012).

Furthermore, the small quantity of organic solvents in PLE allows this technique to acquire broad recognition as a "green extraction" technique (Ibanez et al., 2012). PLE is highly effective in extracting bioactive compounds from various plant materials. Essential oils in *Thymus munbyanus* subsp. coloratus and *Thymus munbyanus* subsp. munbyanus were extracted without degradation by PLE using optimized conditions (Bendif et al., 2018). As a more efficient process that uses fewer solvents and yields higher reproducibility, PLE is a viable alternative to conventional methods in terms of reproducibility, yield, solvent consumption, and extraction time. At 50°C, ethanol was more effective in extracting antioxidants from buriti fruit than water solvent (Rudke et al., 2019). A study by Luthria (2008) demonstrated that pressure, temperature, flush volume, particle size, solid-to-solvent ratio, and static time impact the extraction of phenolic compounds from parsley (*Petroselinum crispum*) flakes. In the study, the extract of lycorine and galanthamine (Amaryllidaceae alkaloids) from *Narcissus jonquilla* was optimized using PLE. This method was considered more efficient than hot solvent extraction, UAE, and MAE (Mroczek and Mazurek, 2009). Chlorogenic acid, GCT (gallocatechin), caffeic acid, epicatechin gallate, catechin, and myricetin were isolated from *Anatolia* propolis using PLE at 40°C, 1500 psi, for 15 minutes (Erdogan et al., 2011).

3.6.5 Supercritical Fluid Extraction (SFE)

At the end of the 19th century, Hannay and Hogarth discovered supercritical fluid extraction technique. Before Zosel received the first recorded patent for the decaffeination of coffee, SFE remained mainly theoretical. Supercritical fluid technology has gained a wide range of scientific interest in recent years. It has been used successfully in several fields, including environmental, pharmaceutical development, polymer processing, and food analysis (Ishwarya and Nisha, 2021). Supercritical states are unique states reached when a substance is subjected to temperatures and pressures that exceed its critical points. As defined by Gandhi et al. (2017), the critical point is the characteristic pressure (Pc) and temperature (Tc) above which the liquid and gas phases are indistinguishable. When a gas or liquid reaches a supercritical state, its specific properties disappear, so it cannot be liquefied by altering pressure or temperature. The properties of a supercritical fluid, such as diffusion, viscosity, surface tension, density, and solvation power, are similar to those of a gas. Kate et al. (2016) found that these properties made it suitable for separating compounds with high yields in a short period. The SFE system comprises several critical components: a mobile phase tank, usually CO_2; a pump to pressurize the gas; a cosolvent vessel and pump; an oven housing the extraction vessel;a controller to maintain high pressure; and a trapping vessel (Figure 3.5). There are several types of meters that could be attached to the system, including flow meters and dry/wet gas meters. CO_2 has a close critical temperature (70°C) and a low critical pressure (74 bars), allowing for operating pressures usually between 100 and 450 bars (Kate et al., 2016).

Only the low polarity of carbon dioxide makes it suitable for lipids, fats, and nonpolar compounds but unsuitable for most pharmaceuticals and drug samples. A chemical modifier has been successfully used to overcome the low polarity of carbon dioxide (Ghafoor et al., 2010; Azmir et al., 2013). When it comes to enhancing the polarity of carbon dioxide, a small amount of modifier is usually sufficient. It is critically important that certain SFE parameters can be tuned so that bioactive compounds can be successfully extracted from plant material (Khaw et al., 2017).

Optimal results from this technique require precise control of these parameters. Temperature, pressure, particle size, moisture content, extraction time, CO_2 flow rate, and solvent-to-feed ratio all influence extraction efficiency (Ibanez et al., 2012; Da Silva et al., 2016). There are some aspects of supercritical fluids that explain their advantages. Compared to liquid solvents, supercritical fluids have a higher diffusion coefficient, lower viscosity, and lower surface tension, resulting in better penetration

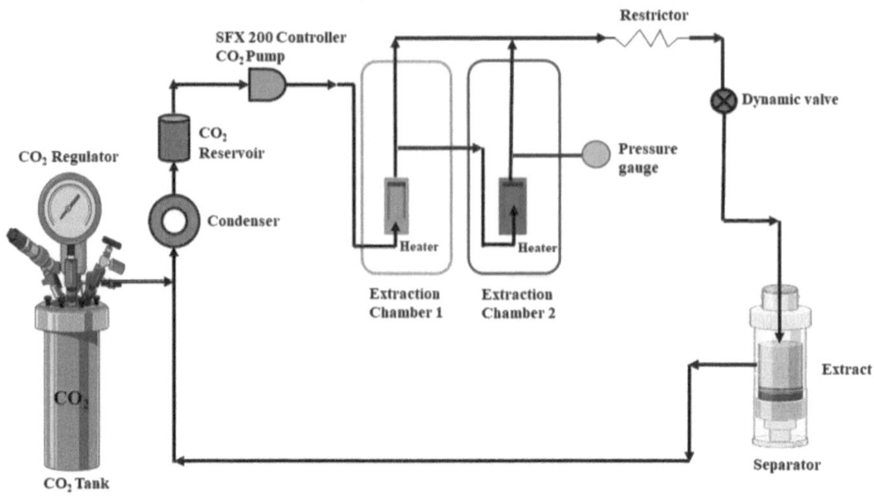

FIGURE 3.5 Schematic diagram of dynamic SFE system.

into the sample matrix. When compared with conventional extraction methods, SFE can significantly reduce extraction time. Complete extraction is achieved by repeatedly refluxing supercritical fluid into the sample. A supercritical fluid has higher selectivity than a liquid solvent since its solvation power can be tuned by changing the pressure or temperature. When supercritical fluid is depressurized, it is possible to bypass the separation of the solute from the solvent, thereby saving time.

The SFE method is ideal for the extraction of thermolabile compounds because it is carried out at room temperature. In contrast to solvent extraction methods, SFE requires a very small sample amount to extract, so the total experiment time is reduced. It is considered eco-friendly because it uses only a small amount of organic solvent. In addition, SFE can be coupled online with chromatographic methods, which is useful when dealing with highly volatile compounds. There is the possibility that supercritical fluid can be recycled and reused, thus reducing waste production. It is possible to arrange SFE measuring scales for specific purposes, whether they are for milligram samples in laboratories or tons of samples in the industrial sector. When employing the SFE process, information about the extraction procedure and mechanism can be used to optimize the extraction process.

In 2019, Gallo-Molina et al. extracted tetrahydrocannabinol from *Cannabis sativa* L. using SFE at 40–80°C and 15–33 MPa. When both temperature and co-solvent (ethanol) concentration were maintained at 60°C and 2%, the raw extracts had the highest THC content at 33 MPa and 36.18% at 15 MPa. Uwineza et al. (2021) extracted total phenolic compounds (TPC) from *Lamium album* using SFE, where methanol was used as a co-solvent. The TPC levels ranged from 234.17 mg GAE/g extract to 650.17 mg GAE/g extract for all extracts tested. Based on SFE extraction of antioxidant compounds, carnosic acid, and carnosol from *Salvia officinalis* L., Pavic et al. (2019) found that pressure significantly affected carnosol extraction, while temperature, CO_2 flow rate, and pressure affected carnosic acid extraction.

Carnosic acid content ranged from 0.29 to 120.0 μg mg^{-1}, and carnosol content varied from 0.46 to 65.5 μg mg^{-1}. Various compounds, including essential oils, phenolic compounds, carotenoids, tocopherols, and tocotrienols, have been extracted using SFE (Wijngaard et al., 2012). In order to optimize the supercritical carbon dioxide extraction of essential oil from *Cyperus rotundus* Linn, the response surface methodology was applied. Compared with soxhlet extraction with n-hexane, SFE produced significantly higher yields of *Cyperus rotundus* Linn. SFE has also been used to extract non-polar components (i.e., fatty acids from *Borago officinalis* L. flower), which was observed to be more effective in preserving fatty acid quality than conventional hydro-distillation. To eliminate the need to use organic solvents, SFE without additives will remain the method of choice for extracting non-polar components from medicinal plants.

3.6.6 Enzyme-Assisted Extraction (EAE)

EAE is considered an eco-friendly, efficient, and mild extraction method, and has recently been applied to the extraction of a wide variety of compounds in the food industry (Puri et al., 2012; Liu et al., 2016; Nadar et al., 2018; Su et al., 2019; Domínguez-Rodríguez et al., 2021; Macedo et al., 2021). Enzymes, enzyme combinations, and EAE are frequently used to extract valuable plant resources by hydrolyzing components, such as cell walls, and binding to target components, such as pectin, based on their inherent properties (specificity and regioselectivity). Enzymes such as pectinases, cellulases, and hemicellulases can improve target compound recovery by degrading and disrupting the structural integrity of cell walls and membranes (Puri et al., 2012; Łubek-Nguyen et al., 2022).

A thorough understanding of enzyme catalysis and mechanism of action is crucial for determining the elite enzyme or enzyme combination for EAE and optimizing conditions. Before selecting suitable enzymes or enzyme combinations, it is also necessary to know the composition of the cell wall of the sample matrix (Nadar et al., 2018). Several factors can influence the degradation and disruption of cell wall structure and the release of the target bioactive compounds, including enzyme composition and concentration, type of extraction solvent, solid-to-liquid ratio, enzyme/substrate ratio, pH, extraction temperature, and extraction time. Regarding enzymatic reactions, temperature is one of the most important factors that can enhance reaction and mass transfer rates. Alternatively, too high a temperature could inactivate enzymes. In addition, some target components might be heat-sensitive and require mild temperatures during processing (Marathe et al., 2019). Furthermore, most enzymes exhibit their activity within a specific pH range; however, there is often an optimal pH for enzymatic activity, and the optimal pH can vary depending on the substrate (Maric et al., 2018; Marathe et al., 2019).

Increasing enzyme/substrate ratios could improve catalytic reaction rates but simultaneously increase enzyme usage and product costs. The enzymes employed may or may not be compatible with solvents with higher solubility than the target components. It is important to note that many enzymes used in the extraction of plant materials are active in water solutions, which are inhibited or even inactivated by polar solutions such as methanol and ethanol. Many bioactive components, such as polyphenols, are more soluble in concentrated methanol/water solutions. It is important to select the solvent carefully or optimize the concentration of polar solvents to achieve the highest yield or bioactivity criteria, for example. Wijesinghe and Jeon (2012) summarize the optimal reaction conditions for many frequently used enzymes. Additionally, specific parameters for the enzymatic reaction, the solvent-to-matrix ratio, particle size, and nature of the extraction solvent can also influence the extraction process.

EAE has been widely used to extract bioactive components from a variety of plant sources, such as fructans from *Dasylirion wheeleri* (Sanchez-Madrigal et al., 2018), total phenolics from pomegranate peels (Nag and Sit, 2018), and *Cassia fistula* pods (Haider et al., 2018), a polysaccharide from *Cedrela sinensis* (Oh et al., 2018). A recombinant α-L-rhamnosidase was used by Kaur et al. (2010) to enhance flavonoid release from citrus peel waste with minimal solvent and energy input. In a similar manner, Puri et al. (2011) also used α-L-rhamnosidase to release naringin from the waste kinnow peel. According to the authors, the enzyme can disrupt the pectin-cellulose complex in the peel, resulting in a higher extraction yield. An EAE study with cellulases and pectinases was carried out on tomato tissues to determine their ability to degrade lycopene. When comparing the EAE process to the conventional extraction process, researchers found that the EAE process increased the extraction yield of lycopene by 206%. For the efficient extraction of phenolic phytochemicals from dried rosemary leaves, EAE is a promising pre-treatment to conventional extractions. Pre-treatment of rosemary leaves for one hour with pectinolytic enzymes significantly increased phenolic yield (Pontillo et al., 2021; Khare et al., 2017).

EAE is an effective method for obtaining bioactive compounds, but it has some potential commercial and technical drawbacks. It is still not possible to completely hydrolyze matrix cell walls with currently available enzyme preparations, which limits the yields of phyto-bioactive compounds (Maric et al., 2018; Vora et al., 2023). In large-scale or industrial production, enzymes may be comparatively expensive if they are processed with a high enzyme-to-substrate ratio. Several factors, including temperature and pH, can influence enzyme behavior, making scaling up EAE a challenge in the industry. Despite this, EAE could offer the possibility of increasing extraction yields, reducing extraction time, and improving extract quality through the use of milder extraction conditions (e.g., lower extraction temperatures).

In Table 3.2, a list of bioactive compounds extracted through different green extraction techniques is provided.

3.7 COMBINATION TECHNIQUES

In recent years, extraction methods have undergone rapid development and improvement (Parveen et al., 2023). As environmental demands increase, several industries seek to increase production rates or yield while reducing production costs. Using a combination of these technologies provides a better balance that is primarily aimed at saving energy and resources (Mustafa and Turner, 2011). Some studies

TABLE 3.2
List of Bioactive Compounds Extracted Using Various Green Extraction Methods

Medicinal Plant Species	Bioactive Compound Extracted	Green Extraction Technique	Solvent Used	Optimum Extraction Parameters	References
Apium graveolens L.	Fixed oils (lipids)	UAE, SFE	Chloroform/methanol	400 W (amplitude 0.5 and rotation 70 cycles)	Shams et al. (2015)
Andrographis echioides L. Nees	Flavonoid	UAE	Ethanol and methanol	Frequency of 40 kHz and 230 W	Ramasamy et al. (2019)
Annona squamosa L.	Phenols and flavonoids	MAE, UAE	Ethanol	400–600 W and 27 ± 2°C	Dahmoune et al. (2015); Donga and Chanda (2020)
Ammodaucus leucotrichus	Phenolics and carbohydrates	PLE	Ethanol and water	180°C, 10.34 MPa, and 10 min	Abderrezag et al., (2021)
Citrus reticulata	Polyphenols	UAE	Acetone (100%, w/v)	—	Saini et al. (2019)
Citrus limetta	Polyphenols	UAE	Acetone (100%, w/v)	—	Saini et al. (2019)
Carica papaya Linn.	Phenols and flavonoids	MAE, UAE	Ethanol	400–600 W and 27 ± 2°C	Dahmoune et al. (2015); Donga and Chanda (2020)
Calendula officinalis	Polyphenols	PEF	Double distilled water	Electric field intensities ranging from 1.2 to 2.0 kV cm^{-1}	Lakka et al. (2021)
Crocus sativus	Polyphenols and anthocyanins	UAE	Ethanol/water mixture (59:41 v/v)	66°C and 800 W	Da Porto and Natolino (2018)
Castanea sativa	Polyphenols	PEF	Double distilled water	Electric field intensities ranging from 1.2 to 2.0 kV cm^{-1}	Lakka et al. (2021)
Ceiba pentandra L. Gaertn.	Phenols and flavonoids	MAE, UAE	Ethanol	400–600 W and 27 ± 2°C	Dahmoune et al. (2015); Donga and Chanda (2020)

(Continued)

TABLE 3.2 (CONTINUED)
List of Bioactive Compounds Extracted Using Various Green Extraction Methods

Medicinal Plant Species	Bioactive Compound Extracted	Green Extraction Technique	Solvent Used	Optimum Extraction Parameters	References
Citrus paradisi L.	Polyphenols	EAE, UAE	Ethanol	For EAE, incubation conditions of pH 4.8 and 60°C were selected, and for UAE, an ultrasonicator frequency was fixed at 20 kHz	Nishad et al. (2019)
Foeniculum vulgare L.	Fixed oils (lipids)	UAE, SFE	Chloroform/methanol	400 W (amplitude 0.5 and rotation 70 cycles)	Shams et al. (2015)
Hibiscus sabdariffa	Total phenolic and organic acids content	SFE	CO_2 and ethanol	Pressure (from 150 to 350 bars), temperature (from 40 to 60°C) and different solvent combinations (CO_2 + ethanol from 7 to 15 %)	Pimentel-Moral et al. (2019)
Myrtus communis L.	Polyphenols	MAE	Ethanol	400–600 W and 27 ± 2°C	Dahmoune et al. (2015)
Moringa stenopetala	Phenolic compound	UAE	Ethanol	30, 40, and 50°C	Dadi et al. (2019)
Medicago sativa L	Polyphenols	SFE and EAE	96% ethanol	50–70°C and pressure from 100 to 300 bar. For EAE, kemzyme is used	Krakowska-Sieprawska et al. (2021)
Petroselinum crispum Mill	Fixed oils (lipids)	UAE, SFE	Chloroform/methanol	400 W (amplitude 0.5 and rotation 70 cycles)	Shams et al. (2015)

(Continued)

TABLE 3.2 (CONTINUED)
List of Bioactive Compounds Extracted Using Various Green Extraction Methods

Medicinal Plant Species	Bioactive Compound Extracted	Green Extraction Technique	Solvent Used	Optimum Extraction Parameters	References
Rheum moorcroftianum	Phenolics and antioxidant compounds	UAE	40% acetone	50 KHz at 37°C	Pandey et al. (2018)
Rosa canina	Polyphenols	PEF	Double distilled water	Electric field intensities ranging from 1.2 to 2.0 kV cm^{-1}	Lakka et al. (2021)
Sideritis raiseri	Polyphenols	PEF	100% water, 25% ethanol in water, 50% ethanol in water, 75% ethanol in water, and 100% ethanol	Pulse durations were 10μs and 100μs, and frequency was 1000 Hz	Athanasiadis et al. (2022)
Trachyspermum ammi L	Phenols and flavonoids	MAE, UAE	Ethanol	400–600 W and 27 ± 2°C	Dahmoune et al. (2015); Donga and Chanda (2020)
Trigonella foenum-graecum L.	Phenols and flavonoids	MAE, UAE	Ethanol	400–600 W and 27 ± 2°C	Dahmoune et al. (2015); Donga and Chanda (2020)
Vigna radiata L.	Phenolic compound	PLE	95% ethanol	1500 psi	Supasatyankul et al. (2022)
Stevia rebaudiana	Steviol glycosides	PLE	Ethanol	120°C, 100 bar, and 60 min	Raspe et al. (2021)

have examined ultrasound-assisted enzymatic extraction (UAEE), a combination of ultrasound-assisted extraction and enzyme-assisted extraction, for the extraction of bioactive compounds (Amiri-Rigi A et al., 2016; Chen et al., 2016; Luo et al., 2019). In UAEE, two complementary extraction methods are combined, bringing additional advantages. In EAE, enzymes work to facilitate recovery by degrading cell walls and disrupting membranes. The cavitation created by ultrasound in UAE has a complementary effect on EAE, as it can disrupt and break down the matrix, facilitating enzymatic reactions and the consequent release of target compounds. In studies with wheat bran (Wang et al., 2014), corn silk (Chen et al., 2014), Ginkgo biloba leaves (Zhang et al., 2015), and blackcurrant (Xu et al., 2015), UAEE was shown to improve yields. According to a report by Amiri-Rigi A et al. (2016), UAEE was also used in the tomato industry to extract lycopene from residual tomato material. Their results showed that the UAEE process was more effective in extracting lycopene than conventional extraction with UAE and EAE, followed by the extraction of tomato residuals in microemulsions. The UAEE process was studied by Tchabo et al. (2015) to isolate phytochemical compounds from mulberry must (phenolics, flavonoids, anthocyanins, etc.). UAEE proved to be more effective than UAE, EAE, or traditional extraction methods in producing phytochemical compounds from mulberry must.

It is possible to combine ultrasound extraction with other techniques, such as extraction by heat-reflux, supercritical CO_2, and microwave. Yang and Wei. (2015) combined heat reflux extraction (conventional, solvent: ethanol) with ultrasound-assisted extraction (40 kHz, 185 W, with agitation) to extract bioactive compounds from *Rabdosia rubescens.* Based on their results, the combination of extraction techniques reduced the processing time and increased the yield of bioactive compounds. The effects of ultrasound and supercritical fluid extraction on the bioactive compounds in red pepper (*Capsicum baccatum* L. var. Pendulum) were investigated by Dias et al. (2016). In the process, 200 to 600 W ultrasound powers were applied for 40 to 80 minutes at 15 to 25 MPa, temperatures of 40 to 60°C, and ultrasound pressures of 15 to 25 MPa. In studies using these techniques, phenolic compounds were found to be more prevalent. MAEs and EAEs have recently gained popularity in phytobioactive compound extraction due to their high extraction efficiency, ease of handling, low solvent consumption, and energy efficiency. When microwave irradiation and enzymes are combined, cell wall integrity can be disrupted, and membrane and wall permeability can be increased. This facilitates the transfer of target molecules into the solvent from the matrix cell (Cheng et al., 2015). The combination of MAE and EAE, known as MAEE, has been studied for oil extraction (Jiao et al., 2012; Li et al., 2013a; Jiao et al., 2014). MAEE can alleviate regulatory challenges regarding the use and disposal of organic solvents, which can negatively impact oil quality and create additional costs. In 2013, Li et al. developed a green MAEE system without organic solvents for high-performance oil extraction from yellow horn seed kernels. After optimization, the MAEE oil extraction yield reached 55.8% at 60°C for 30 minutes at a low temperature. In comparison with conventional OSE, under optimized conditions, they observed a similar extraction yield of 90 min at 90°C.

The UMAE process, which combines UAE and MAE, has been extensively studied and has shown great potential. Due to its shorter extraction time, high efficiency, and yield, UMAE is a cost-effective method with the advantages of fast preparation and intensive extraction (Sun et al., 2019). When cell walls are disrupted, ultrasound can improve the penetration of solvent into the sample matrix, promote the solvation of soluble compounds, and enhance mass transfer by increasing the contact surface area. A microwave accelerates the mass transfer and solubility of the solute, enhancing the desorption of relevant compounds from the sample matrix, and thereby improving the extraction efficiency (Vinatoru and Calinescu, 2017). Liew et al. (2016) compared sequential UMAE, MUAE, MAE, and UAE on grapefruit peel. The highest yield was obtained by UMAE under optimal conditions, followed by MUAE, MAE, and then UAE. MUAEE is a microwave/ultrasonic-assisted enzymatic extraction technique that combines UAE, MAE, and EAE and is applied sequentially or simultaneously to extract bioactive compounds (Yin et al., 2018). In a study by Wu et al. (2015b), antioxidant components were extracted from *Nitraria tangutorum* juice by-products (NJB) using MUAEE. As an optimal extraction condition, 800 W US power, 420 W MW power, 70% ethanol concentration, 0.6% addition of cellulase, pH 4.5 phosphate buffer, a solvent-to-solvent ratio of 20:1, extraction temperature at 66°C, and extraction time of 43 min were used, which led to a total antioxidant capacity of 219.73 mg TE/g, which was 27.62–190.23% higher than SE, SEE, UAE, UAEE, MAE, and MAEE. Based on the chemical composition, the extracts from MUAEE contained the highest levels of TPC, total flavonoid content, and total anthocyanin content, at 141.48%, 133.52%, and 92.42%, respectively. In addition, the study demonstrated that MUAEE could be a powerful and efficient technique for extracting phyto-bioactive compounds.

3.8 CONCLUSIONS AND FUTURE PERSPECTIVES

An extraction process according to the six principles of green extraction of natural products is a novel concept that can address the challenges of the 21st century, protecting both consumers and the environment while also improving the competitiveness of industries to be more innovative, economically beneficial, and ecologically friendly. As part of this approach, we introduce the concept of green extraction, which allows a substance to be extracted in such a way that it uses the least number of resources and can be recycled (co-products, biodegradability). The green extract should be the result of an entire value chain: economic and responsible, beginning with production and harvesting, moving to extraction and separation, and finishing with formulation and marketing. Currently, research is increasingly focused on combining novel green extraction technologies. There is still much to be done to develop novel extraction methods, and optimal conditions are needed to scale up, but it may be a major step toward the sustainable development and use of bioactive compounds from natural products. In the future, research priorities for high-pressure processing technologies should focus on the use of ionic liquids as cosolvents, as this has been proven to be effective on an industrial scale for the extraction/separation of high-value compounds.

ACKNOWLEDGMENT

We are grateful to the Director of Dayalbagh Educational Institute, Dayalbagh, Agra, for encouragement and kind support.

REFERENCES

Abderrezag N, Sánchez-Martínez JD, Louaer O, et al. Optimization of pressurized liquid extraction and in vitro neuroprotective evaluation of ammodaucus leucotrichus. Untargeted metabolomics analysis by UHPLC-MS/MS. *Molecules*. 2021;26(22):6951.

Aitta E, Marsol-Vall A, Damerau A, et al. Enzyme-assisted extraction of fish oil from whole fish and by-products of Baltic herring (Clupea harengus membras). *Foods*. 2021;10(8):1811.

Alara OR, Abdurahman NH, Ukaegbu CI. Extraction of phenolic compounds: A review. *Current Research in Food Science*. 2021;4:200–214.

Ali A, Lim XY, Chong CH, et al. Ultrasound-assisted extraction of natural antioxidants from betel leaves (Piper betle): Extraction kinetics and modeling. *Separation Science and Technology*. 2018;53(14):2192–2205.

Alupului A, Calinescu I, Lavric V. Microwave extraction of active principles from medicinal plants. *UPB Science Bulletin, Series B*. 2012;74(2):129–142.

Amiri-Rigi A, Abbasi S, Scanlon MG. Enhanced lycopene extraction from tomato industrial waste using microemulsion technique: Optimization of enzymatic and ultrasound pretreatments. *Innovative Food Science & Emerging Technologies*. 2016;35:160–167.

Andreou V, Sigala A, Limnaios A, et al. Effect of pulsed electric field treatment on the kinetics of rehydration, textural properties, and the extraction of intracellular compounds of dried chickpeas. *Journal of Food Science*. 2021;86(6):2539–2552.

Andreu V, Picó Y. Pressurized liquid extraction of organic contaminants in environmental and food samples. *TrAC Trends in Analytical Chemistry*. 2019;118:709–721.

Asghari J, Bernd O, Mohsen M. Extraction of bioactive chemical compounds from the medicinal Asian plants by microwave irradiation. *Journal of Medicinal Plants Research*. 2011;5(4):495–506.

Athanasiadis V, Pappas VM, Palaiogiannis D, et al. Pulsed electric field-based extraction of total polyphenols from sideritis raiseri using hydroethanolic mixtures. *Oxygen*. 2022;2(2):91–98.

Azmir J, Zaidul IS, Rahman MM, et al. Techniques for extraction of bioactive compounds from plant materials: A review. *Journal of Food Engineering*. 2013;117(4):426–436.

Bagade SB, Patil M. Recent advances in microwave assisted extraction of bioactive compounds from complex herbal samples: A review. *Critical Reviews in Analytical Chemistry*. 2021;51(2):138–149.

Barba FJ, Zhu Z, Koubaa M, et al. Green alternative methods for the extraction of antioxidant bioactive compounds from winery wastes and by-products: A review. *Trends in Food Science & Technology*. 2016;49:96–109.

Belwal T, Pandey A, Bhatt ID, et al. Optimized microwave assisted extraction (MAE) of alkaloids and polyphenols from Berberis roots using multiple-component analysis. *Scientific Reports*. 2020;10(1):1–10.

Bendif H, Adouni K, Miara MD, et al. Essential oils (EOs), pressurized liquid extracts (PLE) and carbon dioxide supercritical fluid extracts (SFE-CO2) from Algerian Thymus munbyanus as valuable sources of antioxidants to be used on an industrial level. *Food Chemistry*. 2018;260:289–298.

Boussetta N, Vorobiev E. Extraction of valuable biocompounds assisted by high voltage electrical discharges: A review. *Comptes Rendus Chimie*. 2014;17(3):197–203.

Chaturvedi AK. Extraction of neutraceuticals from plants by microwave assisted extraction. *Systematic Reviews in Pharmacy.* 2018;9(1):31–35.

Chaves JO, De Souza MC, Da Silva LC, et al. Extraction of flavonoids from natural sources using modern techniques. *Frontiers in Chemistry.* 2020;8:507887.

Chemat F, Rombaut N, Meullemiestre A, et al. Review of green food processing techniques. Preservation, transformation, and extraction. *Innovative Food Science & Emerging Technologies.* 2017;41:357–377.

Chemat F, Vian MA, Cravotto G. Green extraction of natural products: Concept and principles. *International Journal of Molecular Sciences.* 2012;13(7):8615–8627.

Chemat F, Vian MA, Fabiano-Tixier AS, et al. A review of sustainable and intensified techniques for extraction of food and natural products. *Green Chemistry.* 2020;22(8):2325–2353.

Chen F, Zhang Q, Gu H, et al. An approach for extraction of kernel oil from Pinus pumila using homogenate-circulating ultrasound in combination with an aqueous enzymatic process and evaluation of its antioxidant activity. *Journal of Chromatography A.* 2016;1471:68–79.

Chen S, Chen H, Tian J, et al. Enzymolysis-ultrasonic assisted extraction, chemical characteristics and bioactivities of polysaccharides from corn silk. *Carbohydrate Polymers.* 2014;101:332–341.

Cheng Z, Song H, Yang Y, et al. Optimization of microwave-assisted enzymatic extraction of polysaccharides from the fruit of Schisandra chinensis Baill. *International Journal of Biological Macromolecules.* 2015;76:161–168.

Chiremba C, Rooney LW, Beta T. Microwave-assisted extraction of bound phenolic acids in bran and flour fractions from sorghum and maize cultivars varying in hardness. *Journal of Agricultural and Food Chemistry.* 2012;60(18):4735–4742.

Chouhan PN, Azimov A, Jabborova D, & Singh S. Review on the role of indigenous plants and plant derivatives of Ayurveda in bone healing. *Plant Science Today.* 2023;10(sp2):143–151.

Chuo SC, Nasir HM, Mohd-Setapar SH, et al. A Glimpse into the extraction methods of active compounds from plants. *Critical Reviews in Analytical Chemistry.* 2022;52(4):667–696.

Cravotto G, Bicchi C, Mantegna S, et al. Extraction of kiwi seed oil: Soxhlet versus four different non-conventional techniques. *Natural Product Research.* 2011;25(10):974–981.

Cravotto G, Binello A. *Low-frequency, high-power ultrasound-assisted food component extraction. InInnovative food processing technologies.* Woodhead Publishing. 2016;3–29.

Ćujić N, Šavikin K, Janković T, et al. Optimization of polyphenols extraction from dried chokeberry using maceration as traditional technique. *Food Chemistry.* 2016;194:135–142.

Da Porto C, Natolino A. Extraction kinetic modelling of total polyphenols and total anthocyanins from saffron floral bio-residues: Comparison of extraction methods. *Food Chemistry.* 2018;258:137–143.

Da Silva RP, Rocha-Santos TA, Duarte AC. Supercritical fluid extraction of bioactive compounds. *TrAC Trends in Analytical Chemistry.* 2016;76:40–51.

Dadi DW, Emire SA, Hagos AD, et al. Effect of ultrasound-assisted extraction of Moringa stenopetala leaves on bioactive compounds and their antioxidant activity. *Food Technology and Biotechnology.* 2019;57(1):77.

Dahmoune F, Nayak B, Moussi K. Optimization of microwave-assisted extraction of polyphenols from Myrtus communis L. leaves. *Food Chemistry.* 2015;166:585–595.

Delsart C, Ghidossi R, Poupot C, et al. Enhanced extraction of phenolic compounds from Merlot grapes by pulsed electric field treatment. *American journal of Enology and Viticulture.* 2012;63(2):205–211.

Dias ALB, Arroio Sergio CS, Santos P, et al. Effect of ultrasound on the supercritical CO2 extraction of bioactive compounds from dedo de moca pepper (Capsicum baccatum L. var. pendulum). *Ultrasonics Sonochemistry.* 2016;31:284.

Domínguez-Rodríguez G, Marina ML, Plaza M. Enzyme-assisted extraction of bioactive non-extractable polyphenols from sweet cherry (Prunus avium L.) pomace. *Food Chemistry.* 2021;339:128086.

Donga S, Chanda S. Evaluation of phytochemical profile and antioxidant activity of some medicinal plants seed extracts obtained by traditional and modern (Green) extraction methods. *International Journal of Chemical Studies.* 2020;8(4):3295–3309.

Elez Garofulić I, Kruk V, Martić A, et al. Evaluation of polyphenolic profile and antioxidant activity of Pistacia lentiscus L. leaves and fruit extract obtained by optimized microwave-assisted extraction. *Foods.* 2020;9(11):1556.

Erdogan S, Ates B, Durmaz G, et al. Pressurized liquid extraction of phenolic compounds from Anatolia propolis and their radical scavenging capacities. *Food and Chemical Toxicology.* 2011;49(7):1592–1597.

Gallo-Molina AC, Castro-Vargas HI, Garzón-Méndez WF, et al. Extraction, isolation and purification of tetrahydrocannabinol from the Cannabis sativa L. plant using supercritical fluid extraction and solid phase extraction. *The Journal of Supercritical Fluids.* 2019;146:208–216.

Gandhi K, Arora S, Kumar A. Industrial applications of supercritical fluid extraction: A review. *International Journal of Chemical Studies.* 2017;5(3):336–340.

Gangwar P, Johri P, Datta R, Singh S, & Trivedi M. Overview on 1 phytopharmaceuticals and biotechnology of herbal plants. In *Phytopharmaceuticals and Biotechnology of Herbal Plants.* CRC Press. 2023; 1–17.

Ghafoor K, Hui T, Choi YH. Optimization of ultrasonic-assisted extraction of total anthocyanins from grape peel using response surface methodology. *Journal of Food Biochemistry.* 2011;35(3):735–746.

Ghafoor K, Park J, Choi YH. Optimization of supercritical fluid extraction of bioactive compounds from grape (Vitis labrusca B.) peel by using response surface methodology. *Innovative Food Science & Emerging Technologies.* 2010;11(3):485–490.

Gupta G, Paul S, Singh S, et al. Exploring functional diversity and community structure of diazotrophic endophytic bacteria associated with Pennisetum glaucum growing under field in a semi-arid region. *Land.* 2022;11(7):991.

Haider W, Sultana B, Mushtaq M, et al. Multi-response optimization of enzyme-assisted maceration to enhance the yield and antioxidant activity of Cassia fistula pods extracts. *Journal of Food Measurement and Characterization.* 2018;12(4):2685–2694.

Hannay JB, Hogarth J. I. On the solubility of solids in gases. *Proceedings of the Royal Society of London.* 1880;30(200–205):178–188.

Heleno SA, Diz P, Prieto MA, et al. Optimization of ultrasound-assisted extraction to obtain mycosterols from Agaricus bisporus L. by response surface methodology and comparison with conventional Soxhlet extraction. *Food Chemistry.* 2016;197:1054–1063.

Ibañez E, Herrero M, Mendiola JA, et al. Extraction and characterization of bioactive compounds with health benefits from marine resources: Macro and micro algae, cyanobacteria, and invertebrates. In *Marine bioactive compounds.* Springer. 2012;55–98.

Ishwarya SP, Nisha P. *Headway in supercritical extraction of fragrances and colors.* Elsevier. 2021;20–639.

Jabborova D, Davranov K, Jabbarov Z, et al. Dual inoculation of plant growth-promoting Bacillus endophyticus and Funneliformis mosseae improves plant growth and soil properties in ginger. *ACS Omega.* 2022;7(39):34779–34788.

Jacotet-Navarro M, Rombaut N, Deslis S, et al. Towards a "dry" bio-refinery without solvents or added water using microwaves and ultrasound for total valorization of fruit and vegetable by-products. *Green Chemistry.* 2016;18(10):3106–3115.

Jiao J, Fu YJ, Zu YG, et al. Enzyme-assisted microwave hydro-distillation essential oil from Fructus forsythia, chemical constituents, and its antimicrobial and antioxidant activities. *Food Chemistry.* 2012;134(1):235–243.

Jiao J, Li ZG, Gai QY, et al. Microwave-assisted aqueous enzymatic extraction of oil from pumpkin seeds and evaluation of its physicochemical properties, fatty acid compositions and antioxidant activities. *Food Chemistry.* 2014;147:17–24.

Joannes C, Sipaut CS, Dayou J, et al. The potential of using pulsed electric field (PEF) technology as the cell disruption method to extract lipid from microalgae for biodiesel production. *International Journal of Renewable Energy Research.* 2015; 5(2):598–621.

Kate AE, Singh A, Shahi NC, et al. Novel eco-friendly techniques for extraction of food based lipophilic compounds from biological materials. *Natural Products Chemistry & Research.* 2016;4(5):231–237.

Kaur A, Singh S, Singh RS, et al. Hydrolysis of citrus peel naringin by recombinant α-L-rhamnosidase from Clostridium stercorarium. *Journal of Chemical Technology & Biotechnology.* 2010;85(10):1419–1422.

Khare P, Kishore K, & Sharma DK. A study on the standardization parameters of Cassia angustifolia. *Asian Journal of Pharmaceutical and Clinical Research.* 2017;10(7):329–332.

Khaw KY, Parat MO, Shaw PN, et al. Solvent supercritical fluid technologies to extract bioactive compounds from natural sources: A review. *Molecules.* 2017;22(7):1186.

Krakowska-Sieprawska A, Rafińska K, Walczak-Skierska J, et al. Promising green technology in obtaining functional plant preparations: Combined enzyme-assisted supercritical fluid extraction of flavonoids isolation from Medicago sativa leaves. *Materials.* 2021;14(11):2724.

Kumar K, Srivastav S, Sharanagat VS. Ultrasound assisted extraction (UAE) of bioactive compounds from fruit and vegetable processing by-products: A review. *Ultrasonics Sonochemistry.* 2021;70:105325.

Lakka A, Bozinou E, Stavropoulos G, et al. Enhancement of polyphenols recovery from Rosa canina, Calendula officinalis and Castanea sativa using pulsed electric field. *Beverages.* 2021;7(3):63.

Leong SY, Burritt DJ, Oey I. Evaluation of the anthocyanin release and health-promoting properties of Pinot Noir grape juices after pulsed electric fields. *Food Chemistry.* 2016;196:833–841.

Li J, Zu YG, Luo M, et al. Aqueous enzymatic process assisted by microwave extraction of oil from yellow horn (Xanthoceras sorbifolia Bunge.) seed kernels and its quality evaluation. *Food Chemistry.* 2013a;138(4):2152–2158.

Li Y, Fabiano-Tixier AS, Vian MA, et al. Solvent-free microwave extraction of bioactive compounds provides a tool for green analytical chemistry. *TrAC Trends in Analytical Chemistry.* 2013b;47:1–1.

Liew SQ, Ngoh GC, Yusoff R, et al. Sequential ultrasound-microwave assisted acid extraction (UMAE) of pectin from pomelo peels. *International Journal of Biological Macromolecules.* 2016;93:426–435.

Liu JJ, Gasmalla MA, Li P, et al. Enzyme-assisted extraction processing from oilseeds: Principle, processing and application. *Innovative Food Science & Emerging Technologies.* 2016;35:184–193.

Llompart M, Celeiro M, Dagnac T. Microwave-assisted extraction of pharmaceuticals, personal care products and industrial contaminants in the environment. *TrAC Trends in Analytical Chemistry.* 2019;116:136–150.

Łubek-Nguyen A, Ziemichód W, Olech M. Application of enzyme-assisted extraction for the recovery of natural bioactive compounds for nutraceutical and pharmaceutical applications. *Applied Sciences.* 2022;12(7):3232.

Luo X, Bai R, Zhen D, et al. Response surface optimization of the enzyme-based ultrasound-assisted extraction of acorn tannins and their corrosion inhibition properties. *Industrial Crops and Products*. 2019;129:405–413.

Luthria DL. Influence of experimental conditions on the extraction of phenolic compounds from parsley (Petroselinum crispum) flakes using a pressurized liquid extractor. *Food Chemistry*. 2008;107(2):745–752.

Macedo GA, Santana ÁL, Crawford LM, et al. Integrated microwave-and enzyme-assisted extraction of phenolic compounds from olive pomace. *Lwt*. 2021;138:110621.

Maddala VKS, Singh S, Bandi J. Soil properties influencing nutrient signaling in plant soil systems. In *Essential minerals in plant-soil systems*. Elsevier. 2024;373–393.

Marathe SJ, Jadhav SB, Bankar SB, et al. Improvements in the extraction of bioactive compounds by enzymes. *Current Opinion in Food Science*. 2019;25:62–72.

Marić M, Grassino AN, Zhu Z, et al. An overview of the traditional and innovative approaches for pectin extraction from plant food wastes and by-products: Ultrasound-, microwaves-, and enzyme-assisted extraction. *Trends in Food Science & Technology*. 2018;76:28–37.

Mashuni, Hamid FH, Muzuni, et al. The determination of total phenolic content of cocoa pod husk based on microwave-assisted extraction method. In *AIP Conference Proceedings*. AIP Publishing LLC. 2020;2243:030013.

Mroczek T, Mazurek J. Pressurized liquid extraction and anticholinesterase activity-based thin-layer chromatography with bioautography of Amaryllidaceae alkaloids. *Analytica Chimica Acta*. 2009;633(2):188–196.

Mustafa A, Turner C. Pressurized liquid extraction as a green approach in food and herbal plants extraction: A review. *Analytica Chimica Acta*. 2011;703(1):8–18.

Nadar SS, Rao P, Rathod VK. Enzyme assisted extraction of biomolecules as an approach to novel extraction technology: A review. *Food Research International*. 2018;108:309–330.

Nag S, Sit N. Optimization of ultrasound assisted enzymatic extraction of polyphenols from pomegranate peels based on phytochemical content and antioxidant property. *Journal of Food Measurement and Characterization*. 2018;12(3):1734–1743.

Nieto A, Borrull F, Pocurull E, et al. Pressurized liquid extraction: A useful technique to extract pharmaceuticals and personal-care products from sewage sludge. *TrAC Trends in Analytical Chemistry*. 2010;29(7):752–764.

Nishad J, Saha S, Dubey AK, et al. Optimization and comparison of non-conventional extraction technologies for Citrus paradisi L. peels: A valorization approach. *Journal of Food Science and Technology*. 2019;56(3):1221–1233.

Oh MJ, Lee J, Seo S, et al. Vortex pinning in artificially layered Ba (Fe, Co) 2As2 film. *Cryogenics*. 2018;92:1–4.

Pandey A, Belwal T, Sekar KC, et al. Optimization of ultrasonic-assisted extraction (UAE) of phenolics and antioxidant compounds from rhizomes of Rheum moorcroftianum using response surface methodology (RSM). *Industrial Crops and Products*. 2018;119:218–225.

Parveen R, Khare N, Singh S, & Johri P. An In silico outlook for the detection and surveillance of evolving and persistent plant pathogens. In *Detection, diagnosis and management of soil-borne phytopathogens*. Springer Nature Singapore. 2023;17–40.

Patani A, Patel A, Prajapati D, Khare N, Singh S. Various agriculture crop plant-based bioactive compounds and their use in nanomaterial synthesis and applications. In *Nanomaterials from agricultural and horticultural products*. Springer Nature Singapore. 2023;223–241.

Pavić V, Jakovljević M, Molnar M, et al. Extraction of carnosic acid and carnosol from sage (Salvia officinalis L.) leaves by supercritical fluid extraction and their antioxidant and antibacterial activity. *Plants*. 2019;8(1):16.

Pérez JP, Muñoz AA, Figueroa CP, et al. Current analytical techniques for the characterization of lipophilic bioactive compounds from microalgae extracts. *Biomass and Bioenergy*. 2021;149:106078.

Petigny L, Périno S, Minuti M, et al. Simultaneous microwave extraction and separation of volatile and non-volatile organic compounds of boldo leaves. From lab to industrial scale. *International Journal of Molecular Sciences*. 2014;15(5):7183–7198.

Pico Y. Ultrasound-assisted extraction for food and environmental samples. *TrAC Trends in Analytical Chemistry*. 2013;43:84–99.

Pimentel-Moral S, Borrás-Linares I, Lozano-Sánchez J, et al. Supercritical CO2 extraction of bioactive compounds from Hibiscus sabdariffa. *The Journal of Supercritical Fluids*. 2019;147:213–221.

Pontillo AR, Papakosta-Tsigkri L, Lymperopoulou T, et al. Conventional and enzyme-assisted extraction of rosemary leaves (Rosmarinus officinalis L.): Toward a greener approach to high added-value extracts. *Applied Sciences*. 2021;11(8):3724.

Puértolas E, López N, Saldaña G, et al. Evaluation of phenolic extraction during fermentation of red grapes treated by a continuous pulsed electric fields process at pilot-plant scale. *Journal of Food Engineering*. 2010;98(1):120–125.

Puri M, Kaur A, Schwarz WH, et al. Molecular characterization and enzymatic hydrolysis of naringin extracted from kinnow peel waste. *International Journal of Biological Macromolecules*. 2011;48(1):58–62.

Puri M, Sharma D, Barrow CJ. Enzyme-assisted extraction of bioactives from plants. *Trends in Biotechnology*. 2012;30(1):37–44.

Ramasamy J, Kandasamy R, Palanisamy S, et al. Optimization of ultrasonic-assisted extraction of flavonoids and anti-oxidant capacity from the whole plant of Andrographis echioides (L.) nees by response surface methodology and chemical composition analysis. *Pharmacognosy Magazine*. 2019;15(65):547.

Raspe DT, Ciotta SR, Mello BT, et al. Pressurized liquid extraction of steviol glycosides from Stevia rebaudiana leaves. *Chemical Engineering Transactions*. 2021;87:301–306.

Richter BE, Jones BA, Ezzell JL, et al. Accelerated solvent extraction: A technique for sample preparation. *Analytical Chemistry*. 1996;68(6):1033–1039.

Rodríguez-Pérez C, Quirantes-Piné R, Fernández-Gutiérrez A, et al. Optimization of extraction method to obtain a phenolic compounds-rich extract from Moringa oleifera Lam leaves. *Industrial Crops and Products*. 2015;66:246–254.

Roselló-Soto E, Galanakis CM, Brnčić M, et al. Clean recovery of antioxidant compounds from plant foods, by-products and algae assisted by ultrasounds processing. Modeling approaches to optimize processing conditions. *Trends in Food Science & Technology*. 2015;42(2):134–149.

Rudke AR, Mazzutti S, Andrade KS, et al. Optimization of green PLE method applied for the recovery of antioxidant compounds from buriti (Mauritia flexuosa L.) shell. *Food Chemistry*. 2019;298:125061.

Saini A, Panesar PS, Bera MB. Comparative study on the extraction and quantification of polyphenols from citrus peels using maceration and ultrasonic technique. *Current Research in Nutrition and Food Science*. 2019;7(3):678.

Sánchez-Madrigal MÁ, Viesca-Nevárez SL, Quintero-Ramos A, et al. Optimization of the enzyme-assisted extraction of fructans from the wild sotol plant (Dasylirion wheeleri). *Food Bioscience*. 2018;22:59–68.

Sao P, Parveen R, Khattri A, Sharma S, Tiwari N, Singh S. Understanding the metabolomics of medicinal plants under environmental pollution: Challenges and opportunities. *Environmental Pollution and Medicinal Plants*. 2022;193–205.

Shams KA, Abdel-Azim NS, Tawfik WA, et al. Green extraction techniques: Effect of extraction method on lipid contents of three medicinal plants of Apiaceae. *Journal of Chemical and Pharmaceutical Research*. 2015;7(4):1080–1088.

Shingare MS, Shingate BB. 10 ultrasound in synthetic applications and organic chemistry. In *Handbook on applications of ultrasound. Sonochemistry for sustainability (Chap. 10).* CRC Press. 2012;213–261.

Singh A, Sengar RS, Singh AK, Singh S, Kumar A, Ahmad K. Impact of genetic engineering on agriculture food crops. *Biotech Today: An International Journal of Biological Sciences.* 2019; 9(1):17–24.

Su CH, Nguyen HC, Bui TL, et al. Enzyme-assisted extraction of insect fat for biodiesel production. *Journal of Cleaner Production.* 2019;223:436–444.

Sun H, Li C, Ni Y, et al. Ultrasonic/microwave-assisted extraction of polysaccharides from Camptotheca acuminata fruits and its antitumor activity. *Carbohydrate Polymers.* 2019;206:557–564.

Supasatyankul B, Saisriyoot M, Klinkesorn U, et al. Extraction of phenolic and flavonoid compounds from Mung Bean (Vigna radiata L.) seed coat by Pressurized Liquid Extraction. *Molecules.* 2022;27(7):2085.

Tchabo W, Ma Y, Engmann FN, et al. Ultrasound-assisted enzymatic extraction (UAEE) of phytochemical compounds from mulberry (Morus nigra) must and optimization study using response surface methodology. *Industrial Crops and Products.* 2015;63:214–225.

Timmermans RA, Mastwijk HC, Berendsen LB, et al. Moderate intensity Pulsed Electric Fields (PEF) as alternative mild preservation technology for fruit juice. *International Journal of Food Microbiology.* 2019;298:63–73.

Tiwari BK. Ultrasound: A clean, green extraction technology. *TrAC Trends in Analytical Chemistry.* 2015;71:100–109.

Tyagi P and Ranjan R. Phytochemical screening of all active components and Thin Layer Chromatography (TLC) profiling of diosgenin of all parts of tribulus terrestris and pedalium murex. *Zeichen Journal.* 2022;8:387–403.

Uwineza PA, Gramza-Michałowska A, Bryła M, et al. Antioxidant activity and bioactive compounds of lamium album flower extracts obtained by supercritical fluid extraction. *Applied Sciences.* 2021;11(16):7419.

Verma N, Sao P, Srivastava A, Singh S. Physiological and molecular responses to drought, submergence and excessive watering in plants. In *Harsh environment and plant resilience: Molecular and functional aspects.* Springer. 2021;305–321.

Vinatoru M, Mason TJ, Calinescu I. Ultrasonically assisted extraction (UAE) and microwave assisted extraction (MAE) of functional compounds from plant materials. *TrAC Trends in Analytical Chemistry.* 2017;97:159–178.

Vishwakarma SK, Ilyas T, Shahid M, et al. Bacillus spp.: Nature's gift to agriculture and humankind. In *Applications of bacillus and bacillus derived genera in agriculture, biotechnology and beyond.* Springer Nature Singapore. 2024;1–36.

Vora Z, Pandya J, Vaikuntapu PR, Singh S, Asha S, Purama RK. Genome-wide association of defense hormone crosstalk in plants. In *Hormonal cross-talk, plant defense and development.* Academic Press. 2023;353–371.

Vorobiev E, Lebovka N. Pulsed-electric-fields-induced effects in plant tissues: Fundamental aspects and perspectives of applications. In *Electrotechnologies for extraction from food plants and biomaterials.* Springer. 2009;39–81.

Vorobiev E, Lebovka NI. Extraction of intercellular components by pulsed electric fields. In *-Pulsed electric fields technology for the food industry.* Springer. 2006;153–193.

Vuong QV, Thanh DT, Bhuyan DJ, et al. Optimization of ultrasound-assisted extraction conditions for euphol from the medicinal plant, Euphorbia tirucalli, using response surface methodology. *Industrial Crops and Products.* 2015;63:197–202.

Wang J, Sun B, Liu Y, et al. Optimisation of ultrasound-assisted enzymatic extraction of arabinoxylan from wheat bran. *Food Chemistry.* 2014;150:482–488.

Wianowska D, Gil M. Critical approach to PLE technique application in the analysis of secondary metabolites in plants. *TrAC Trends in Analytical Chemistry.* 2019;114:314–325.

Wijesinghe WA, Jeon YJ. Enzyme-assistant extraction (EAE) of bioactive components: A useful approach for recovery of industrially important metabolites from seaweeds: A review. *Fitoterapia*. 2012;83(1):6–12.

Wijngaard HH, Ballay M, Brunton N. The optimisation of extraction of antioxidants from potato peel by pressurised liquids. *Food Chemistry*. 2012;133(4):1123–1130.

Wu C, Wang F, Liu J, et al. A comparison of volatile fractions obtained from Lonicera macranthoides via different extraction processes: Ultrasound, microwave, Soxhlet extraction, hydrodistillation, and cold maceration. *Integrative Medicine Research*. 2015a;4(3):171–177.

Wu D, Gao T, Yang H, et al. Simultaneous microwave/ultrasonic-assisted enzymatic extraction of antioxidant ingredients from Nitraria tangutorun Bobr. juice by-products. *Industrial Crops and Products*. 2015b;66:229–238.

Xhaxhiu K, Korpa A, Mele A, et al. Ultrasonic and soxhlet extraction characteristics of the orange peel from "Moro" Cultivars grown in Albania. *Journal of Essential Oil Bearing Plants*. 2013;16(4):421–428.

Xu Y, Zhang L, Yang Y, et al. Optimization of ultrasound-assisted compound enzymatic extraction and characterization of polysaccharides from blackcurrant. *Carbohydrate Polymers*. 2015;117:895–902.

Yahya NA, Attan N, Wahab RA. An overview of cosmeceutically relevant plant extracts and strategies for extraction of plant-based bioactive compounds. *Food and Bioproducts Processing*. 2018;112:69–85.

Yang L, Wang H, Zu YG, et al. Ultrasound-assisted extraction of the three terpenoid indole alkaloids vindoline, catharanthine and vinblastine from Catharanthus roseus using ionic liquid aqueous solutions. *Chemical Engineering Journal*. 2011;172(2–3):705–712.

Yang YC, Wei MC. Kinetic and characterization studies for three bioactive compounds extracted from Rabdosia rubescens using ultrasound. *Food and Bioproducts Processing*. 2015;94:101–113.

Yang Y, Zhang, F. Ultrasound-assisted extraction of rutin and quercetin from *Euonymus alatus* (Thunb.) Sieb. *Ultrasonics Sonochemistry*, 2008;15(4):308–313.

Yin C, Fan X, Fan Z, et al. Optimization of enzymes-microwave-ultrasound assisted extraction of Lentinus edodes polysaccharides and determination of its antioxidant activity. *International Journal of Biological Macromolecules*. 2018;111:446–454.

Zderic A, Zondervan E. Polyphenol extraction from fresh tea leaves by pulsed electric field: A study of mechanisms. *Chemical Engineering Research and Design*. 2016;109:586–592.

Zhang L, Guo S, Wang M, et al. PEG-based ultrasound-assisted enzymatic extraction of polysaccharides from Ginkgo biloba leaves. *International Journal of Biological Macromolecules*. 2015;80:644–650.

Zu G, Zhang R, Yang L, et al. Ultrasound-assisted extraction of carnosic acid and rosmarinic acid using ionic liquid solution from Rosmarinus officinalis. *International Journal of Molecular Sciences*. 2012;13(9):11027–11043.

4 Green Technologies for Isolating Bioactives from Plant-Based Food and Their Applications in Human Health and Longevity

Vibha Gajbe, Azza. S. Naik, Roji Waghmare, and Shuchi Midha*

4.1 INTRODUCTION

"Bioactive" is a term used to define biologically active molecules that elicit a significant effect in organisms, tissues, or cells. These molecules can be employed in human nutrition, disease alleviation, and other physiological applications (Banwo et al., 2021). They belong to different classes of compounds, including macromolecules like carbohydrates, lipids, and proteins, or occur as smaller molecules such as omega-3 fatty acids, peptides, short-chain fatty acids, nucleosides and nucleotides, polyphenols, pigments such as carotenoids, lycopene, and vitamins (Singh et al., 2022). They can be derived from animal or plant sources and from terrestrial as well as marine ecosystems. Bioactive molecules in the natural plant material must be extracted, concentrated, stabilized, and formulated for specific human applications (Figure 4.1). Potent plant-derived molecules have been utilized against several diseases, such as microalgal omega-3 fatty acids for joint health and anti-inflammation, and other plant-derived antioxidants for promoting longevity (Ferguson et al., 2020; Fernandes et al., 2019, Trivedi et al., 2024).

Bioactive molecules are extracted from plant raw materials, transitioning from conventional methods that use hazardous chemicals to novel and greener technologies (Prado et al., 2021) such as supercritical extraction, ultrasound, and microwave-assisted extraction. These greener methods increase the yield and recovery of bio-actives (Marić et al., 2018). Non-thermal techniques minimize the thermal degradation of bioactives. With increased emphasis on attaining UN sustainable

* Corresponding author

 DOI: 10.1201/9781003392910-4

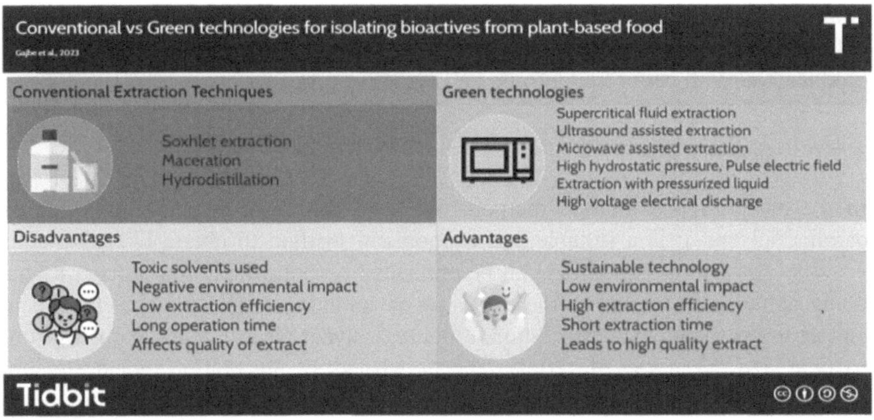

FIGURE 4.1 Graphical abstract.

development goals and a push by several global policy initiatives, it is imperative to move towards greener methods (Pavez et al., 2018).

A detailed investigation through sustainability assessment tools such as life cycle assessment (LCA), life cycle cost (LCC), and cost-benefit analysis (CBA) needs to be undertaken to judiciously select the best technology for extraction (Hoogmartens et al., 2014). Green methods do not necessarily imply complete avoidance of chemicals and solvents; the use of ionic liquids, deep eutectic solvents, bio-based solvents, and solvent modifiers in supercritical extraction systems signifies the role these chemical agents play in enhancing solvent properties and improving extraction capabilities (Lim et al., 2022; Sportiello et al., 2023). Enzyme-assisted extraction is another promising technique used for extracting bio-actives (Das et al., 2021). Since novel processing technologies for extraction might result in significantly different properties, in Europe, authorization under Regulation (EC) No. 258/97 is required before extracts can be utilized for human consumption (Crevel, 2023)

4.2 CONVENTIONAL EXTRACTION METHODS FOR BIOACTIVE COMPOUNDS

4.2.1 Soxhlet Extraction

The Soxhlet extraction method serves as a benchmark for evaluating newer extraction techniques (Azmir et al., 2013; Khare et al., 2017; Parveen et al., 2023). In this method, material without moisture is positioned in a thimble. This packed thimble and material are then positioned in a distillation beaker (Osorio-Tobón, 2020). Bioactive compounds are separated from the solid sample with the application of heated and condensed solvent. This method is not suitable for the extraction of thermally sensitive compounds (Santos & Santana, 2022). The fact that the sample is packed in a thimble during Soxhlet extraction allows the matrix to be continually in contact with new solvents and eliminates the need for extract filtering. Fuente et al. (2022) compared the microwave-assisted and Soxhlet extraction methods for obtaining bioactive oils from salmon side streams. Microwave-assisted extraction produced oil with better lipid profile and bioactive properties.

4.2.2 Maceration

In the maceration extraction method, the solid substance is combined with the organic solvent. It is a suitable and economical method to obtain bioactive compounds. The rate of extraction can be enhanced by optimizing the process by utilizing agitation to accelerate the diffusion of components into the solvent, raising temperatures to enhance dissolution, reducing solvent viscosity, and reducing solid particle size (Verma et al., 2021). The drawbacks of maceration include lengthy extraction durations, heavy reliance on solvents, incomplete yield of extract, less energy efficiency, and less recovery (Santos & Santana, 2022). For extraction, maceration comprises numerous steps. Raw samples are ground into fine particles to improve the surface area; then, a suitable solvent is added to a packed vessel and filtered to remove contaminants (Azmir et al., 2013; Gupta et al., 2022). The benefits of maceration include lower temperatures and simple equipment (Das et al., 2019). Maceration requires longer extraction periods and results in poorer recovery of bioactive compounds (Osorio-Tobón, 2020). Meregalli et al. (2020) found 12% more recovery and 25% reduced extraction time during ultrasound-assisted extraction of bioactive compounds from red araca´ peel in comparison to the maceration method (116.81 Eq. mg of cyanidin-3-glycoside/100 g of peel using 90% ethanol, for 2 hrs, at 40°C, and mass-to-volume ratio of 1 g/10 mL).

4.2.3 Hydrodistillation

Water serves as the solvent in the hydrodistillation extraction method (Siddique et al., 2020). The solid material and water are combined in this method using an electric heating mantle. The volatile chemicals are then condensed and collected. Due to the application of increased temperatures of water, hydrodistillation has the drawback of losing certain components, such as natural color compounds and thermally sensitive bioactive compounds (Naik et al., 2021; Santos & Santana, 2022). In this method, the raw samples are packed in a still chamber, then enough water is added and heated to a boiling temperature. Alternatively, direct steam can be introduced into the sample (Tungare et al., 2022). The major parameters that liberate the bioactive chemicals from plant tissue are hot water and steam. The vaporized combination of water and oil is condensed by indirect water cooling (Singh et al., 2019). Oil and bioactive chemicals separate from the water when the condensed mixture moves from the condenser to the separator (Azmir et al., 2013). Drinić et al. (2021) found that supercritical carbon dioxide followed by the conventional method was more effective than the Soxhlet extraction and hydrodistillation methods for extracting bioactive compounds from hemp. Another study (Costa et al., 2012; Bala et al., 2011) demonstrated that supercritical fluid extraction isolated more bioactive compounds from *Lavandula viridis* than hydrodistillation. However, higher bioactive compounds were extracted in the hydrodistillation method.

4.3 GREEN TECHNOLOGIES FOR EXTRACTION OF BIOACTIVE COMPOUNDS

Conventional techniques consume more time, energy, and large quantities of organic solvents, with low yield and selectivity (Maddala et al., 2023). To overcome these limitations, novel green extraction methods are being explored.

4.3.1 Supercritical Fluid Extraction

Supercritical fluid extraction uses non-toxic organic solvents and is safe for human health and the environment. It lowers extraction time, speeds up processing, reduces waste generation, lowers energy consumption, and enhances efficiency and product quality (Soquetta et al., 2018).

4.3.1.1 CO_2 as Supercritical Fluid

Carbon dioxide is used as an organic solvent in the superfluid extraction process. The U.S. Food and Drug Administration and the European Food Safety Authority have declared CO_2 a safe solvent (Khaw et al., 2017) due to its chemical inertness, non-toxic and inflammable nature, and environmental safety. Some common chemical solvents in the extraction of bioactive compounds are depicted in Table 4.1.

4.3.1.2 Parameters Influencing Supercritical Solvent Extraction

4.3.1.2.1 Temperature and Pressure

The temperature influences solute solubility in two opposite ways. First, the increase in temperature decreases solvent density, causing a reduction in solvation power. As

TABLE 4.1
Common Supercritical Solvents Used in Supercritical Fluid Extraction

Chemical Solvent	Molar Mass (g/mol)	Critical Temperature (°C)	Critical Pressure (bar)	Density (kg/m^3)
CO_2	44.01	31.1	73.8	470.0
Ethane	30.07	48.2	48.8	200.0
Ethylene	28.05	10.1	50.40	215.0
Ethanol	46.07	63	62.10	280.0
Propane	44.097	42	42.49	220.0
Propylene	42.08	91.9	46.0	232.0
Acetone	58.08	235	46.5	278.0
Water	18.02	373.94		348.0

Source: Uwineza & Waśkiewicz (2020)

a result, the rate of vapor formation of solute molecules increases, which increases the solubility of solute and produces a higher extraction yield. Second, at high temperature, solvent density decreases, lowering the solvation and solubility of the solute, leading to an undesired yield (Cvjetko Bubalo et al., 2018; Espinosa-Pardo et al., 2017). These two opposite effects cause a phenomenon of retrogradation or crossover effect (Mazzutti et al., 2012). Hence, the temperature and pressure should be optimized to avoid degradation of the targeted compound. A study by Bimakr et al. (2011) observed that increased temperature led to improved extraction yield, with the highest extraction yield of 60.57 mg/g obtained at 60°C. They also reported an increase in extraction yield with increased pressure ranging between 100 and 200 bar and above 200 bar.

4.3.1.2.2 Time of Extraction

The extraction time should be short enough to maintain the quality of the product, but not so short that it results in inadequate extraction (Johri & Dwivedi, 2022). A longer extraction time uses more solvent and time, and can degrade bioactive compounds. Bimakr et al. (2011) stated that the high extraction yield of spearmint leaves was achieved at 90 min at 100 bar pressure. Molino et al. (2019) extracted 55.4% of β-carotene at 550 bars and 65°C with varying extraction times between 30 and 110 min, and a CO_2 flow rate of 14.48 g/min from microalgae *Dunaliella salina*.

4.3.1.2.3 The Type of Solvent

The solvent nature improves bioactive compounds' extraction yield (Gangwar et al., 2023). For example, a study by Ngo et al. (2017) showed that 50% acetone resulted in a higher content of phenolic compounds and saponins than water and absolute methanol from roots of S. chinensis. In another study, propane was found to be the best extraction solvent for extracting sunflower seed oil (Nimet et al., 2011). Bonthuys et al. (2011) reported ethane to be more effective in separating alkane and alcohol.

4.3.1.2.4 The Type of Co-Solvent

SC-CO_2 is non-polar as it has zero dipole moment and hence it is inefficient in extracting polar solutes. To enhance the extraction capabilities small amounts of polar co-solvents (Raventós et al., 2002) like water, methanol, and ethanol (Uwineza & Waśkiewicz, 2020) are often added. Klein et al. (2019) reported that adding co-solvents such as ethanol and ethyl acetate improved extraction rates and increased α-/β-amyrin yields. The highest yield of α-/β-amyrin was obtained using 5% ethyl acetate and global extraction yield was achieved using 15% ethanol.

4.3.2 Ultrasound-Assisted Extraction

Ultrasound-assisted extraction (UAE) uses less solvent, energy, and time, consequently enhancing the yield of the final product with high purity (Yusoff et al., 2022). The technique is environmentally sustainable, as it generates green products and consumes less solvent (Kumar Kshitij, 2021).

4.3.2.1 Working Principle of Ultrasound-Assisted Extraction

Ultrasound-assisted extraction uses 20 kHz to 100 MHz ultrasound waves. When ultrasound waves transmit through the liquid solvent, it results in cavitation phenomena. During cavitation, bubbles expand and implode at a temperature of 4500°C and pressure of nearly 100 MPa. As a result, turbulence and shock wave effects are observed. The high heat and pressure, combined with elevated turbulence levels, escalate the mass transport of solute molecules. When ultrasound waves propagate through the solvent, expansion and compression effects can be observed (Vardanega et al., 2014).

Ultrasound waves can cause mechanical effects in pure liquid. The uniform surrounding environment keeps the bubble globular during mechanical actions. The bubble implodes on the surface of the solid sample, developing a high-speed jet of solvent that hits the plant cell membrane. This results in cell membrane fragmentation and cell disruption. The cell disruption improves solvent penetration in plant cells, mass transfer, and extraction of bioactive compounds. When ultrasound waves pass through the solvent, they cause the pores of the solid plant matrix to expand, allowing the matrix to absorb water and swell. Swelling and hydration accelerate solvent-solute diffusion and increase mass transport. The combined effects of expansion, compression, mechanical, hydration, and swelling contribute to increased extraction yield in all ultrasound-assisted extraction (Diego T. Santos et al., 2014).

4.3.2.2 Parameters Influencing Ultrasound-Assisted Extraction

Ultrasound-assisted extraction is influenced by the frequency and intensity of waves, temperature, time for extraction, and choice of solvent.

Low-frequency waves improve extraction yield, but high-frequency waves decrease mass transfer and produce a low yield. The low frequency may form fewer cavities, resulting in a high cavitation effect (Howard et al., 2011). However, high-frequency sound waves impede bubble implosion due to the compression effects (Vardanega et al., 2014). Mason et al. (2011) found that a shorter compression-rarefaction cycle at high frequencies delays bubble implosion (Figure 4.2).

Guandalini et al. (2019) reported a high yield of pectin from mango peels at when treated at 20 kHz. González-Centeno et al. (2014) extracted phenolics from grape

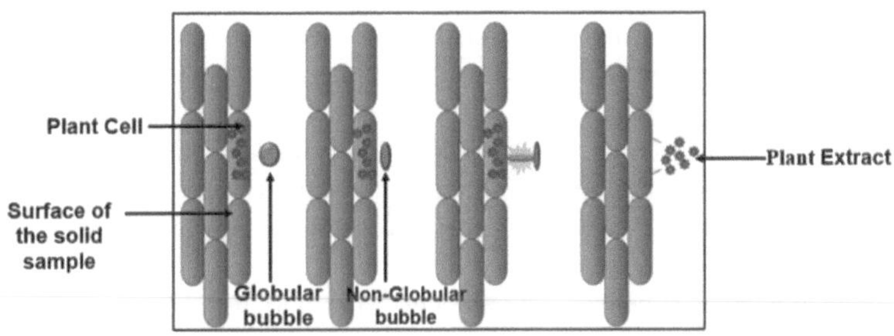

FIGURE 4.2 Cavitation bubble collapse releasing plant extract (credit: Vardanega et al., 2014).

pomace at 40, 80, and 120 kHz, and reported that the 40 kHz frequency produced a high yield.

The intensity of ultrasound waves affects extraction yield by enhancing diffusivity and enlarging the solute-solvent contact area. Very high intensity can lead to greater bubble formation, which decreases the cavitation effect. This further degrades the targeted compounds and reduces the yield. Al-Dhabi et al. (2017), found that increasing power from 100 to 200 W increased the yield of phenolic compounds extracted from waste spent coffee grounds. Gonzalez-Centeno et al. (2014) and Bruno Romanini et al. (2021) extracted high yields of phenolics and flavonoids from grape pomace at 40 kHz, 150 W/L, and 25 minutes. Furthermore, Hosseini et al. (2019) extracted pectin from orange peel and observed that the yield improved under optimized conditions of 150 W power, and 10 min irradiation time.

Ultrasound-assisted extraction produces a good extraction yield at high temperatures. Possible reasons are that i) high temperatures reduce the viscosity of the solvent and enhance the solubility of the solute. Due to reduced cavitation, increasing temperature reduces yield. A similar trend has been reported (Yuting et al., 2014) for pectin extraction from grapefruit peel. ii) High temperatures generate more bubbles, and shear stress degrades the product. iii) The solvent develops maximum vapor pressure and forms a greater number of bubbles at high temperatures. This reduces the pressure gradient around the bubble. The larger bubbles formed at high temperatures collapse mildly without damaging plant cells, eventually reducing the extraction yield.

The increased extraction time leads to a higher extraction yield, but further increases in time can reduce the yield. Longer time enhances pores and fragments on the plant material's surface, improving solvent diffusivity and penetration into the cell wall, ultimately increasing extraction yield. Kazemi et al. (2019) applied an irradiation time of 30 min for pectin extraction from eggplant peel. Kaur et al. (2021) observed the highest yield of bioactive compounds from kinnow mandarin peel in a shorter extraction time of 15 min.

4.4 CONCLUSION

This chapter presents a detailed overview of the extraction of bioactive compounds from plant-based food by traditional and green extraction methods. In conclusion, green extraction methods offer numerous benefits over traditional extraction methods, including the use of eco-friendly and non-toxic solvents, the production of bioactive compounds in a high purity state, shorter extraction time, improved extraction yield and quality of extract, decreased energy expenditure, cost-effectiveness, mitigating environmental impact, and minimizing exposure to human health. Bioactives have the potential to be developed into new drugs and supplements and functional foods, and can help to preserve plant biodiversity. This versatility and growing need for bioactive molecules encourages researchers to explore novel technologies for the extraction of bioactive molecules.

REFERENCES

Al-Dhabi, N. A., Ponmurugan, K., & Maran Jeganathan, P. (2017). Development and validation of ultrasound-assisted solid-liquid extraction of phenolic compounds from waste spent coffee grounds. *Ultrasonics Sonochemistry*, 34, 206–213. https://doi.org/10.1016/j.ultsonch.2016.05.005

Alonso-Riaño, P., Sanz Diez, M. T., Blanco, B., Beltrán, S., Trigueros, E., & Benito-Román, O. (2020). Water Ultrasound-Assisted Extraction of Polyphenol Compounds from Brewer's Spent Grain: Kinetic Study, Extract Characterization, and Concentration. *Antioxidants*, 9(3), 265. https://doi.org/10.3390/antiox9030265

Azmir, J., Zaidul, I. S. M., Rahman, M. M., Sharif, K. M., Mohamed, A., Sahena, F., Jahurul, M. H. A., Ghafoor, K., Norulaini, N. A. N., & Omar, A. K. M. (2013). Techniques for extraction of bioactive compounds from plant materials: A review. *Journal of Food Engineering*, 117(4), 426–436. https://doi.org/10.1016/j.jfoodeng.2013.01.014

Bala, J. M., Vennila, J. J., Ganesh, S., & Sachidanand Singh, S. S. (2011). Analysis of active components in Acanthus ilicifolius leaves for their potential chemoprotective activity against hepatocellular carcinoma: An in silico approach. *Journal of Pharmacy Research*, 4(6), 1809–1812.

Banwo, K., Olojede, A. O., Adesulu-Dahunsi, A. T., Verma, D. K., Thakur, M., Tripathy, S., Singh, S., Patel, A. R., Gupta, A. K., Aguilar, C. N., & Utama, G. L. (2021). Functional importance of bioactive compounds of foods with Potential Health Benefits: A review on recent trends. *Food Bioscience*, 43, 101320. https://doi.org/10.1016/j.fbio.2021.101320

Bimakr, M., Rahman, R. A., Taip, F. S., Ganjloo, A., Salleh, L. M., Selamat, J., Hamid, A., & Zaidul, I. S. M. (2011). Comparison of different extraction methods for the extraction of major bioactive flavonoid compounds from spearmint (Mentha spicata L.) leaves. *Food and Bioproducts Processing*, 89(1), 67–72. https://doi.org/10.1016/j.fbp.2010.03.002

Bonthuys, G. J. K., Schwarz, C. E., Burger, A. J., & Knoetze, J. H. (2011). Separation of alkanes and alcohols with supercritical fluids. Part I: Phase equilibria and viability study. *The Journal of Supercritical Fluids*, 57(2), 101–111. https://doi.org/10.1016/j.supflu.2011.02.012

Bruno Romanini, E., Misturini Rodrigues, L., Finger, A., Perez Cantuaria Chierrito, T., Regina da Silva Scapim, M., & Scaramal Madrona, G. (2021). Ultrasound assisted extraction of bioactive compounds from BRS Violet grape pomace followed by alginate-Ca2+ encapsulation. *Food Chemistry*, 338, 128101. https://doi.org/10.1016/j.foodchem.2020.128101

Costa, P., Grosso, C., Gonçalves, S., Andrade, P. B., Valentão, P., Gabriela Bernardo-Gil, M., & Romano, A. (2012). Supercritical fluid extraction and hydrodistillation for the recovery of bioactive compounds from Lavandula viridis L'Hér. *Food Chemistry*, 135(1), 112–121. https://doi.org/10.1016/j.foodchem.2012.04.108

Crevel, R. (2023). Novel foods and ingredients: Laws and regulations Europe. In *Reference module in food science*. Elsevier. https://doi.org/10.1016/B978-0-12-823960-5.00047-0

Cvjetko Bubalo, M., Vidović, S., Radojčić Redovniković, I., & Jokić, S. (2018). New perspective in extraction of plant biologically active compounds by green solvents. *Food and Bioproducts Processing*, 109, 52–73.https://doi.org/10.1016/j.fbp.2018.03.001

Das, S., Johri, P., Sharma, R., Kashyap, M., Singh, S., & Singh, S. (2019). Comparative modeling, characterization and energy minimization studies of Aquaporin 9: An exclusive target protein for rheumatoid arthritis. *International Journal of Pharmaceutical Investigation*, *9*(2), 43–46. https://doi.org/10.5530/ijpi.2019.2.9

Das, S., Nadar, S. S., & Rathod, V. K. (2021). Integrated strategies for enzyme assisted extraction of bioactive molecules: A review. *International Journal of Biological Macromolecules*, 191, 899–917. https://doi.org/10.1016/j.ijbiomac.2021.09.060

De la Fuente, B., Pinela, J., Mandim, F., Heleno, S. A., Ferreira, I. C. F. R., Barba, F. J., Berrada, H., Caleja, C., & Barros, L. (2022). Nutritional and bioactive oils from salmon (Salmo salar) side streams obtained by Soxhlet and optimized microwave-assisted extraction. *Food Chemistry*, 386, 132778. https://doi.org/10.1016/j.foodchem.2022.132778

Drinić, Z., Vladic, J., Koren, A., Zeremski, T., Stojanov, N., Tomić, M., & Vidović, S. (2021). Application of conventional and high-pressure extraction techniques for the isolation of bioactive compounds from the aerial part of hemp (Cannabis sativa L.) assortment Helena. *Industrial Crops and Products*, 171, 113908. https://doi.org/10.1016/j.indcrop.2021.113908

Espinosa-Pardo, F. A., Nakajima, V. M., Macedo, G. A., Macedo, J. A., & Martínez, J. (2017). Extraction of phenolic compounds from dry and fermented orange pomace using supercritical CO_2 and cosolvents. *Food and Bioproducts Processing*, 101, 1–10. https://doi.org/10.1016/j.fbp.2016.10.002

Ferguson, J. J. A., Stojanovski, E., MacDonald-Wicks, L., & Garg, M. L. (2020). High molecular weight oat β-glucan enhances lipid-lowering effects of phytosterols. A randomised controlled trial. *Clinical Nutrition*, 39(1), 80–89. https://doi.org/10.1016/j.clnu.2019.02.007

Fernandes, S. S., Coelho, M. S., & Salas-Mellado, M. de las M. (2019). Bioactive compounds as ingredients of functional foods. In *Bioactive -compounds* (pp. 129–142). Elsevier. https://doi.org/10.1016/B978-0-12-814774-0.00007-4

Gangwar, P., Johri, P., Datta, R., Singh, S., & Trivedi, M. (2023). Overview on 1 Phytopharmaceuticals and biotechnology of herbal plants. In *Phytopharmaceuticals and biotechnology of herbal plants* (pp. 1–17). CRC Press.

González-Centeno, M. R., Knoerzer, K., Sabarez, H., Simal, S., Rosselló, C., & Femenia, A. (2014). Effect of acoustic frequency and power density on the aqueous ultrasonic-assisted extraction of grape pomace (Vitis vinifera L.) – A response surface approach. *Ultrasonics Sonochemistry*, 21(6), 2176–2184. https://doi.org/10.1016/j.ultsonch.2014.01.021

Guandalini, B. B. V., Rodrigues, N. P., & Marczak, L. D. F. (2019). Sequential extraction of phenolics and pectin from mango peel assisted by ultrasound. *Food Research International*, 119, 455–461. https://doi.org/10.1016/j.foodres.2018.12.011

Gupta, G., Paul, S., Singh, S., Pietramellara, G., Pathan, S. I., Danish, S., & Jha, P. N. (2022). Exploring functional diversity and community structure of diazotrophic endophytic bacteria associated with Pennisetum glaucum growing under field in a semi-arid region. *Land*, 11(7), 991.

Hoogmartens, R., van Passel, S., van Acker, K., & Dubois, M. (2014). Bridging the gap between LCA, LCC and CBA as sustainability assessment tools. *Environmental Impact Assessment Review*, 48, 27–33. https://doi.org/10.1016/j.eiar.2014.05.001

Hosseini, S. S., Khodaiyan, F., Kazemi, M., & Najari, Z. (2019). Optimization and characterization of pectin extracted from sour orange peel by ultrasound assisted method. *International Journal of Biol Macromolecular*, 125, 621–629. https://doi.org/10.1016/j.ijbiomac.2018.12.096

Howard, C., Maddern, A., Privopoulos, E., Leong, T., Ashokkumar, M., & Kentish, S. (2011). PAPERS Acoustic characteristics for effective ambulance sirens The fundamentals of power ultrasound-a review. In *Acoustics Australia* (Vol. 39, Issue 2). www.acoustics.asn.au

Johri, P., & Dwivedi, A. (2022). Covid19 and the rise of telemedicine. *International Journal of Biology, Pharmacy and Allied Sciences*, *11*(4), 1653–1662. https://doi.org/10.31032/ijbpas/2022/11.4.5992

Kaur, S., Panesar, P. S., & Chopra, H. K. (2021). Standardization of ultrasound-assisted extraction of bioactive compounds from kinnow mandarin peel. *Biomass Conversion and Biorefinery*. https://doi.org/10.1007/s13399-021-01674-9

Kazemi, M., Khodaiyan, F., & Hosseini, S. S. (2019). Eggplant peel as a high potential source of high methylated pectin: Ultrasonic extraction optimization and characterization. *LWT*, 105, 182–189. https://doi.org/10.1016/j.lwt.2019.01.060

Khare, P., Kishore, K., & Sharma, D. K. (2017). A study on the standardization parameters of Cassia angustifolia. *Asian Journal of Pharmaceutical and Clinical Research*, 10(7), 329–332.

Khaw, K.-Y., Parat, M.-O., Shaw, P. N., & Falconer, J. R. (2017). Solvent supercritical fluid technologies to extract bioactive compounds from natural sources: A review. *Molecules*, 22(7), 1186. https://doi.org/10.3390/molecules22071186

Klein, E. J., Náthia-Neves, G., Vardanega, R., Meireles, M. A. A., da Silva, E. A., & Vieira, M. G. A. (2019). Supercritical CO2 extraction of α-/β-amyrin from uvaia (Eugenia pyriformis Cambess.): Effects of pressure and co-solvent addition. *The Journal of Supercritical Fluids*, 153, 104595. https://doi.org/10.1016/j.supflu.2019.104595

kumar kshitij. (2021). Ultrasound assisted extraction of bioactive compounds from fruit and vegetable. *Ultrasonics Chemistry*. https://doi.org/10.1016/j.ultsonch.2020.105325

Lim, J. R., Chua, L. S., & Mustaffa, A. A. (2022). Ionic liquids as green solvent and their applications in bioactive compounds extraction from plants. *Process Biochemistry*, 122, 292–306. https://doi.org/10.1016/j.procbio.2022.10.024

Maddala, V. K. S., & Singh, S. (2023). Essential oils from plants and their role in nanomaterial synthesis characterization and applications. In *Secondary metabolites based green synthesis of nanomaterials and their applications* (pp. 191–207). Springer Nature Singapore.

Marić, M., Grassino, A. N., Zhu, Z., Barba, F. J., Brnčić, M., & Rimac Brnčić, S. (2018). An overview of the traditional and innovative approaches for pectin extraction from plant food wastes and by-products: Ultrasound-, microwaves-, and enzyme-assisted extraction. *Trends in Food Science & Technology*, 76, 28–37. https://doi.org/10.1016/j.tifs.2018.03.022

Mason, T. J., Cobley, A. J., Graves, J. E., & Morgan, D. (2011). New evidence for the inverse dependence of mechanical and chemical effects on the frequency of ultrasound. *Ultrasonics Sonochemistry*, 18(1), 226–230. https://doi.org/10.1016/j.ultsonch.2010.05.008

Mazzutti, S., Ferreira, S. R. S., Riehl, C. A. S., Smania, A., Smania, F. A., & Martínez, J. (2012). Supercritical fluid extraction of Agaricus brasiliensis: Antioxidant and antimicrobial activities. *The Journal of Supercritical Fluids*, 70, 48–56. https://doi.org/10.1016/j.supflu.2012.06.010

Meregalli, M. M., Puton, B. M. S., Camera, F. D., Amaral, A. U., Zeni, J., Cansian, R. L., Mignoni, M. L., & Backes, G. T. (2020). Conventional and ultrasound-assisted methods for extraction of bioactive compounds from red araçá peel (Psidium cattleianum Sabine). *Arabian Journal of Chemistry*, 13(6), 5800–5809. https://doi.org/10.1016/j.arabjc.2020.04.017

Molino, A., Larocca, V., di Sanzo, G., Martino, M., Casella, P., Marino, T., Karatza, D., & Musmarra, D. (2019). Extraction of bioactive compounds using supercritical carbon dioxide. *Molecules*, 24(4), 782. https://doi.org/10.3390/molecules24040782

Naik, A. S., Suryawanshi, D., Kumar, M., & Waghmare, R. (2021). Ultrasonic treatment: A cohort review on bioactive compounds, allergens and physico-chemical properties of food. *Current Research in Food Science*, 4, 470–477. https://doi.org/10.1016/j.crfs.2021.07.003

Ngo, T., Scarlett, C. J., Bowyer, M. C., Ngo, P. D., & Vuong, Q. van. (2017). Impact of different extraction solvents on bioactive compounds and antioxidant capacity from the root of Salacia chinensis L. *Journal of Food Quality*, 2017, 1–8. https://doi.org/10.1155/2017/9305047

Nimet, G., da Silva, E. A., Palú, F., Dariva, C., Freitas, L. dos S., Neto, A. M., & Filho, L. C. (2011). Extraction of sunflower (Heliantus annuus L.) oil with supercritical CO2 and subcritical propane: Experimental and modeling. *Chemical Engineering Journal*, 168(1), 262–268. https://doi.org/10.1016/j.cej.2010.12.088

Osorio-Tobón, J. F. (2020). Recent advances and comparisons of conventional and alternative extraction techniques of phenolic compounds. *Journal of Food Science and Technology*, 57(12), 4299–4315. https://doi.org/10.1007/s13197-020-04433-2

Parveen, R., Khare, N., Singh, S., & Johri, P. (2023). An in silico outlook for the detection and surveillance of evolving and persistent plant pathogens. In *Detection, diagnosis and management of soil-borne phytopathogens* (pp. 17–40). Springer Nature Singapore.

Pavez, P., Honores, J., Millán, D., & Isaacs, M. (2018). UN sustainable development goals: How can sustainable/green chemistry contribute? *Current Opinion in Green and Sustainable Chemistry*, 13, 154–157. https://doi.org/10.1016/j.cogsc.2018.06.013

Prado, J. M., Vardanega, R., Debien, I. C. N., Meireles, M. A. A., Gerschenson, L. N., Sowbhagya, H. B., Chemat, S., Prado, J. M., Vardanega, R., Debien, I. C. N., Meireles, M. A. A., Gerschenson, L. N., Sowbhagya, H. B., & Chemat, S. (2021). Conventional extraction. In *Food waste recovery* (pp. 109–127). Elsevier. https://doi.org/10.1016/B978-0-12-820563-1.00015-9

Quiroz, J. Q., Torres, A. C., Ramirez, L. M., Garcia, M. S., Gomez, G. C., & Rojas, J. (2019). Optimization of the microwave-assisted extraction process of bioactive compounds from annatto seeds (Bixa orellana L.). *Antioxidants*, 8(2), 37. https://doi.org/10.3390/antiox8020037

Raventós, M., Duarte, S., & Alarcón, R. (2002). Application and possibilities of supercritical CO2 extraction in food processing industry: An overview. *Food Science and Technology International*, 8(5), 269–284. https://doi.org/10.1106/108201302029451

Santos, D. T., Vardanega, R., & de Almeida, M. A. (2014). Intensification of bioactive compounds extraction from medicinal plants using ultrasonic irradiation. *Pharmacognosy Reviews*, 8(16), 88. https://doi.org/10.4103/0973-7847.134231

Santos, T. R. J., & Santana, L. C. L. de A. (2022). Conventional and emerging techniques for extraction of bioactive compounds from fruit waste. *Brazilian Journal of Food Technology*, 25. https://doi.org/10.1590/1981-6723.13021

Siddique, M. A. B., Tzima, K., Rai, D. K., & Brunton, N. (2020). Conventional extraction techniques for bioactive compounds from herbs and spices. In *Herbs, spices and medicinal plants* (pp. 69–93). Wiley. https://doi.org/10.1002/9781119036685

Singh, S., Datta, R., Johri, P., & Trivedi, M. (Eds.). (2022). *Phytopharmaceuticals and biotechnology of herbal plants*. CRC Press.

Singh, A., Sengar, R. S., Singh, A. K., Singh, S., Kumar, A., & Ahmad, K. (2019). Impact of genetic engineering on agriculture food crops. *Biotech Today: An International Journal of Biological Sciences*, 9(1), 17–24.

Soquetta, M. B., Terra, L. de M., & Bastos, C. P. (2018). Green technologies for the extraction of bioactive compounds in fruits and vegetables. *CyTA - Journal of Food*, 16(1), 400–412. https://doi.org/10.1080/19476337.2017.1411978

Sportiello, L., Favati, F., Condelli, N., di Cairano, M., Caruso, M. C., Simonato, B., Tolve, R., & Galgano, F. (2023). Hydrophobic deep eutectic solvents in the food sector: Focus on their use for the extraction of bioactive compounds. *Food Chemistry*, 405, 134703. https://doi.org/10.1016/j.foodchem.2022.134703

Trivedi, M., Singh,, S., Johri, P., & Sanchez, P. L. (Eds.). (2024). *Nutraceuticals inspiring the contemporary therapy for lifestyle diseases*. CRC Press.

Tungare, K., Shahu, R., Zambare, V., Agarwal, P., Jobby, R., Nisar, N., Alabdallah, N. M., Al-Saeed, F. A., Johri, P., Singh, S., Saeed, M., & Jha, P. (2022). Toxicity mitigation of textile dye reactive blue 4 by hairy roots of helianthus annuus and testing its effect in in vivo model systems. *BioMed Research International*, 2022. https://doi.org/10.1155/2022/195893

Uwineza, P. A., & Waśkiewicz, A. (2020). Recent advances in supercritical fluid extraction of natural bioactive compounds from natural plant materials. *Molecules*, 25(17), 3847. https://doi.org/10.3390/molecules25173847

Vardanega, R., Santos, D. T., & de Almeida, M. A. (2014). Intensification of bioactive compounds extraction from medicinal plants using ultrasonic irradiation. In *Pharmacognosy reviews* (Vol. 8, Issue 16, pp. 88–95). Medknow Publications. https://doi.org/10.4103/0973-7847.134231

Verma, N., Sao, P., Srivastava, A., & Singh, S. (2021). Physiological and molecular responses to drought, submergence and excessive watering in plants. In *Harsh environment and plant resilience: Molecular and functional aspects* (pp. 305–321). Springer.

Yusoff, I. M., Mat Taher, Z., Rahmat, Z., & Chua, L. S. (2022). A review of ultrasound-assisted extraction for plant bioactive compounds: Phenolics, flavonoids, thymols, saponins and proteins. *Food Research International*, 157, 111268. https://doi.org/10.1016/j.foodres.2022.111268

Yuting. (2014). Effects of ultrasound and/or heating on the extraction of pectin from grapefruit peel. *Journal of Food Engineering*, 126, 72–81. https://doi.org/10.1016/j.jfoodeng.2013.11.004

5 Green Extraction of Bioactive Compounds from Plant Biomass and Their Application in Meat as Natural Antioxidants

*Ritesh Pandey, Parul Johri, Rajesh K. Tiwari, and Mala Trivedi**

5.1 INTRODUCTION

It is well known to everyone that free radicals are molecular species that may exist independently and have an unpaired electron in their atomic orbits. For instance, a superoxide molecule can react with a fatty acid and lose one of its electrons. After this, the fatty acid transforms into a free radical that can interact with adjacent fatty acids. As a result of this series of events, cell membrane fluidity and permeability shift, membrane-bound proteins become less active, and receptor proteins encounter structural alterations that either affect or abolish their function (Alkadi,2018). Free radicals are a threat to macromolecules such as DNA, RNA, proteins, and fatty acids (Augustyniak, 2010). Cells can get damaged as a result of chain reactions that free radicals might start. Several processes in our body produce free radicals. Although the generation of free radicals is inevitable, human bodies have evolved by establishing and maintaining defensive systems that lessen their influence. One major part of the defensive system is antioxidants. Antioxidants are molecules that can block free radicals from stealing electrons and can help combat oxidative stress. Aside from these physiological advantages, antioxidants are compounds known to prevent oxidation. Antioxidants have become an integral class of food additives due to their unique ability to enhance the shelf life of food items without compromising their sensory or nutritional properties. Proteins, fats, vitamins, and minerals are abundant in meat (Uzombah, T. A., 2022). However, the presence of unsaturated fat in membrane phospholipids in meat leads fat to oxidise during processing and storage. Due to fat oxidation and the development of off-flavours, the quality of meat and meat products deteriorates. Oxidation of lipids, degradation of proteins, and

* Corresponding author

DOI: 10.1201/9781003392910-5

loss of other valuable molecules are the reasons for the meat spoilage process. The breakdown of proteins and lipids can lead to the creation of new chemicals that alter the flavour, tenderness, juiciness, odour, and texture of the meat. Lipid oxidation accelerates the conversion of oxymyoglobin (bright red colour) to metmyoglobin (brown discolouration), affecting the physical appearance of meat and meat products (Lu and Pharm-Mondala, 2018). Meat is also prone to deterioration and microbial development during slaughter and post-slaughter handling. As a result, meat providers employ a variety of food additives to increase the shelf life of meat and meat products. Synthetic antioxidants such as BHA (butylated hydroxyanisole), BHT (butylated hydroxytoluene), propyl gallate, and TBHQ (tert-butylhydroquinone) are used (Xu et al., 2021). However, synthetic phenolic antioxidants can cause endocrine disruption, cytotoxicity, carcinogenicity, and oxidative stress when used improperly or in excess. Therefore, the meat industry's attention is moving to antioxidants that are new, efficient, cost-effective, environmentally friendly, and natural alternatives to potentially dangerous synthetic antioxidants. Meat industries have been working hard to discover new, effective, and edible antioxidants and antibacterial substances derived from natural sources. Natural ingredients (plant extracts, herbs, spices) have recently sparked fresh interest in food preservation. Because of the antioxidant, bacteriostatic, or bactericidal qualities of polyphenolic compounds and essential oils, these additives have been added to meals since ancient times as flavourings/seasonings, traditional medicine, and food preservatives (Karre et al., 2013). Natural antioxidants are formed by natural molecular formations in plants, animals, mushrooms, and microorganisms such as algae and bacteria, and are thus derived directly from organic sources such as fruits, vegetables, cereals, and meat. In addition to their preservation function, these substances are known to provide health benefits when consumed. These herbs, spices, essential oils, and plant extracts have varied disease resistance properties and are thus frequently utilised as potential green alternatives to food additives, preservatives, and nutritional supplements. Natural antioxidants can be extracted and used in the meat, are safe to digest, and are considerably more affordable than their synthetic counterparts. Fruits and other plant materials are an excellent alternative to conventional antioxidants due to their high phenolic component concentration (Santos-Sánchez et al., 2017). Plum, grape seed extract, cranberry, pomegranate, bearberry, pine bark extract, rosemary, oregano, and other spices act as antioxidants in meat and poultry products. Pomegranate, pine bark extract, cinnamon, and cloves have been shown to have higher antioxidant levels than certain synthetic alternatives. Plum products, grape seed extract, pine bark extract, rosemary, and some spices have all been found to have an effect on the colour of final meat or poultry products (Ahmad et al., 2013). According to the research, the use of antioxidants as food preservatives is dependent on the type or class of food being preserved as well as the overall price consideration. For example, using synthetic antioxidants to preserve foods with high rancidity levels is preferable because they are more potent and thus have higher performance levels, whereas natural antioxidants with lower potency and performance levels can suffice for hydrogenated oils with lower rancidity levels (Uzombah, 2022). However, the effectiveness of antioxidants can be measured by the extent to which their use promotes food safety, consumer health, disabling occurrences of degenerative diseases in vivo, and preventing or minimising lipid peroxidation, which produces toxic compounds that enhance deterioration of flavour, colour, texture, and nutritional values, which leads to overall depreciation of food quality. Natural antioxidants have been shown to be equally effective as synthetic antioxidants at stabilising food items and contributing to meat products having a longer shelf life.

5.2 SYNTHETIC ANTIOXIDANTS: THE CONSEQUENCES

Synthetic antioxidants are created artificially or made by utilising a variety of processes. These synthetic polyphenolic chemicals are primarily responsible for capturing free radicals and halting chain reactions. Polyphenolic derivatives often have many hydroxyl or methoxy groups (Xu et al., 2021). The only heterocyclic, N-containing molecule documented to be utilised as an antioxidant in food, particularly animal feed, is ethoxy quinine (Xu et al., 2021). The majority of documented synthetic phenolic antioxidants are p-substituted, whereas the majority of natural phenolic compounds are o-substituted (Xu et al., 2021). Because of their decreased toxicity, p-substituted compounds are chosen. To increase solubility in fats and oils and to lessen toxicity, synthetic phenolic antioxidants are always replaced with alkyl groups. Synthetic phenolic antioxidants (SPAs) are commonly utilised to delay oxidative processes and extend product shelf life in a variety of industrial and commercial items (Lu and Mabury, 2020). Synthetic antioxidants are employed in the food industry to prevent lipid peroxidation and oxidation of food ingredients. Tert-butyl hydroxyanisole (BHA), tert-butyl hydroxytoluene (BHT), tert-butyl hydroquinone (TBHQ), and propyl gallate (PG) are the most popular synthetic phenolic antioxidants used to prevent oxidative rancidity in edible oils or lipid-based foods (Shahidi et al., 1992). The safety and toxicity of synthetic antioxidants, on the other hand, have prompted serious concerns since such materials may cause liver swelling, affect liver enzyme activity, and be carcinogenic (Sood and Venkatesh 2011). BHA and BHT may potentially be harmful, according to reports (Table 5.1). Furthermore, because synthetic antioxidants have low solubility and low antioxidant activity, their uses have been severely limited (Sood and Venkatesh 2011).

5.3 PLANT-BASED NATURAL ANTIOXIDANTS

Due to the high antioxidant potential of these chemicals in the meat matrix, numerous plant extracts are increasingly being investigated in meat products as possible natural antioxidants for enhancing oxidative stability. The antioxidant attributes of these plant extracts are due to flavonoids (flavanol, flavones, anthocyanins), phenolic acids (hydroxybenzoic, hydroxycinnamic acids), diterpenes, tannins (hydrolysable and condensed tannins), stilbenes, coumarins, lignans, quinones, curcuminoids, and others including phenolic alkaloids, phenolic glycosides, phenolic terpenoids, and essential oils) (Awad et al., 2021). Tea phenolics, such as epigallocatechin gallate (EGCG), exhibit antioxidant potential equivalent to alpha-tocopherols or vitamin E in terms of electron reduction (550 mV) (Awad et al., 2021). Plant-derived

TABLE 5.1
Classification of Synthetic Antioxidants

Synthetic Antioxidants	Properties	Limit in Foods
BHA	A lipid-soluble, monophenolic antioxidant that works better against lipid oxidation in animal fat than it does in vegetable oil.	<200 mg/kg*
BHT	A monophenolic fat-soluble antioxidant. More durable at high temperatures than BHA, but both function synergistically.	<100 mg/kg*
TBHQ	A powerful diphenolic antioxidant. It is utilised as a preservative in foods such as unsaturated vegetable oils and various edible animal fats. Even in the presence of iron, it causes no discolouration and has no effect on the flavour or odour of the item to which it is added.	120 mg/kg*
Propyl gallate	An ester produced by the condensation of gallic acid with propanol. It functions as an antioxidant and is used as a food additive to preserve oils and fats in food items.	<200 mg/kg*
EDTA	A water-soluble antioxidant that is added to foods. It binds with trace minerals including copper, iron, and nickel that may be present in the foodstuff. These minerals, if not inactivated, can cause discolouration, rancidity, and textural degradation. When used as an antioxidant, EDTA inhibits oxygen from generating colour changes and rancidity.	75 ppm*
Ethoxyquin	Primarily utilised as an antioxidant to prevent carotenoid oxidation in animal feeds, vegetables, and fruits during storage.	<100 ppm*

* 21 CFR Part 172

antioxidants are found in abundance in fruits, vegetables, spices, herbs, cereals, grains, and seeds. It is also important to keep in mind that the by-products formed during the processing of these foods can be a source of antioxidants, whose usage would allow their reutilisation while avoiding the significant environmental and economic losses connected with this waste (Table 5.2).

5.3.1 Phenols

Phenols are secondary metabolites of plant metabolism produced from the aromatic amino acid phenylalanine via the acetic acid (simple phenols) and shikimic acid (phenylpropanoids) pathways. Plants' colour, flavour, and astringency are all influenced by phenols (Manessis et al., 2020). Phenolic compounds function as hydrogen donors, quenching free radicals, or transferring single electrons to decrease chemical compounds with oxidative activity (Wade, 2018). Aside from their antioxidant capacity, phenolic compounds play an important functional and biological role in plant physiology, acting as structural polymers, UV radiation screeners, pollination and soil bacterium attractants, non-specific defence mechanisms, phytoalexins, and

TABLE 5.2
Classification of Phenolic Compounds Based on Their Carbon Chains

Class	Structure	Plant Sources
Phenols and benzoquinones	C6	*Primula obconica*, sorghum, berries, fruit wines, olive oil.
Phenolic acids	C6-C1	Blueberries, blackberries, persimmon, apple juice, cider, cherry laurels, canola meal, oranges, rye.
Acetophenones and phenylacetic acids	C6-C2	Root bark of *Derris indica*, Ageratina pichinchensis, spruce, pine, pisum, nicotiana, phaseolus, triticum, trapoleum, lycopersicon, avena, balsamic vinegar, cocoa powder.
Hydroxycinnamic acids, phenylpropenes, coumarins-isocoumarins, and chromones	C6-C3	Berries, pomes, herbs, seeds, cereal grains, leafy greens, asparagus, cinnamon, cloves, and potatoes
Naphthoquinones	C6-C4	Oranges, Plumbaginaceae, Droseraceae, Ebenaceae.
Xanthones	C6-C1-C6	*Gentianna ottonis*, Guttiferae, Moraceae, Clusiaceae, and Polygalaceae.
Stilbenes and anthraquinones	C6-C2-C6	Grapes, pine, peanuts, sorghum, rheum, Rubiaceae, black pepper.
Flavonoids (flavones, flavonols, flavanones, flavan-3-ols, isoflavones, anthocyanidin compounds)	C6-C3-C6	Celery, parsley, red prickly pears, olives, acerola, lychees, avocados, green and black tea, cherries, raspberries, strawberries, grapes, and red wine.

potentially signaling compounds for systemic acquired resistance (Manessis et al., 2020).

5.3.2 Tannins

Tannins are astringent, polyphenolic molecules that are formed by the polymerisation of phenylpropanoid compounds (Swanson 2003). Because of their capacity to bind with at least two protein molecules and form insoluble, cross-linked tannin-protein complexes, tannins have traditionally been utilised for protein precipitation. Tannins are classified into (i) condensed tannins (proanthocyanidins), which are polymers of catechin, epicatechin, prodelphinidins, profisetinidins, and prorobinetidins, and (ii) hydrolysable tannins, which can be hydrolysed by weak acids or bases, with the latter being mixtures of carbohydrates with gallic and ellagic acid (gallotannins and ellagitannins) (Shirmohammadli et al., 2018). Proanthocyanidins transfer hydrogen and electrons (primary antioxidant activity), chelate ferrum, and inhibit cyclooxygenase activity (secondary antioxidant action) (Shirmohammadli et al., 2018). Tannins have favourable effects on lipid metabolism and immune response control, lower blood pressure, and promote anticarcinogenic and antimutagenic action when consumed in tiny amounts. However, because tannins have the tendency to flocculate proteins, diets high in tannins have reduced nutritional value, hence reducing their digestibility (Chung et al., 1998).

5.3.3 Flavonoids

Flavonoids and isoflavonoids are present in plants and are derived from the aromatic amino acids, phenylalanine and tyrosine, as well as malonate. The flavan nucleus, which consists of three rings of carbon atoms, is the fundamental structure of flavonoids (C6-C3-C6). Flavonoids are classified according to their degree of oxidation and the pattern of carbon atom ring substitution (flavones, flavanones, isoflavones, flavonols, flavanonols, flavan-3-ols, anthocyanidins, biflavones, chalcones, aurones, and coumarins) (Panche et al., 2016). The two main ways that flavonoids work as antioxidants are (i) by preventing the enzymatic processes that lead to the creation of free radicals and (ii) by scavenging them (Panche et al., 2016). Flavonoids have demonstrated a strong capacity as antioxidants in in vitro studies and are thought to lower the chance of developing cardiovascular illnesses, hypertension, Alzheimer's disease, and several forms of cancer. These advantages may still not be fully utilised since flavonoids have a limited capacity for enteric absorption in humans.

5.3.4 Stilbenes

Stilbenes have a structure that is 1,2-diphenylethylene (Manessis et al., 2020). Its primary sources in human diets include grapes and wine, peanuts, and some varieties of berries. They are divided into monomeric and oligomeric stilbenes. By regulating redox status, cell proliferation, mitochondrial activity, and the generation of inflammatory indicators, stilbenes have a protective impact against chronic illnesses such as cancer, and cardiovascular and neurodegenerative disorders (Manessis et al., 2020). With its positive impact on heart health, resveratrol, which is found in red wine, is the stilbene that has been researched the most. Resveratrol's bioavailability is decreased through metabolic processes and the influence of the intestinal microflora, although its metabolites are still linked to a number of health advantages.

5.3.5 Vitamin E

Tocopherols and tocotrienols are compounds that belong to the vitamin E group. These are phenolic substances produced by plants (e.g., cabbage, oregano, and paprika) and are important components of both human and animal diets. Compounds from the vitamin E group contain a chroman head, consisting of a heterocyclic and phenolic ring coupled with a phytyl chain (Kamal-Eldin and Appelqvist, 1996). Tocopherols occur in four homologues, α, β, γ, and δ, discriminated by the number of methyl substituents of the saturated phytyl chain and the substitution patterns of the phenolic ring. The three trans-double bonds in their hydrocarbon tails distinguish tocotrienols from tocopherols. Tocopherols' antioxidant action is based on their ability to donate hydrogen to lipid free radicals, which has been established both in vitro and in vivo. The strongest antioxidant effect was seen in both types of investigations for α-tocopherol, followed by the homologues β, γ, and δ. The highest antioxidant capacity of α-tocopherol, in vivo, is exhibited due to its preferential retention and distribution in animal species (Manessis et al., 2020). By scavenging peroxy-radicals and quenching or reacting with singlet oxygen $^1O_2$ or ROS, vitamin E compounds safeguard membrane lipids. Tocopherols promote protein kinase C and protein phosphatase 2 activity, as well as gene expression and cell proliferation, whereas tocotrienols are neuroprotective, anti-cancer, and cholesterol-lowering.

5.3.6 Carotenoids

Carotenoids are lipophilic, pigmented compounds that are produced naturally by microbes and plants (e.g., acerola, cabbage, carrots, and paprika) but not by vertebrates.[16] Carotenoids have a polyisoprenoid structure made up of a carbon chain with conjugated carbon bonds and a near-bilateral symmetry around the central double bond. Cyclic end-groups are connected to the central chain and can be replaced with oxygen-containing functional groups. Depending upon the type of substitution, carotenoids are further divided into carotenes (such as β-carotene and lycopene), which only include carbon and hydrogen atoms, and oxycarotenoids (like xanthophylls), which contain at least one oxygen atom. Lycopene (with eleven conjugated double bonds and two acyclic end-groups) and β-carotene (with eleven conjugated double bonds and two cyclohexene type groups) are the most researched carotenoids (Manessis et al., 2020). Carotenoids' ability to physically quench singlet oxygen ($^1O_2$) and peroxyl radicals indicates how effective they are as antioxidants (Manessis et al., 2020). This ability is based on the quantity of conjugated double bonds present in the compound. Besides this, carotenoids have provitamin A action and are linked to the control of lipoxygenases and the expression of the connexin 43 gene. Carotenoids have been found to lower the risk of several malignancies (e.g., lung cancer) in the general population and to help prevent cataracts and macular degeneration (Manessis et al., 2020).

5.4 GREEN EXTRACTION OF BIOACTIVE COMPOUNDS FROM PLANT BIOMASS

The efficiency of plant extracts is primarily dependent on the extraction process, which includes the initial material, the extraction method, the solvent, and the processing parameters (temperature and extraction time) that affect their antioxidant activity. If we talk about the extraction of polyphenols from vegetable materials, many procedures are required, including feed preparation, extraction, purification, and drying. Feed preparation is a critical step in the extraction process. The feed is prepared before being fed into the extractor to eliminate undesired elements and to maximise extraction efficiency (Awad et al., 2021). The unwanted components are eliminated depending on the portion of the sample to be retrieved. For example, before a fruit rind extraction, the seeds and pulp must be removed. This allows the extractor's available capacity to be utilised more efficiently. Another advantage is that the solvent-to-feed ratio may be increased while using the same amount of solvent. To improve the surface area for mass transmission, the feed must sometimes be ground. It is advised that the food be cleaned before being mixed with the solvent to eliminate foreign contaminants such as dirt particles, tiny stones, and so on (Panja, 2017). Prior to extraction, the feed may be frozen in some situations. Typically, the material must be softened. The most critical stage in isolating polyphenols from vegetable components is extraction. The method and solvent are critical aspects in maximising selective recovery while minimising unwanted compounds (Manessis et al., 2020; Panja, 2017). Traditional techniques involve basic equipment, and the quality, amount, and duration of extraction are frequently questioned. These techniques have been refined over time to boost extraction efficiency and decrease extraction time. To extract BACs from plant material, traditional extraction procedures are still routinely employed. Nevertheless, the harsh processing conditions utilised may have harmed the quality and biological activity of the compounds produced, prompting researchers to look for other approaches that are more in line with the notion of "green" technology (Panja, 2017). Eco-innovative technologies emerge in this regard with the goal of developing techniques characterised by the synthesis of safe products (minimising or eliminating the use of non-toxic solvents), faster extraction rates, more effective energy use, increase in mass and heat transfer, decrease in equipment size and number of processing steps, and preservation of the natural environment and resources (Pateiro et al., 2021). Some of the most remarkable techniques include accelerated solvent extraction (ASE), enzyme-assisted extraction (EAE), high hydrostatic pressure extraction (HHPE), infrared-assisted extraction (IAE), microwave-assisted extraction (MAE), pulsed electric field extraction (PEF), subcritical fluid extraction (SFE), and ultrasound-assisted extraction (UAE), or combinations of some of these (Panja, 2017). Thus, the recent trend is to substitute toxic organic solvents (such as acetone, hexane, benzene, methanol, and chloroform) with potentially safer ones. For extraction processes, green solvents including water, aqueous ethanol solutions, natural deep eutectic solvents, and supercritical fluids are favoured. The utilisation of these solvents for extracting phenolic compounds from plants is outlined in Table 5.3.

After being extracted, BACs are stabilised using a variety of methods to preserve both their quality and bioactive characteristics during storage. In order to best protect delicate components, freeze-drying or encapsulation is optimised for the creation of powders and emulsions from extracts (Rezvankhah, 2019). In addition to protecting them from hazardous conditions, encapsulation techniques like spray-drying can help maintain the nutritional content, bioavailability, solubility, and effectiveness of BACs by masking off-flavors and odours, regulating their release and handling in foodstuffs (Pateiro et al., 2021).

TABLE 5.3
Extraction of Phenolic Compounds from Plant Sources Utilising Green Solvents and Technology

Extraction Technique	Plant Source	BACs	Extraction Conditions	Reference
SFE	Lavender (Lavandula *angustifolia Mill.*)	Oxygenated monoterpenes, coumarin, herniarin	40–60°C, 100–300 bar, 1–3 kg/h CO_2, 90 min	(Busatta et al., 2017)
	Marjoram (*Origanummajorana*)	*cis*-Sabinene hydrate, linalyl acetate, terpinen-4-ol,	40°C and 200 atm, 180 min	
	Oregano (*Origanum vulgare*)	α-terpineol, cis-Sabinene hydrate, thymol, carvacrol, terpinen-4-ol, geraniol		
SWE	Ginger (Zingiber officinale)	Gingerol	130–140° C, 10–40 min, 2 bar	(Yulianto et al., 2017)
	Turmeric rhizomes (*Curcuma longa* L)	Curcumin	120–160°C, 6–22 min, 10 bar	(Kiamahalleh et al., 2016)
	Wild garlic (*Allium ursinum* L.)	Phenols and flavonoids	120–200°C, 10–30 min, HCL: 0–1.5%	(Tomšik et al., 2017)
NADES	Onion, olive, tomato and pear industrial by-products	Gallic acid, 3-hydroxytyrosol, tyrosol, catechin, caffeic acid, rutin, coumaric acid, trans-ferulic acid, ooleuropein, cinnamic acid, quercetin, luteolin, naringenin, apigenin	LaGlc UAE (0–60 min, 40°C, 200 W)	(Pateiro et al., 2021)
	Rosemary (*Rosmarinus officinalis* L.)	*Phenolic acids*: rosmarinic acid and ferulic acid *Flavonoids*: 7-methylrosmanol, rutin, naringin	ChCl: ProOH UAE (120 min, 40°C)	(Pateiro et al., 2021)

ChCl: choline chloride; Glc: glucose; La: lactic acid; Pro: L-proline; ProOH: 1,2-propanediol

TABLE 5.4
Antioxidants from Natural Sources with Possible Uses in Meat and Meat Products

Natural Source	Active Compound	Application
Oregano essential oil	Thymol, para-cymene, gamma-terpinene, carvacrol	Raw and cooked porcine/bovine meat
Sage essential oil	Eucalyptol, camphor, alpha-pinene	Raw and cooked porcine/bovine meat
Curry and mint leaves	Phenolics	Raw ground pork meat
Rosemary and oregano leaves extracts	Phenolics, rosmarinic acid, carnosic acid, carnosol	Raw pork batter
Grape seeds and peels extract	Phenolics	Raw and cooked ground chicken
Sea buckthorn (berry residues)	Polyphenols	Raw and cooked mechanically deboned chicken and turkey
Red grape pomace extract	Total polyphenolics, total anthocyanins	Pork burger
Defatted canola meal	Sinapic, ferulic, para-hydrobenzoic acid	Chicken meat homogenate
Borage leaves extract	Phenolics	Dry fermented sausage enriched with omega-3 polyunsaturated fatty acids
Pomegranate peel powder and extract	Anthocyanins	Chicken patties

5.4.1 Application of Plant-Derived Natural Antioxidants in Meat

The traditional approach of integrating natural antioxidants into meat products is more efficient than spraying them on their surface. Direct contact of antioxidant compounds with meat is preferred for efficient action; however, exposure to external factors such as bright light, oxygen, high pH, and temperature may cause their antioxidant potential to be deactivated. Innovative methods, such as active packaging and edible coatings/films, take advantage of antioxidants' benefits while minimising their negative side effects. These methods must be followed because meat and meat products are vulnerable to oxidation processes due to their composition, especially those that include the breakdown of lipids. Along with this, antioxidants are crucial to maintaining the quality of these items. Using natural antioxidants, especially those containing polyphenols, may be a reliable approach due to their effectiveness in scavenging free radicals and the added health advantages they provide to meat products (Table 5.4).

5.5 CONCLUSION

Natural extracts are derived from numerous plant sources and are rich in many bioactive components containing powerful antioxidants. Several of these plant extracts display equivalent or even superior antioxidant activities compared to routinely used synthetic antioxidants in the meat industry. Many in vitro studies have proven the favourable effects of natural antioxidants on lipid and protein oxidation, the lengthening of shelf-life, the antioxidant profiles of functionalised meat products, and the corresponding potential health advantages. However, there are several challenges regarding the application of natural antioxidants, such as (i) the increased expense of preparing and storing plants/fruits and the antioxidants they produce, and that (ii) meat's organoleptic features, including flavour, colour, and aroma, may change as a result of the higher levels needed to exhibit effective antioxidant action, which might reduce marketability. Apart from this, time, storage, and treatment-dependent variations in antioxidant profiles restrict their universal and unquestionable usage. It has not been comprehensively proven that adding antioxidants to meat has health benefits. As a result, the meat industry is primarily driven by the advantages of antioxidants with regard to shelf-life, safety, and quality assurance of meat and meat products, and only secondarily by advertising the possible health benefits for customers. The use of different combinations of extraction technologies has shown to have a synergistic or auxiliary effect. There is a need for efficient identification, segregation, and purification of the bioactive components in plant extracts for their optimal application as natural antioxidants in the meat sector.

REFERENCES

Ahmad, S. R., Gokulakrishnan, P., Giriprasad, R., & Yatoo, M. A. (2013). Fruit-based natural antioxidants in meat and meat products: A review. *Critical Reviews in Food Science and Nutrition*, *55*(11), 1503–1513. https://doi.org/10.1080/10408398.2012.701674

Alkadi, H. (2018). A review on free radicals and antioxidants. *Infectious Disorders – Drug Targets*, *18*. https://doi.org/10.2174/1871526518666180628124323

Augustyniak, A., Bartosz, G., Čipak, A., Gunars Duburs, Lubica Horakova, Wojciech Łuczaj, Majekova, M., Odysseos, A. D., Račková, L., Elżbieta Skrzydlewska, Stefek, M., Strosova, M., G. Tirzitis, Maggi, F., Viskupicova, J., Vraka, P. S., & Neven Zarkovic. (2010). Natural and synthetic antioxidants: An updated overview. *Free Radical Research*, *44*(10), 1216–1262. https://doi.org/10.3109/10715762.2010.508495

Awad, A. M., Kumar, P., Ismail-Fitry, M. R., Jusoh, S., Ab Aziz, M. F., & Sazili, A. Q. (2021). Green extraction of bioactive compounds from plant biomass and their application in meat as natural antioxidant. *Antioxidants*, *10*(9), 1465. https://doi.org/10.3390/antiox10091465

Busatta, C., Barbosa, J., Cardoso, R. I., Paroul, N., Rodrigues, M., Oliveira, D. de, Oliveira, J. V. de, & Cansian, R. L. (2017). Chemical profiles of essential oils of marjoram (Origanum majorana) and oregano (Origanum vulgare) obtained by hydrodistillation and supercritical CO2. *Journal of Essential Oil Research*, *29*(5), 367–374. https://doi.org/10.1080/10412905.2017.1340197

Chung, K.-T., Wong, T. Y., Wei, C.-I., Huang, Y.-W., & Lin, Y. (1998). Tannins and human health: A review. *Critical Reviews in Food Science and Nutrition*, *38*(6), 421–464. https://doi.org/10.1080/10408699891274273

Kamal-Eldin, A., & Appelqvist, L.-Å. (1996). The chemistry and antioxidant properties of tocopherols and tocotrienols. *Lipids*, *31*(7), 671–701. https://doi.org/10.1007/bf02522884

Karre, L., Lopez, K., & Getty, K. J. K. (2013). Natural antioxidants in meat and poultry products. *Meat Science*, *94*(2), 220–227. https://doi.org/10.1016/j.meatsci.2013.01.007

Kiamahalleh, V., Najafpour-Darzi, M., Rahimnejad, G., Moghadamnia, M., A. A., & Valizadeh, K. M. (2016). High performance curcumin subcritical water extraction from turmeric (Curcuma longa L.). *Journal of Chromatography B*, *1022*, 191–198. https://doi.org/10.1016/j.jchromb.2016.04.021

Liu, R., & Mabury, S. A. (2020). Synthetic phenolic antioxidants: A review of environmental occurrence, fate, human exposure, and toxicity. *Environmental Science & Technology*, *54*(19), 11706–11719. https://doi.org/10.1021/acs.est.0c05077

Lu, F. S. H., & Pham-Mondala, A. (2018). *Trends in synthetic and natural antioxidants for shelf life extension of meat and meat products* (n.d.). Www.aocs.org. https://www.aocs.org/stay-informed/inform-magazine/featured-articles/trends-in-synthetic-and-natural-antioxidants-for-shelf-life-extension-of-meat-and-meat-products-november/december-2018?SSO=True

Manessis, G., Kalogianni, A. I., Lazou, T., Moschovas, M., Bossis, I., & Gelasakis, A. I. (2020). Plant-derived natural antioxidants in meat and meat products. *Antioxidants*, *9*(12), 1215. https://doi.org/10.3390/antiox9121215

Panche, A. N., Diwan, A. D., & Chandra, S. R. (2016). Flavonoids: An overview. *Journal of Nutritional Science*, *5*(e47). https://doi.org/10.1017/jns.2016.41

Panja, P. (2017). *Green extraction methods of food polyphenols from vegetable materials.* https://www.semanticscholar.org/paper/Green-extraction-methods-of-food-polyphenols-from-Panja/ab3661a39904c14e7c10aaf89ff1a510070b0341

Pateiro, M., Gómez-Salazar, J. A., Jaime-Patlán, M., Sosa-Morales, M. E., & Lorenzo, J. M. (2021). Plant extracts obtained with green solvents as natural antioxidants in fresh meat products. *Antioxidants*, *10*(2), 181. https://doi.org/10.3390/antiox10020181

Rezvankhah, A., Emam-Djomeh, Z., & Askari, G. (2019). Encapsulation and delivery of bioactive compounds using spray and freeze-drying techniques: A review. *Drying Technology*, *38*(1-2), 235–258. https://doi.org/10.1080/07373937.2019.1653906

Santos-Sánchez, N. F., Salas-Coronado, R., Valadez-Blanco, R., Hernández-Carlos, B., & Guadarrama-Mendoza, P. C. (2017). Natural antioxidant extracts as food preservatives. *Acta Scientiarum Polonorum Technologia Alimentaria*, *16*(4), 361–370. https://doi.org/10.17306/j.afs.0530

Shahidi, F., Janitha, P. K., & Wanasundara, P. D. (1992). Phenolic antioxidants. *Critical Reviews in Food Science and Nutrition*, *32*(1), 67–103. https://doi.org/10.1080/10408399209527581

Shirmohammadli, Y., Efhamisisi, D., & Pizzi, A. (2018). Tannins as a sustainable raw material for green chemistry: A review. *Industrial Crops and Products*, *126*, 316–332. https://doi.org/10.1016/j.indcrop.2018.10.034

Sood, D., & Venkatesh, R. (2011). *A review of the physiological implications of antioxidants in food.* Digital WPI. Worcester Polytechnic Institute. https://digital.wpi.edu/concern/student_works/z316q182d?locale=en

Swanson, B. G. (2003, January 1). *Tannins and polyphenols.* Elsevier eBooks. https://doi.org/10.1016/b0-12-227055-x/01178-0

Tomšik, A., Pavlić, B., Vladić, J., Cindrić, M., Jovanov, P., Sakač, M., Mandić, A., & Vidović, S. (2017, October 1). Subcritical water extraction of wild garlic (Allium ursinum L.) and process optimization by response surface methodology. *The Journal of Supercritical Fluids/the Journal of Supercritical Fluids.* https://doi.org/10.1016/j.supflu.2017.05.012

Uzombah, T. A. (2022). The implications of replacing synthetic antioxidants with natural ones in the food systems. In *www.intechopen.com.* IntechOpen. https://www.intechopen.com/chapters/81679

Wade, L. G. (2018). phenol | Definition, structure, uses, & facts. In *Encyclopædia Britannica.* https://www.britannica.com/science/phenol

Xu, X., Liu, A., Hu, S., Ares, I., Martínez-Larrañaga, M.-R., Wang, X., Martínez, M., Anadón, A., & Martínez, M.-A. (2021). Synthetic phenolic antioxidants: Metabolism, hazards and mechanism of action. *Food Chemistry*, *353*, 129488. https://doi.org/10.1016/j.foodchem.2021.129488

Yulianto, M., Kusumo, P., & Hartati, I. (2017, January 1). Subcritical water extraction of gingerol from zingiber officinale. *Rasayan Journal of Chemistry/Rasayan Journal of Chemistry*. https://doi.org/10.7324/rjc.2017.1031619

6 Endophytes and Plant Metabolites
Inter-Relation and Ecological Roles

Zaryab Shafi, Krishna Nanda Dhal, Amrita Gupta, Mala Trivedi, and Pramod Kumar Sahu

6.1 INTRODUCTION

In the contemporary landscape of global challenges, the search for novel and functional bioactive compounds has intensified, aiming to address a multitude of pressing issues (Verma et al., 2021). These include the rise of drug-resistant bacteria, the emergence of life-threatening viruses, the increasing incidence of fungal infections worldwide, and the persistent struggle to combat food scarcity in specific regions. In this context, endophytes have emerged as a valuable source of innovative natural compounds with far-reaching applications in business, agriculture, and medicine (Gupta et al., 2022).

Endophytes, often residing within the living tissues of host plants without causing apparent harm, form a fascinating component of the plant's microbiome population (Jabborova et al., 2022). What makes them particularly intriguing is their ability to produce a diverse array of metabolites under culture conditions, some of which closely resemble those generated by their respective host plants. These metabolites, such as taxol, camptothecin, vinblastine, hypericin, rohitukine, capsaicin, podophyllotoxin, emodin, α-irone, β-irone, toosendanin, phillyrin, piperine, sanguinarine, and others, hold immense potential for various applications (Naik et al., 2019; Bala et al., 2011).

Among the fascinating topics within endophyte biology is the question of why endophytes produce metabolites that mimic those of their host plants. This phenomenon sparks inquiries into the mechanisms and purpose behind this intriguing convergence (Bielecka et al., 2022; Patani et al., 2023). The interactions between plants and endophytes are complex, encompassing parasitic, symbiotic, and mutualistic relationships. These interactions are influenced by several factors, including the infection pattern of the endophyte, its mode of transmission, the genetic background of the host, and environmental conditions (Wani et al., 2022). Endophytes can spread

DOI: 10.1201/9781003392910-6

both horizontally and vertically, the latter indicating mutualistic relationships and the former often indicating host hostility (Samreen et al., 2021).

Host plants derive numerous benefits from their associations with endophytes, including enhanced nutrient access, protection against desiccation, and defense against surface-feeding insects and parasitic fungi, among others (Mathur et al., 2023). In some instances, endophytes alter host tissues to facilitate the extraction of nutrients. This intricate interplay can include changes in the structure of epidermal cells, promoting the flow of nutrients to the stromatal mycelium (Šantrůček et al., 2022). Furthermore, endophytes establish endosymbiotic relationships by dispersing their mycelium within plant tissue cells, optimizing nutrient flow (Palomares-Rius et al., 2021). Beyond the realm of agriculture, endophytes are also exploited in the pharmaceutical industry. These remarkable microorganisms are a source of bioactive compounds, including antibiotics, anti-cancer agents, and antiviral substances, which can aid host plants in developing systemic resistance to infections.

The aim of this chapter is to delve into the multifaceted world of endophytes, exploring their interactions with host plants, the intriguing convergence of metabolites, and the wide-ranging applications in agriculture and medicine. It is through understanding these remarkable associations that we can unlock innovative solutions to address the multifarious challenges facing humanity today.

6.2 ENDOPHYTE ESTABLISHMENT IN HOST PLANTS

Endophytes have been discovered in a wide range of environments, including temperate, aquatic, tropical, and xerophytic environments, deserts, Antarctica, rainforests, mangrove swamps, geothermal soils, and coastal forests (Yadav et al., 2021; Gangwar et al., 2023). Endophyte-plant interaction is governed by both organisms' genes and is influenced by environmental factors (Todd et al., 2022; Sao et al., 2022). When it comes to growth, obligate bacterial endophytes are totally dependent on the host plant, whereas facultative endophytes are biphasic and alternate between plants and the soil. Endophytes are either recruited from the soil rhizosphere or transferred through the seeds. According to (Xu et al., 2022), they enter the host plant through wounds created by microbial or nematode phytopathogens or fissures developed in the lateral root junction. They then promptly spread to the endorhizosphere. Additionally, root hairs and the spaces between epidermal cells serve as entry points for endophytic bacteria in plant roots (Verma et al., 2021). Enzymatic activity, such as the synthesis of endoglucanases, pectinases, and cellulases, contributes to the destruction of plant cell walls during the invasive phase and helps the invaders inhabit the endorhizosphere. Larger cellulose fibers are loosened by endoglucanases, which may facilitate access to the plant. Exoglucanases may further aid in the colonization process (Pandey et al., 2021). In addition to lignin-degrading enzymes, several strains of Streptomyces have been shown to generate hydrolytic cell wall-degrading enzymes such as chitinases, cellulases, hemicellulases, amylases, and glucanases (Maddala et al., 2024). Glucoamylase was generated by the maize endophytic *Streptosporangium* sp. It appears that the cell wall-degrading enzymes endoglucanase and polygalacturonase are necessary for *Burkholderia* sp. to infect *Vitis vinifera* (Singh et al., 2017).

Endophytes are thought to be in symbiotic relationships with all plant species, including non-vascular plants, conifers, ferns, and angiosperms (Sharma et al., 2020; Meddya et al., 2023). In seemingly healthy tissues, many plants can contain hundreds of endophytes (Martínez-Romero et al., 2021). It's possible that fungi and terrestrial plants first co-evolved considerably earlier than in the early Devonian era. It has been suggested that the early mutualistic symbiosis between an alga and fungus catalyzed the evolution of modern land plants from their aquatic forebears. The fungal partner helped the ancestors of land plants survive against mineral deficiency, UV radiation, drought, and temperature fluctuations, which were characteristics of early Paleozoic land (Benton et al., 2018). In fact, endophytes are known to help their host plants adapt to challenges including salt, drought tolerance, metals, disease, heat, and herbivory, as well as boost growth through the production of plant hormones and nutrient acquisition (Koza et al., 2022; Chouhan et al., 2023). For example, *Dichanthelium lanuginosum* plants are dependent on the endophytic *Curvularia* sp. in a symbiotic relationship. This has been reported to provide thermoprotection, while the grass *Leymus mollis* and its endophyte *Fusarium culmorum* have been stated to provide salt tolerance (Venugopalan et al., 2015), with the endophyte in both cases also gaining advantages from the symbiotic relationship. The generation of secondary metabolites is one of the methods employed by endophytic fungi to assist in the survival of their host plants (Parveen et al., 2023). There are two key concepts put forth to explain how endophytes could aid the host plants in chemical defense (Mattoo et al., 2021). The mosaic theory postulates that endophytes alter the genetically uniform and otherwise homogeneous chemical composition of plant organs, changing their value to herbivores or their susceptibility to pathogen infection (Iason et al., 2018). In contrast, a parallel theory put forth by (Arnold et al., 2003) proposes that endophytes may function as "acquired immune systems" in their host plants. Ergot and indole diterpene alkaloids, lolines, and swainsonine—a trihydroxyindolizidine alkaloid—are examples of secondary metabolites generated by endophytes that have been shown to inhibit herbivory (Panaccione et al., 2014). A recent study discovered a direct link between the amount of a resident taxol-producing fungal endophyte and the variation in taxol content among different parts of a *Taxus* plant. Depleting the endophyte through the use of fungicides caused the plant to express its taxol biosynthesis genes to be less active, which in turn led to less taxol accumulation (Mohammadi et al., 2022).

6.3 BIOACTIVE COMPOUND PRODUCTION POTENTIAL OF ENDOPHYTES

Applications of bacterial endophytes in pharmaceutical and drug development are numerous (Gouda et al., 2016; Vora et al., 2023). As novel bioactive agents and a possible source of natural products for use in oxidative stress, endophytes linked with ethnomedicinal plants are useful (Adeleke et al., 2021). Multi-drug resistance

(MDR) in pathogenic microbes is combated by antimicrobial drugs. There is a growing interest in natural pathogen control techniques using novel, environmentally friendly chemicals as a result of the fact that many bacteria of agricultural relevance have developed resistance to routinely employed antimicrobial compounds. Common endophyte metabolites include amines and amides, which are poisonous to insects but not to mammals. Additionally, endophytes create extracellular hydrolases, such as cellulases, proteinases, lipases, and esterases, to build a defense against plant invasions (Kachroo et al., 2022).

Ziziphora capitata and *Hypericum perforatum* both have endophytic bacteria that are members of the *Arthrobacter*, *Achromobacter*, *Bacillus*, *Enterobacter*, *Pantoea*, *Serratia*, and *Stenotrophomonas* genera. In comparison to *Z. capitata*, *H. perforatum* promoted the colonization of more bacteria with antagonistic activity. These isolates were effective in stopping tomato root rot brought on by *F. oxysporum* (Singh et al., 2020). Secondary metabolites have an adaptive function in ecological interactions and environmental obstacles by serving as defense chemicals or signalling molecules, even though they are not required for an organism's growth. Endophytic microbes create low-molecular-weight secondary metabolites such as antibacterial compounds, phytohormones or their precursors, and vitamins such as B1, B12, and bioprotectants (Pandey et al., 2021). Steroids, flavonoids, terpenoids, peptides, alkaloids, polyketones, quinols, and phenols are examples of secondary metabolites. These chemicals are also useful in medicinal applications, such as those that are anti-cancer, anti-inflammatory, antioxidant, and antibacterial (Hamidpour et al., 2017).

6.4 PLANT VERSUS ENDOPHYTE METABOLISM

Plant/endophyte metabolism interacts on multiple levels: (a) the endophyte promotes host metabolism, (b) the host stimulates endophyte metabolism, (c) the host and endophyte share sections of a specific pathway and contribute partially, (d) the host can metabolize endophyte metabolites, and (e) the endophyte can metabolize secondary substances from the host. The latter two options can be applied to one, many, or all enzymatic steps in biochemical transformation.

6.4.1 Changes in Host Metabolism

Endophytes have been demonstrated in certain cases to stimulate host secondary metabolism; however, this aspect of the relationship has received far less attention than endophyte metabolism induction, despite the fact that defense-related phenolic chemicals in endophyte-infected roots have been identified previously (Pacifico et al., 2019; Khare et al., 2017). An increase in phenols was detected in grasses (particularly *Lolium perenne*) after infection with endophytes from the *Clavicipitaceae* family (Malinowski et al., 2019). Chlorogenic acid was the main component implicated in an elevation of the antioxidative activity of plant extracts following endophyte colonization. Other significant compounds were glycosylated flavonoids and hydroxycinnamic acids (Ancheeva et al., 2020). Since some endophytes may have herbicidal activity, a new situation arises when plants are exposed to poisonous endophyte substances. The endophyte-derived peptide leucinostatin A can be glycosylated by a *Taxus* species, yet this chemical may still be poisonous to other, non-host plants (Ludwig-Müller et al., 2019). Such toxic substances may be employed to ascertain the particularity of a relationship between a fungus and a plant.With regard to microbial metabolism and the role of the plant host, endophytes can affect their host plant's metabolism, as previously mentioned. However, it is possible to hypothesize that the established host range may also change or have an impact on the pattern of secondary metabolites in endophytes. In pathogenic fungi, host plants have a greater impact on the metabolite pattern. After performing an LC-MS study, principal component analysis was used to separate the two pine-infecting *Heterobasidium* species and the three non-pine-infecting *Heterobasidium* species into five unique groups (Vallet et al., 2020). The synthesis of the phytotoxic benzohydrofuran fomannoxin by the two pine-infecting fungi was the principal difference between them and the non pine-infecting fungi (Hansson et al., 2014). However, because this chemical was found in another *Heterobasidium* sp. that was not infecting pine, the variations were considered to be related to differential gene regulation.

6.4.2 Synthesis of Plant Host Compounds by Endophytes

The most exciting issue is either the synthesis or metabolism of host plant compounds by endophytes; however, only a few examples can be described here. Bolivar-Anillo et al. (2021), in a recent study, stated that "The biosynthetic potential of endophytes has received momentum in recent years owing to the continuous finding of endophytes capable of producing plant chemicals (Khare et al., 2024) (Figure 6.1). However, utilizing endophytes to produce the needed plant metabolites has yet to be successful.

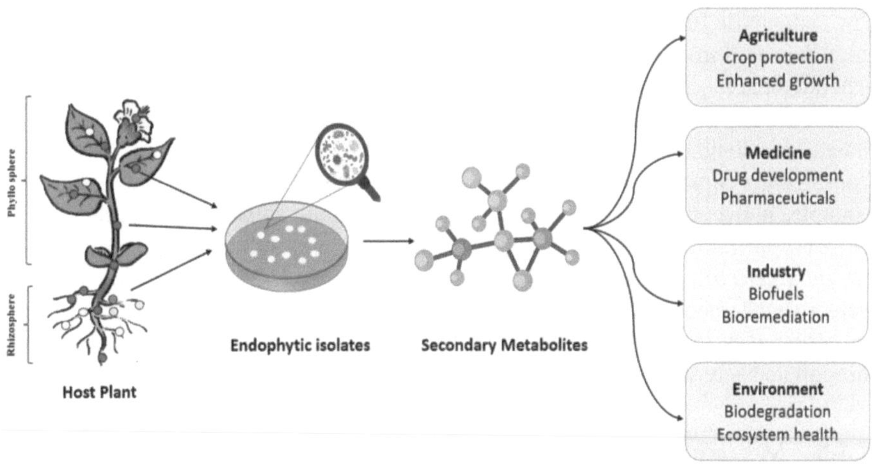

FIGURE 6.1 Diagram illustrating the application of endophytic secondary metabolites in various fields.

Therefore, it is crucial to look at the interactions that endophytes have with other endophytes, host plants, insect pests, and other particular herbivores." This suggests that thorough research must be done to determine the compound's nature, including whether it originates from a plant, an endophyte, or both. In the latter situation, the endophyte may contribute to the whole route, but there is also a chance that it just contributes to some portions of the biosynthesis. It has been proposed that horizontal gene transfer is at least one plausible scenario for how endophytes obtain their genes encoding biosynthetic enzymes and that this may account for the reported ability to produce these plant compounds. The synthesis of paclitaxel (Taxol) by endophytes for medical purposes may be the most well-known example of (partial) plant metabolite synthesis yet. Its synthesis in cell cultures has been highly optimized (Venugopalan et al., 2015). However, because there was such a high demand for the molecule as an anti-cancer medicine, sources other than plants were sought. The isolation of further fungal endophytes from various Taxus spp. followed the discovery of Taxol synthesis in the endophytic fungus *Taxomyces andreanae*, which was identified in *Taxus brevifolia* (El-Sayed et al., 2020). *Pestalotiopsis microspora*, an endophytic fungus isolated from another species of the *Taxaceae*, was shown to have additional bioactive compounds (Rana et al., 2019). Torreyanic acid is a substance with anti-carcinogenic properties. Similar findings were made with plant species from different families, such as *Berberidaceae*, where an endophyte was able to produce podophyllotoxin, or *Meliaceae*, where an endophytic fungus, *Fusarium* sp., was found to produce the compound rohitukine, a precursor for another anticancer medication (Uzma et al., 2018). *Trametes hirsuta*, a new endophytic fungus that also produces similar metabolites, has been proposed as a novel and promising biological system for the production of podophyllotoxin (Yadav et al. 2019). In an in vitro culture system, an endophyte associated with a *Chaetomium* species was isolated from *Hypericum perforatum* and shown to be capable of producing hypericin and the probable biosynthetic precursor emodin (Henzelyová et al., 2020). The new endophytic fungus *Eupenicillium parvum* was isolated by Chutulo et al., (2018) from the neem tree, *Azadirachta indica*, and it was discovered that the fungus also generated the insecticidal chemicals azadirachtin A and B. Of course, none of these instances rule out the possibility of other endophytes coexisting with the abovementioned host plants (Castronovo et al., 2021) on the variety of microbial endophytes from *Azadirachta indica*, as an example. As endophytes, fungi are included in the examples above, but endophytic bacteria can also share substances with their host. The *oxylipin* family of antimicrobial compounds is present in stem preparations of *Alternanthera brasiliana* (*Amaranthaceae*) (Teixeira et al., 2019). A number of them were also detected in *Bacillus* bacteria isolated from Alternanthera plants, prompting researchers to conclude that the host plant received the antimicrobial oxylipins through its bacterial endophytes (Tanvir et al., 2019).

6.4.3 Metabolism of Plant Host Compounds by Endophytes

The host plant's metabolome can be metabolized by the fungal leaf endophyte *Paraconiothyrium variabile* (Liu et al., 2021). It is yet unclear how these metabolic processes serve a function. As a result, metabolome alterations were observed, and the compounds where the most changes were seen that had the potential for helping the fungus were investigated. It was discovered that the endophyte specifically biotransformed glycosylated flavonoids (Mattoo et al., 2021). In every instance, this caused the medicinal host plant *Cephalotaxus harringtonia* to undergo deglycosylation, which produced the appropriate aglycone. The hyphal development of spores that had germinated turned out to be significantly benefited by the deglycosylated flavonoids (Qin et al., 2023). These findings demonstrated a tight biochemical relationship between the two species. Another occurrence was found for symbiotic fungi that changed the metabolism of their hosts in such a manner that leaf-cutting ants preferred non-colonized plants or their extracts. Although the specific leaf components that caused this response have not been identified, they are considered to be low or non-volatile compounds (Saia et al., 2021). There are several instances where endophytes can make their host less enticing to herbivorous predators. The host's defensive compounds are detoxified as a result of the endophyte's alteration of plant metabolites. For example, maize plants release benzoxazinoids to protect themselves from numerous pests. The poisonous substances are present in a glycosylated form, and deglycosylation produces a metabolite that is not harmful. Therefore, if the target enzyme is present, enzymatic deglycosylation will produce tolerant bacteria (Li et al., 2022). Additionally, it was found in this study that fungi, which can break down hazardous compounds, can help in leading the path for the colonization of other endophytes that are less tolerant. Other endophytic fungi employ various metabolic processes such as acylation, oxidation, reduction, hydrolysis, and nitration to detoxify these plant compounds (Rodriguez et al., 2016).

6.5 POSSIBILITIES IN THE ENDOPHYTE-MEDIATED PRODUCTION OF BIOACTIVE COMPOUNDS

The ability of endophytes to imitate the secondary metabolite repertory of the host plant has only recently been discovered, despite the fact that they have been studied for more than a century. According to the "xenohormesis" hypothesis (Ogbe et al., 2020), "heterotrophs" have evolved the ability to sense signaling and stress-induced molecules from plants as a result of evolutionary selection pressures. This ability has been preserved over time even though the ability to biosynthesize these compounds has been lost. This supports the theory that endophytes may receive stress signals from plants and respond by mounting defenses that are comparable to those of the host plant. Therefore, it stands to reason that conserved gene clusters between plants and their endophytes would exist. These gene clusters might be triggered by appropriate interactions, resulting in the synthesis of secondary metabolites that are comparable in both plants and fungi (Kiesewalter et al., 2021). We use the example of endophytes that produce camptothecin to exemplify how interactions with the host plant alter the metabolic machinery of endophytes. Camptothecin and its derivatives have evolved to defend plants from insect and disease attacks as well as against heat stress by interacting with and blocking the DNA topoisomerase I enzyme (Chagas et al., 2018). Many of the rate-limiting enzymes have been found, even though the camptothecin biosynthesis route in plants has not yet been fully defined (Gao et al., 2022). In response, Venugopalan et al. (2015) proposed a host-dependent cross-species biosynthetic pathway for the production of camptothecin in endophytic fungi. Some of the fungal biosynthetic enzymes were shown to be homologous to their host counterparts, suggesting that HGT during coevolution may have occurred. In order to defend themselves from endogenously generated camptothecin, the host plant and its endophytes both created a number of intrinsic and co-evolutionary resistance mechanisms including mutations in the Top1 gene (Venugopalan et al., 2015). Thus, every aspect of endophyte growth, development, and survival, as well as those of their plant hosts, is dependent on their co-evolution, which is defined by balanced antagonism, the characteristic of endophytism (Kusari et al., 2014). There is a chance that these plant metabolites might actually be the result of their individual endophytes because endophytic fungi have been discovered to mimic the secondary metabolite profile of the host plant (Kusari et al., 2012). This has been shown in the example of ansamitocin, an effective anti-tumor maytansinoid that was first isolated from plants in the *Rhamnaceae* and *Euphorbiaceae* families before being found later in the actinomycetes *Actinosynnema pretiosum* ssp. *pretiosum* and *A. pretiosum* ssp. *auranticum*. Despite being isolated from the shrub *Maytenus serrata*, comprehensive analysis of the ansamitocin biosynthetic gene clusters has been carried out in the actinomycete *A. pretiosum* ssp. *auranticum* rather than the plant (Staniek et al., 2008). The absence of a crucial gene for the biosynthesis of maytansinoids in plants and the lack of evidence for the production of maytansinoids from cultured plant cells (callus and suspension) further supported the idea that a bacterial endophyte may actually be the true source of the maytansinoid backbone and that symbiosis may have caused the production in plants (Ghasemnezhad et al., 2021).

6.6 TYPES OF BIOACTIVE COMPOUNDS PRODUCED BY ENDOPHYTES

Endophytes have primarily gained more attention in the past ten years as a result of their capacity to produce a variety of beneficial secondary metabolites (Hilário et al., 2022). Compounds synthesized from endophytes include steroids, quinones, enniatins, isocoumarins, benzopyranones, terpenoids, tetralones, xanthones, cytochalasins, and phenols among other structural classes (Abo Nouh et al., 2021) (Table 6.1). They occasionally represent variants of already recognized structures, such as the plant hormone indole-3-acetic acid or the fungal steroid ergosterol (Salvatore et al., 2020). When the endophyte is competing with other plant-adjacent organisms, the biosynthesis of such metabolites is crucial (Khoshru et al., 2020). As a result, these substances frequently have antibacterial properties. Some of the compounds present in endophytes are frequently responsible for the defense of plants against pests (e.g., Boro et al., 2022). An effective method for finding (novel) organisms with high antibacterial potential is to isolate endophytes from the vicinity of their hosts, and in many cases, new compounds could be isolated and analyzed (e.g.,

TABLE 6.1
Bioactive Compounds from Various Plant Sources, Their Bioactivities, and Associated Endophyte Strains

Compound	Bioactivity	Host Plant	Endophyte Strain	Reference
Azadirachtin A and B	Natural pesticide	*Azadirachta indica* L. (neem)	*Eupenicillum parvum*	(Chutulo et. al., 2018)
Berberine	Antibiotic, anti-cancer	*Curcuma long* L. (Turmeric)	*S6*	(McCubrey et al., 2017)
Camptothecin	Anti-cancer	*Camptotheca acuminate*	Several strains	(Kaur et al., 2020)
Cytochalasin 1 Cytochalasin 2 Cytochalasin 3 Cytochalasin E	Anti-cancer, antibiotic	*Tripterygium wilfordii*	*Rhinocladiella* sp.	(Khiralla et al., 2016)
Digoxin glycoside	Cardiotonic, anti-epileptic	*Digitalis grandiflora*	Several strains	(Pandita et al., 2021)
Emodin	Anti-diabetic, antiviral	*Hypericum perforatum*	*Thielavia subthermophila INFU/Hp/KF/34B*	(Meshram et al., 2019)
Gentiopicrin	Anti-malarial fungicide and larvicide	*Gentiana macrophylla*	*QJ16 and QJ18*	(Venugopalan et al., 2015)
Ginkgolide B	Used in the treatment of cardiovascular diseases and migraine	*Psidium guajava*	*Fusarium oxysporum* SYP0056	(Das et al., 2022)
Gymnemagenin	Anti-diabetic	*Gymnema sylvestre* R.Br	*Penicillum oxalicum*	(Parthasarathy et al., 2014)
Huperzine A	Potential treatment for neurodegenerative diseases	*Huperzia serrata* (Thunb. exMurray) Trev.	*Phlegmariurus taxifolius*	(Thi Minh Le et al., 2019)
Hypericin	Anti-depressant, anti-inflammatory, antibacterial, antioxidant, antiviral	*Hypericum perforatum*	*T.subthermophila* INFU/Hp/KF/34B	(Venugopalan et al., 2015)
Kaempferol	Antioxidant, anti-cancer	*Lantana camara*	*Fusarium chlamydosprum*	(Imran et al., 2019)
Lapachol	Anti-cancer	*Tabebuia argentea*	*Aspergillus niger* *Alternaria niger*	(Srinivasa et al., 2022)

(Continued)

TABLE 6.1 (CONTINUED)
Bioactive Compounds from Various Plant Sources, Their Bioactivities, and Associated Endophyte Strains

Compound	Bioactivity	Host Plant	Endophyte Strain	Reference
Piperine	Antimicrobial, anti-depressant, anti-inflammatory, antioxidative, anti-tumor	*Rosa chinensis Jacq.*	*Periconia* sp.	(Srivastava et al., 2017)
Podophyllotoxin	Anti-cancer	*Juniperus recurva*	*Fusarium oxysporum*	(Kour et al., 2008)
Quercetin glycoside	Anti-hypertensive, anti-cancer, anti-inflammatory, antioxidant	*Avicennia marina*	*Penicillium* sp. Gh01	(Toghueo et al., 2020)
Rhein	Anti-tumor, anti-inflammatory, antimicrobial, and hemostatic	*Centella asiatica Linn.*	*Fusarium saloni R13*	(Sharma et al., 2020)
Rohitukine	Anti-inflammatory, immune-modulatory, anti-cancer	*Dysoxylum binectariferum* (Meliaceae) and *Schumanniophyton problematicum* (Rubiaceae)	*Fusarium proliferatum* (MTCC 9690)	(Toghueo et al., 2020)
Sanguinarine	Anti-cancer	*Macleaya cordata*	*F.proliferatum* BLH51	(Wang et al., 2014)
Sipeimine	Antitussive and expectorant	*Fritillaria ussuriensis*	*Fritillaria ussuriensis Fu7*	(An et al., 2022)
Subglutinols A and B	Immuno-suppressants	*Tripterygium wilfordii*	*Fusarium subglutinans*	(Tiwari et al., 2022)
Tanshinone I and IIA	Cardiotonic, anti-inflammatory	*Salvia miltiorrhiza*	*Emericella foeniculicola TR21*	(Nicoletti et al., 2015)
Taxol	Anti-cancer	Several plants	Several strains	(Gallego et al., 2017)
Toosendanin	Digestive tract-parasiticide, selective presynaptic blocker, agricultural insecticide	*Melia toosendan* and *M. azedarach*	O-L-5,O-SC II-4, O-RC-3	(Zhang et al., 2019)

Caruso et al., 2022). However, there are other studies that describe antibacterial properties in endophyte extracts without further identifying the active principle (e.g., Techaoei et al., 2020); therefore there is still a significant possibility for new compounds within these findings. Endophytes can also release substances that are toxic to plants, according to a screening for endophytes that included antimicrobial and herbicidal compounds (Yusupov et al., 2020). Additionally, this study discovered that there were twice as many endophytes manufacturing these herbicides as there were plant pathogenic bacteria and three times as many soil microbes. Additionally, endophytic fungi that produce bioactive compounds and host plants may be influenced by insects (Kaur et al., 2020). As a result, among the compounds produced by endophytes are also insecticidal substances. However, studies showing that fungi have their own symbionts may complicate the scenario. For example, the plant pathogenic fungus *Rhizopus* does not synthesize rhizoxin as previously thought but rather a symbiotic bacterium (*Burkholderia*) within the fungus (Newman et al., 2015). The endophytic fungus can also impact their metabolism in a complex interaction if there is more than one in a particular host plant (Yan et al., 2019).

Strong physiological actions are displayed by the bioactive substances that plants generate. However, a number of problems affect their manufacture, including uneven quality and a low level of productivity. In a regulated environment, microorganisms create primary and secondary metabolites with the highest level of efficiency and consistency (Landi et al., 2020). It has been demonstrated that the connection of plants with endophytic bacteria is a source of materials and products with higher therapeutic potential than plants alone. Bioactive compounds produced by microorganisms, such as secondary metabolites and enzymes, have been widely used in the production of these industrial products, including pharmaceuticals, biofuels, biopesticides, detergents, food, and food supplements (Singh et al., 2017). Since their first significance was understood, methods for producing metabolites and enzymes have been improved. After fermentation, the presence of soluble phenolic compounds and antioxidant activity were increased by *B. subtilis* and *Lactobacillus plantarum* in kidney bean extracts (Verni et al., 2020).

6.6.1 Antimicrobial Compounds

The bioactive natural compounds synthesized by endophytes are antimicrobial metabolites (Paramanantham et al., 2019). Endophytes have created secondary metabolites as a resistant defense against pathogenic invasion (Naik et al., 2019). There are several structural groups of antimicrobial compounds produced by endophytes, including peptides, alkaloids, steroids, quinines, terpenoids, phenols, and flavonoids (Manganyi et al., 2020). The emerging alternative strategy to combat the rising levels of drug resistance is the use of new antibacterial metabolites from endophytes (Pandey et al., 2020). Endophytic actinomycetes with a broad spectrum of antibacterial activity were identified from 26 medicinal plants from the Panxi plateau, making them attractive repositories of new bioactive compounds (Narayanan et al., 2022). Elite Citrus nobilis fruit-derived Endophytic *Streptomyces* sp. TQR12-4 exhibited antibacterial activity and inhibited test pathogens *Colletotrichum truncatum*,

Geotrichum candidum, and *F. oxysporum* (Hong-Thao et al., 2016). As antibacterial agents, nanoparticles may be very important. The therapy of several cancers has been significantly impacted by nanoparticles. Endophytic bacteria produce a variety of nanoparticles, opening up a brand-new topic of study in the field of pharmaceutical engineering (Oladipo et al., 2022). With regard to HIV-1, hepatitis B, respiratory syncytial virus, and herpes simplex virus, silver nanoparticles also exhibit antibacterial and antiviral activities (Hamouda et al., 2021). Silver nanoparticles (AgNPs) were produced by the reduction of silver nitrate ($AgNO_3$) by endophytic cultures of *Bacillus* sp. obtained from the therapeutic plants *Adhatoda beddomei* (Malabar nut) and *Garcinia xanthochymus* (Egg tree) (Tariq et al., 2020).

6.6.2 Anti-Cancerous Compounds

Several endophyte-produced bioactive substances have been recognized as anticancer medicines (Basha et al., 2019). A powerful anti-angiogenic activity was displayed by the endophytic bacterial strain EML-CAP3, which was isolated from a *C. annuum* L. (red pepper) leaf. This endophytic bacterial strain synthesized lipophilic peptides that displayed anti-angiogenic potential in the development of tumors and reduced the proliferation of human umbilical vein endothelial cells (Sawant et al., 2020). Ginsenosides, which are found in ginseng (*Panax ginseng*), are renowned for having anti-cancerous properties. Ginsenoside content was high in the modified *Paenibacillus polymyxa*, an endophytic bacterium of the ginseng leaf. Inoculating ginseng plants with this endophytic bacterial strain by foliar sprays and irrigation improved plant growth and the concentration of ginsenosides (Murthy et al., 2019). Exopolysaccharides (EPS) generate morphological irregularities in the treated cells, which are linked to mitochondrial malfunction and operate as anti-tumor modes of action. As a novel anti-cancer agent, *Bacillus* is the source of the first anti-tumoral EPS, a natural substance with significant therapeutic efficacy (Wu et al., 2021). Some neoplastic cells, such as lymphoblasts, require the conversion of L-asparagine, which is catalyzed by the enzyme L-asparaginase. For both children and adults with acute lymphoblastic leukemia, L-asparaginase added to the multi-drug treatment caused a significant improvement and full remission in the majority of patients (Abdelrazek et al., 2019). L-asparaginase was produced effectively by endophytic *B. licheniformis, B. pseudomycoides, and Paenibacillus denitriformis* (Kumari et al., 2021).

6.6.3 Antibiotics

Bioactive molecules known as antibiotics are created by microbes as secondary metabolites in order to destroy or inhibit other pathogens. In the 20th century, they were crucial in the fight against infectious diseases. Drugs derived from bacteria, fungi, or soil organisms have been the main focus of pharmaceutical firms up until now. Actinobacteria stand out among bacteria because they create antibiotics. Of all antibiotics, 80% are created by productive microbes known as the *Streptomyces* species (Narayanan et al., 2022). *Streptomyces* isolates have shown at least 7,000 distinct secondary metabolites (Mahmood et al., 2022). Streptomyces produce fungicides, antibiotics, immune system modulators, and plant growth promoters (Mun et al., 2020). Endophytic *Streptomyces* sp. LJK109 isolated from *Alpinia galangal* root contains 3-methylcarbazoles, a strong anti-inflammatory component, and inhibits macrophage production of the inflammatory mediators NO, PGE2, TNF-a, IL-1b, IL-6, and IL-10 in a dose-dependent manner (Sinha et al. 2020). Numerous antibiotics are produced by the majority of endophytic bacteria (Table 6.2). The endophytic bacteria that make ecomycin, pseudomycins, and kakadumycins are among the producers of

TABLE 6.2
Endophytic Bacteria-Derived Compounds and Their Biological Activities

Compounds	Endophytic Bacteria	Biological Activity	References
Ecomycin	*Pseudomonas viridiflava*	Antifungal	(Pandey et al., 2017)
Bacilysocin	*B. subtilis 168*	Antifungal	(Kaspar et al., 2019)
Nystatin	*Streptomyces noursei*	Antifungal	(Khoram et al., 2019)
KB425796-A	*Paenibacillus* sp. 530603	Antifungal	(Ambarwati et al., 2019)
Bacillomycin	*B. subtilis, B. amyloliquefaciens*	Antifungal, Hemolytic	(Labiadh et al., 2021)
Munumbicin	*Streptomyces* NRRL-30562	Antibacterial	(Singh et al., 2020)
Harmaomycin	*Streptomyces* sp.	Antibacterial	(Lata et al., 2019)
Subtilin	*B. subtilis*	Antibacterial	(Halami et al., 2019)
Tetracyclin	*Streptomyces remosus and S. aureofaciens*	Antibacterial	(Yin et al., 2019)
Bacteriocins	*B. subtilis*	Antibacterial	(Xiang et al., 2021)
Amicoumacin	*B. subtilis*	Antibacterial, anti-inflammatory	(Wang et al., 2020)
Artemisinin	*Pseudonocardia* sp.	Anti-malarial	(Radiastuti et al., 2022)
Coronamycin	*Streptomyces* sp.	Anti-malarial	(Joo et al., 2021)
Spectinomycin	*Streptomyces spectabilis*	Anti-tuberculosis	(Ogawara et al., 2019)
Treponemycin	*Streptomyces* strain MS-6-6	Anti-tuberculosis	(Sabido et al., 2021)
Camptothecin	*Lysinibacillus* sp. *and B. cereus*	Anti-cancer	(Ruan et al., 2021)
Indolocarbazoles	*Streptomyces* sp.	Anti-cancer	(Kumar et al., 2021)
Doxorubicin	*Streptomyces* sp.	Treatment of breast cancers and tumors	(Chiu et al., 2021)
Anthracyclin	*Streptomyces* sp. YIM66403	Anti-tumor	(Silva et al., 2020)
Daptomycin	*Streptomyces roseoporous*	Bacterial infection of the skin and underlying tissues	(Ghosh et al., 2022)
Strepturidin	*Streptomyces albus* DSM 40763	Immunotherapy	(Aamir et al., 2020)
Xiamycin	*Streptomyces* sp.	Anti-HIV activity	(El-Gendy et al., 2022)

novel antibiotics (Choudhary et al., 2022). Ecomycin, a medication used to treat skin, eye, and gut infections, was synthesized by *Pseudomonas viridiflava*, an epiphyte or endophyte of the leaves of several types of grass. In addition to the typical amino acids alanine, serine, threonine, and glycine, the structure of ecomycin includes certain uncommon amino acids such as homoserine and betahydroxyaspartic acid (Nomani et al., 2020). A chlorine-containing ansamycin known as naphthomycin K showed cytotoxicity against the P388 and A-549 cell lines. *Bafilomycin* family member 24-demethylbafilomycin C1 was synthesized by endophytic *Streptomyces* sp. CS and targets both the autophagy and apoptosis pathways in juvenile B cell acute lymphoblastic leukemia (Pandey et al., 2021). Two novel novobiocin analogues and cedarmycins were generated by endophytic Streptomyces species isolated from Aucuba japonica and *Cryptomeria japonica*. Alnumycin, a brand-new naphthoquinone antibiotic, was also discovered in *Streptomyces* sp., an endophytic strain from *Alnus glutinosa*. A novel peptide antibiotic with a broad spectrum of action against several human diseases, phytopathogenic fungi, and bacteria was synthesized by the endophyte of the snake vine plant, *Streptomyces* sp. NRRL30562. Kakadumycin, a chemically similar compound to echinomycin, was generated by the endophytic *Streptomyces* sp. NRRL30566, which was isolated from a fern-leaved Grevillea tree *(Grevillea pteridifolia)* (Choudhary et al. 2022).

6.7 CONCLUSION AND FUTURE PROSPECTS

In conclusion, the potential of endophytes, both fungal and bacterial, as invaluable sources of bioactive compounds and their integration into sustainable agricultural practices is an exciting frontier in the field of biotechnology. The vast reservoir of knowledge we now possess, combined with ongoing breakthroughs in endophyte research, points to a future where endophytes will play a pivotal role in shaping the landscape of secondary metabolite production.

The promise of high-value natural products derived from endophytic fungi is particularly alluring as we continue to refine and adapt various yield enhancement strategies for these remarkable organisms. To fully unlock the commercial potential of endophytes, we recognize the imperative need for a deeper understanding of host plant-endophyte interactions at the molecular level. Such insights will enable the fine-tuning of bioprocesses, ensuring the sustainable and large-scale production of these precious secondary metabolites. The utilization of endophytic bacteria for bioactive compound production, either independently or in synergy with plants, is an ever-expanding frontier. With each passing day, we are unraveling the unique capabilities of these microorganisms. Bio-fertilizers, which harness the benefits of beneficial microorganisms in agriculture, hold the promise of revolutionizing organic farming. Integrating bacterial endophytes into bio-fertilizer technology is a logical step forward, but it is clear that further research is needed to fully grasp how endophytes influence plant physiology and metabolism. Understanding how they utilize the products of primary and secondary metabolism as sources of nourishment and precursors for the creation and enhancement of significant secondary metabolites remains a fascinating challenge.

As we move forward, it is evident that endophyte research offers immense potential not only for the biotechnology and pharmaceutical industries but also for sustainable agriculture. The future promises a deeper understanding of the intricate interplay between endophytes and their host plants, as well as innovative strategies to maximize their commercial and ecological benefits. By embracing the potential of these extraordinary organisms, we are poised to unlock a world of possibilities for the development of bioactive compounds and sustainable agricultural practices, ensuring a brighter and more environmentally friendly future.

REFERENCES

Aamir M, Rai KK, Zehra A, Dubey MK, Samal S, Yadav M, Upadhyay RS. Endophytic actinomycetes in bioactive compounds production and plant defense system. In *Microbial Endophytes* 2020 Jan 1 (pp. 189–229). Woodhead Publishing.

Abdelrazek NA, Elkhatib WF, Raafat MM, Aboulwafa M. Diverse origins of microbial L-asparaginases and their current miscellaneous applications. *Archives of Pharmaceutical Sciences Ain Shams University.* 2019 Jan 1;3(1):21–36.

Abo Nouh FA, Gezaf SA, Abdel-Azeem AM. Recent advances in fungal antimicrobial molecules. *Industrially Important Fungi for Sustainable Development.* 2021 2:177–203.

Adeleke BS, Babalola OO. Pharmacological potential of fungal endophytes associated with medicinal plants: A review. *Journal of Fungi.* 2021 Feb 17;7(2):147.

Ambarwati A, Wahyuono S, Moeljopawiro S, Sembiring L, Yuwono T. Antimicrobial activities and phylogenetic study of bacteria associated with Cyperus rotundus rhizosphere from Cemoro Sewu Plateau, Indonesia. *Biodiversitas Journal of Biological Diversity.* 2019 Jul 21;20(8):2206–2212.

An YL, Wei WL, Guo DA. Application of analytical technologies in the discrimination and authentication of herbs from fritillaria: A review. *Critical Reviews in Analytical Chemistry.* 2022 Oct 4:1–22. https://doi.org/10.1080/10408347.2022.2132374

Ancheeva E, Daletos G, Proksch P. Bioactive secondary metabolites from endophytic fungi. *Current Medicinal Chemistry.* 2020 Apr 1;27(11):1836–54.

Arnold AE, Herre EA. Canopy cover and leaf age affect colonization by tropical fungal endophytes: ecological pattern and process in Theobroma cacao (Malvaceae). *Mycologia.* 2003 May 1;95(3):388–98.

Bala JM, Vennila JJ, Ganesh S, Singh S. Analysis of active components in Acanthus ilicifolius leaves for their potential chemoprotective activity against hepatocellular carcinoma: An in silico approach. *Journal of Pharmacy Research.* 2011;4(6):1809–12.

Basha J, Rajagopal WK, Meenambiga SS, Arulmathi R, Kumaresan S, Kathiravan G, Prabhakaran R, Rajesh SV, Gayathri G, Kathiresan AK. Analysis of endophytic fungal extracts of therapeutically important hydrophytes for different chemical compounds. *Research Journal of Pharmacy and Technology.* 2019 Oct 12;12(10):5004–6.

Benton MJ. Hyperthermal-driven mass extinctions: Killing models during the Permian–Triassic mass extinction. *Philosophical Transactions of the Royal Society A: Mathematical, Physical and Engineering Sciences.* 2018 Oct 13;376(2130):20170076.

Bielecka M, Pencakowski B, Nicoletti R. Using next-generation sequencing technology to explore genetic pathways in endophytic fungi in the syntheses of plant bioactive metabolites. *Agriculture.* 2022 Jan 28;12(2):187.

Bolivar-Anillo HJ, González-Rodríguez VE, Cantoral JM, García-Sánchez D, Collado IG, Garrido C. Endophytic bacteria Bacillus subtilis, isolated from Zea mays, as potential biocontrol agent against Botrytis cinerea. *Biology.* 2021 Jun;10(6):492.

Boro M, Sannyasi S, Chettri D, Verma AK. Microorganisms in biological control strategies to manage microbial plant pathogens: A review. *Archives of Microbiology*. 2022 Nov;204(11):1–5.

Caruso DJ, Palombo EA, Moulton SE, Zaferanloo B. Exploring the promise of endophytic fungi: A Review of novel antimicrobial compounds. *Microorganisms*. 2022 Oct 8;10(10):1990.

Castronovo LM, Vassallo A, Mengoni A, Miceli E, Bogani P, Firenzuoli F, Fani R, Maggini V. Medicinal plants and their bacterial microbiota: A review on antimicrobial compounds production for plant and human health. *Pathogens*. 2021 Jan 22;10(2):106.

Chagas FO, de Cassia Pessotti R, Caraballo-Rodriguez AM, Pupo MT. Chemical signaling involved in plant–microbe interactions. *Chemical Society Reviews*. 2018;47(5):1652–704.

Chiu CF, Chiu SJ, Bai LY, Feng CH, Hu JL, Lin WY, Huang HY, Weng JR. A macrolide from Streptomyces sp. modulates apoptosis and autophagy through Mcl-1 downregulation in human breast cancer cells. *Environmental Toxicology*. 2021 Jul;36(7):1316–25.

Choudhary J, Mishra SK, Rao Z, Shukla S. Endophytic Microbes: Potential Source of Antibiotic Production. *Microbial Biotechnology: Role in Ecological Sustainability and Research*. 2022 Sep 26:89–106.

Chouhan PN, Azimov A, Jabborova D, & Singh S. Review on the role of indigenous plants and plant derivatives of Ayurveda in bone healing. *Plant Science Today*. 2023;10(sp2):143–51.

Chutulo EC, Chalannavar RK. Endophytic mycoflora and their bioactive compounds from Azadirachta indica: A comprehensive review. *Journal of Fungi*. 2018 Mar 24;4(2):42.

Das PP, Singh KR, Nagpure G, Mansoori A, Singh RP, Ghazi IA, Kumar A, Singh J. Plant-soil-microbes: A tripartite interaction for nutrient acquisition and better plant growth for sustainable agricultural practices. *Environmental Research*. 2022 Nov 1;214:113821.

El-Gendy MM, Yahya SM, Hamed AR, El-Bondkly AM. Assessment of the phylogenetic analysis and antimicrobial, antiviral, and anticancer activities of marine endophytic Streptomyces species of the soft coral Sarcophyton convolutum. *International Microbiology*. 2022 Jan;25(1):133–52.

El-Sayed AS, El-Sayed MT, Rady AM, Zein N, Enan G, Shindia A, El-Hefnawy S, Sitohy M, Sitohy B. Exploiting the biosynthetic potency of taxol from fungal endophytes of conifers plants; genome mining and metabolic manipulation. *Molecules*. 2020 Jun 30;25(13):3000.

Gallego A, Malik S, Yousefzadi M, Makhzoum A, Tremouillaux-Guiller J, Bonfill M. Taxol from Corylus avellana: Paving the way for a new source of this anti-cancer drug. *Plant Cell, Tissue and Organ Culture (PCTOC)*. 2017 Apr;129(1):1–6.

Gangwar P, Johri P, Datta R, Singh S, Trivedi M. Overview on 1 Phytopharmaceuticals and Biotechnology of Herbal Plants. In *Phytopharmaceuticals and Biotechnology of Herbal Plants*. 2023 (pp. 1–17). CRC Press.

Gao B, Yang B, Feng X, Li C. Recent advances in the biosynthesis strategies of nitrogen heterocyclic natural products. Natural Product Reports. 2022.

Ghasemnezhad A, Frouzy A, Ghorbanpour M, Sohrabi O. Microbial Endophytes: New Direction to Natural Sources. In *Endophytes: Mineral Nutrient Management*, Volume 3 2021 (pp. 123–55). Springer.

Ghosh S, Das S. Impact of climate change on microbial endophytes: Novel nanoscale cell factories. In *Microbiome Under Changing Climate* 2022 Jan 1 (pp. 161–185). Woodhead Publishing.

Gouda S, Das G, Sen SK, Shin HS, Patra JK. Endophytes: A treasure house of bioactive compounds of medicinal importance. *Frontiers in Microbiology*. 2016 Sep 29;7:1538.

Gupta G, Paul S, Singh S, Pietramellara G, Pathan SI , Danish S, Jha PN. Exploring functional diversity and community structure of diazotrophic endophytic bacteria associated with Pennisetum glaucum growing under field in a semi-arid region. *Land.* 2022;11(7):991.

Halami PM. Sublichenin, a new subtilin-like lantibiotics of probiotic bacterium Bacillus licheniformis MCC 2512T with antibacterial activity. *Microbial Pathogenesis.* 2019 Mar 1;128:139–46.

Hamidpour R, Hamidpour S, Hamidpour M, Shahlari M, Sohraby M, Shahlari N, Hamidpour R. Russian olive (Elaeagnus angustifolia L.): From a variety of traditional medicinal applications to its novel roles as active antioxidant, anti-inflammatory, anti-mutagenic and analgesic agent. *Journal of Traditional and Complementary Medicine.* 2017 Jan 1;7(1):24–9.

Hamouda T, Ibrahim HM, Kafafy HH, Mashaly HM, Mohamed NH, Aly NM. Preparation of cellulose-based wipes treated with antimicrobial and antiviral silver nanoparticles as novel effective high-performance coronavirus fighter. *International Journal of Biological Macromolecules.* 2021 Jun 30;181:990–1002.

Hansson D, Wubshet S, Olson Å, Karlsson M, Staerk D, Broberg A. Secondary metabolite comparison of the species within the Heterobasidion annosum sl complex. *Phytochemistry.* 2014 Dec 1;108:243–51.

Henzelyová J, Antalová M, Nigutová K, Logoida M, Schreiberová A, Kusari S, Čellárová E. Isolation, characterization and targeted metabolic evaluation of endophytic fungi harbored in 14 seed-derived Hypericum species. *Planta Medica.* 2020 Sep;86(13/14):997–1008.

Hilário S, Gonçalves MF. Endophytic diaporthe as promising leads for the development of biopesticides and biofertilizers for a sustainable agriculture. *Microorganisms.* 2022 Dec 12;10(12):2453.

Hong-Thao PT, Mai-Linh NV, Hong-Lien NT, Van Hieu N. Biological characteristics and antimicrobial activity of endophytic Streptomyces sp. TQR12-4 isolated from elite Citrus nobilis cultivar Ham Yen of Vietnam. *International Journal of Microbiology.* 2016 Oct 4;2016. https://doi.org/10.1155/2016/7207818

Iason GR, Taylor J, Helfer S. Community-based biotic effects as determinants of tree resistance to pests and pathogens. *Forest Ecology and Management.* 2018 May 15;417:301–12.

Imran M, Salehi B, Sharifi-Rad J, Aslam Gondal T, Saeed F, Imran A, Shahbaz M, Tsouh Fokou PV, Umair Arshad M, Khan H, Guerreiro SG. Kaempferol: A key emphasis to its anticancer potential. *Molecules.* 2019 Jun 19;24(12):2277.

Jabborova D, Davranov K, Jabbarov Z, Bhowmik SN, Ercisli S, Danish S, Datta R. Dual inoculation of plant growth-promoting Bacillus endophyticus and Funneliformis mosseae improves plant growth and soil properties in ginger. *ACS Omega.* 2022;7(39):34779–88.

Joo HS, Deyrup ST, Shim SH. Endophyte-produced antimicrobials: A review of potential lead compounds with a focus on quorum-sensing disruptors. *Phytochemistry Reviews.* 2021 Jun;20(3):543–68.

Kachroo H. Endophytes: An Asset for Extracellular Hydrolases Production. In *Endophyte Biology* 2022 Oct 27 (pp. 149–83). Apple Academic Press.

Kaspar F, Neubauer P, Gimpel M. Bioactive secondary metabolites from Bacillus subtilis: A comprehensive review. *Journal of Natural Products.* 2019 Jul 9;82(7):2038–53.

Kaur P, Kumar V, Singh R, Dwivedi P, Dey A, Pandey DK. Biotechnological strategies for production of camptothecin from fungal and bacterial endophytes. *South African Journal of Botany.* 2020 Nov 1;134:135–45.

Khare P, Kishore K, Sharma DK. A study on the standardization parameters of Cassia angustifolia. *Asian Journal of Pharmaceutical and Clinical Research.* 2017;10(7):329–32.

Khare P, Kishore K, Sharma DK, Khare N. Investigation of memory enhancing activity of combination extracts of bauhinia variegata and madhuca longifolia leaves against colchicine: An experimental study and biochemical alterations in Mice. *Indian Journal of Pharmaceutical Education and Research.* 2024;58(1s):s282–s288.

Khiralla A, Spina R, Yagi S, Mohamed I, Laurain-Mattar D. Endophytic fungi: Occurrence, classification, function and natural products. *Endophytic Fungi: Diversity, Characterization and Biocontrol.* 2016 Dec 30:1–9.

Khoram Z, Naine A, Rafieinezha R, Hakimaneh SM, Hakimaneh SM, Shayehg SS, Salari AM. The antifungal effects of two herbal essences in comparison with Nystatin on the Candida Strains Isolated from the edentulous patients. *Journal of Contemporary Dental Practice.* 2019 Jun 21;20(6): 716–719.

Khoshru B, Moharramnejad S, Gharajeh NH, Asgari Lajayer B, Ghorbanpour M. Plant Microbiome and Its Important in Stressful Agriculture. In *Plant Microbiome Paradigm* 2020 (pp. 13–48). Springer.

Kiesewalter HT, Lozano-Andrade CN, Wibowo M, Strube ML, Maróti G, Snyder D, Jørgensen TS, Larsen TO, Cooper VS, Weber T, Kovács ÁT. Genomic and chemical diversity of Bacillus subtilis secondary metabolites against plant pathogenic fungi. *Msystems.* 2021 Feb 23;6(1):e00770–20.

Kour A, Shawl AS, Rehman S, Sultan P, Qazi PH, Suden P, Khajuria RK, Verma V. Isolation and identification of an endophytic strain of Fusarium oxysporum producing podophyllotoxin from Juniperus recurva. *World Journal of Microbiology and Biotechnology.* 2008 Jul;24(7):1115–21.

Koza NA, Adedayo AA, Babalola OO, Kappo AP. Microorganisms in plant growth and development: Roles in abiotic stress tolerance and secondary metabolites secretion. *Microorganisms.* 2022 Jul 28;10(8):1528.

Kumar P, Chauhan A, Kumar M, Kuanr BK, Kundu A, Solanki R, Kapur MK. In vitro and in silico anticancer potential analysis of Streptomyces sp. extract against human lung cancer cell line, A549. *3 Biotech.* 2021 Jun;11(6):1–2.

Kumari P, Kumar S, Patra A, Singh B, Pandey V, Mehta S. Plant-Microbe Symbiosis led synthesis of Bioactive Compounds. In *Antioxidants in Plant-Microbe Interaction* 2021 (pp. 21–40). Springer.

Kusari S, Hertweck C, Spiteller M. Chemical ecology of endophytic fungi: origins of secondary metabolites. *Chemistry & Biology.* 2012 Jul 27;19(7):792–8.

Kusari S, Singh S, Jayabaskaran C. Biotechnological potential of plant-associated endophytic fungi: Hope versus hype. *Trends in Biotechnology.* 2014 Jun 1;32(6):297–303.

Labiadh M, Dhaouadi S, Chollet M, Chataigne G, Tricot C, Jacques P, Flahaut S, Kallel S. Antifungal lipopeptides from Bacillus strains isolated from rhizosphere of Citrus trees. *Rhizosphere.* 2021 Sep 1;19:100399.

Landi M, Zivcak M, Sytar O, Brestic M, Allakhverdiev SI. Plasticity of photosynthetic processes and the accumulation of secondary metabolites in plants in response to monochromatic light environments: A review. *Biochimica et Biophysica Acta (BBA)-Bioenergetics.* 2020 Feb 1;1861(2):148131.

Lata RK, Divjot K, Nath YA. Endophytic microbiomes: Biodiversity, ecological significance and biotechnological applications. *Research Journal of Biotechnology.* 2019 Oct;14:10.

Li LY, Hu YL, Sun JL, Yu LB, Shi J, Wang ZR, Guo ZK, Zhang B, Guo WJ, Tan RX, Ge HM. Resistance and phylogeny guided discovery reveals structural novelty of tetracycline antibiotics. *Chemical Science.* 2022;13(43):12892–8.

Liu R, Bao ZX, Zhao PJ, Li GH. Advances in the study of metabolomics and metabolites in some species interactions. *Molecules.* 2021 Jan;26(11):3311.

Maddala VKS, Singh S, Bandi J. Soil properties influencing nutrient signaling in plant soil systems. In *Essential Minerals in Plant-Soil Systems.* 2024 (pp. 373–93). Elsevier.

Mahmood KI, Najmuldeen HH, Rachid SK. Physiological regulation for enhancing biosynthesis of biofilm-inhibiting secondary metabolites in Streptomyces cellulosae. *Cellular and Molecular Biology*. 2022 May 31;68(5):33–46.

Malinowski DP, Belesky DP. Epichloë (formerly Neotyphodium) fungal endophytes increase adaptation of cool-season perennial grasses to environmental stresses. *Acta Agrobotanica*. 2019;72(2). https://doi.org/10.5586/aa.1767

Manganyi MC, Ateba CN. Untapped potentials of endophytic fungi: A review of novel bioactive compounds with biological applications. *Microorganisms*. 2020 Dec 6;8(12):1934.

Martínez-Romero E, Aguirre-Noyola JL, Bustamante-Brito R, González-Román P, Hernández-Oaxaca D, Higareda-Alvear V, Montes-Carreto LM, Martínez-Romero JC, Rosenblueth M, Servín-Garcidueñas LE. We and herbivores eat endophytes. *Microbial Biotechnology*. 2021 Jul;14(4):1282–99.

Mathur M, Tiwari RK, Johri P, Trivedi M. 2023. Harnessing Rhizosphere Microbiomes in Crop Productivity. In U. B. Singh et al. (eds.), *Detection, Diagnosis and Management of Soil-borne Phytopathogens* (pp. 401–12). Springer Nature Singapore. 2022. https://doi.org/10.1007/978-981-19-8307-8_16

Mattoo AJ, Nonzom S. Endophytic fungi: Understanding complex cross-talks. *Symbiosis*. 2021 Apr;83(3):237–64.

McCubrey JA, Lertpiriyapong K, Steelman LS, Abrams SL, Yang LV, Murata RM, Rosalen PL, Scalisi A, Neri LM, Cocco L, Ratti S. Effects of resveratrol, curcumin, berberine and other nutraceuticals on aging, cancer development, cancer stem cells and microRNAs. *Aging (Albany NY)*. 2017 Jun;9(6):1477.

Meddya S, Meshram S, Sarkar D, Datta R, Singh S, Avinash G, Thulasinathan T. Plant stomata: An unrealized possibility in plant defense against invading pathogens and stress tolerance. *Plants*. 2023;12(19):3380.

Meshram V, Gupta M. Endophytic fungi: A quintessential source of potential bioactive compounds. *Endophytes for a Growing World*. 2019 Mar 21;277:277–309.

Mohammadi Ballakuti N, Ghanati F, Zare-Maivan H, Alipour M, Moghaddam M, Abdolmaleki P. Taxoid profile in endophytic fungi isolated from Corylus avellana, introduces potential source for the production of Taxol in semi-synthetic approaches. *Scientific Reports*. 2022 Jun 7;12(1):1–1.

Mun BG, Lee WH, Kang SM, Lee SU, Lee SM, Lee DY, Shahid M, Yun BW, Lee IJ. Streptomyces sp. LH 4 promotes plant growth and resistance against Sclerotinia sclerotiorum in cucumber via modulation of enzymatic and defense pathways. *Plant and Soil*. 2020 Mar;448(1):87–103.

Murthy HN, Dalawai D, Park SY, Paek KY. Endophytes of. In *Endophytes and Secondary Metabolites* 2019 (pp. 165–86). Springer.

Naik S, Shaanker RU, Ravikanth G, Dayanandan S. How and why do endophytes produce plant secondary metabolites?. *Symbiosis*. 2019 Jul;78(3):193–201.

Narayanan Z, Glick BR. Secondary metabolites produced by plant growth-promoting bacterial endophytes. *Microorganisms*. 2022 Oct 11;10(10):2008.

Newman DJ, Cragg GM. Endophytic and epiphytic microbes as "sources" of bioactive agents. *Frontiers in Chemistry*. 2015 May 22;3:34.

Nicoletti R, Fiorentino A. Plant bioactive metabolites and drugs produced by endophytic fungi of Spermatophyta. *Agriculture*. 2015 Sep 29;5(4):918–70.

Nomani MM, Shirsh S. Endophytes: An asset for their hosts. *Journal of Pharmacognosy and Phytochemistry*. 2020;9(6):1011–6.

Ogawara H. Comparison of antibiotic resistance mechanisms in antibiotic-producing and pathogenic bacteria. *Molecules*. 2019 Sep 21;24(19):3430.

Ogbe AA, Finnie JF, Van Staden J. The role of endophytes in secondary metabolites accumulation in medicinal plants under abiotic stress. *South African Journal of Botany*. 2020 Nov 1;134:126–34.

Oladipo A, Enwemiwe V, Ejeromedoghene O, Adebayo A, Ogunyemi O, Fu F. Production and functionalities of specialized metabolites from different organic sources. *Metabolites*. 2022 Jun 10;12(6):534.

Pacifico D, Squartini A, Crucitti D, Barizza E, Lo Schiavo F, Muresu R, Carimi F, Zottini M. The role of the endophytic microbiome in the grapevine response to environmental triggers. *Frontiers in Plant Science*. 2019 Oct 9;10:1256.

Palomares-Rius JE, Gutiérrez-Gutiérrez C, Mota M, Bert W, Claeys M, Yushin VV, Suzina NE, Ariskina EV, Evtushenko LI, Subbotin SA, Castillo P. 'Candidatus Xiphinematincola pachtaicus' gen. nov., sp. nov., an endosymbiotic bacterium associated with nematode species of the genus Xiphinema (Nematoda, Longidoridae). *International Journal of Systematic and Evolutionary Microbiology*. 2021;71(7).

Panaccione DG, Beaulieu WT, Cook D. Bioactive alkaloids in vertically transmitted fungal endophytes. *Functional Ecology*. 2014 Apr;28(2):299–314.

Pandey AT, Pandey I, Hachenberger Y, Krause BC, Haidar R, Laux P, Luch A, Singh MP, Singh AV. Emerging paradigm against global antimicrobial resistance via bioprospecting of mushroom into novel nanotherapeutics development. *Trends in Food Science & Technology*. 2020 Dec 1;106:333–44.

Pandey KD, Patel AK, Singh M, Kumari A. Secondary metabolites from bacteria and viruses. *Natural Bioactive Compounds*. 2021 Jan 1:19–40.

Pandey PK, Singh S, Singh MC, Singh AK, Pandey P, Pandey AK, Pathak M, Kumar M, Shakywar RC, Patidar RK. Inside the plants: Bacterial endophytes and their natural products. *International Journal of Current Microbiology and Applied Sciences*. 2017;6:33–41.

Pandita D, Pandita A. Secondary Metabolites in Medicinal and Aromatic Plants (MAPs): Potent Molecules in Nature's Arsenal to Fight Human Diseases. In *Medicinal and Aromatic Plants* 2021 (pp. 41–84). Springer.

Paramanantham P, Pattnaik S, Siddhardha B. Natural Products from Endophytic Fungi: Synthesis and Applications. In *Advances in Endophytic Fungal Research* 2019 (pp. 83–103). Springer.

Parthasarathy R, Sathiyabama M. Gymnemagenin-producing endophytic fungus isolated from a medicinal plant Gymnema sylvestre R. Br. *Applied Biochemistry and Biotechnology*. 2014 Mar;172(6):3141–52.

Parveen R, Khare N, Singh S, Johri P. An In Silico Outlook for the Detection and Surveillance of Evolving and Persistent Plant Pathogens. In *Detection, Diagnosis and Management of Soil-borne Phytopathogens* 2023 (pp. 17–40). Springer Nature Singapore.

Patani A, Patel A, Prajapati D, Khare N, Singh S. Various Agriculture Crop Plant-based Bioactive Compounds and Their Use in Nanomaterial Synthesis and Applications. In *Nanomaterials from Agricultural and Horticultural Products* 2023 (pp. 223–41). Springer Nature Singapore.

Qin X, Xu J, An X, Yang J, Wang Y, Dou M, Wang M, Huang J, Fu Y. Insight of endophytic fungi promoting the growth and development of woody plants. *Critical Reviews in Biotechnology*. 2023 Jan 2:1–22.

Radiastuti N, Suryani MJ, Susilowati DN, Purnamaningsih R, Mulya K. The Diversity of Endophitic Fungi on Annua Mutant Plantation (Artemisia annua L.) Based on ITS rDNA Marker. In *IOP Conference Series: Earth and Environmental Science* 2022 Feb 1 (Vol. 976, No. 1, p. 012038). IOP Publishing.

Rana KL, Kour D, Sheikh I, Dhiman A, Yadav N, Yadav AN, Rastegari AA, Singh K, Saxena AK. Endophytic Fungi: Biodiversity, Ecological Significance, and Potential Industrial Applications. In *Recent Advancement in White Biotechnology through Fungi* 2019 (pp. 1–62). Springer.

Rodriguez P, Gonzalez D, Giordano SR. Endophytic microorganisms: A source of potentially useful biocatalysts. *Journal of Molecular Catalysis B: Enzymatic.* 2016 Nov 1;133:S569–81.

Ruan Q, Patel G, Wang J, Luo E, Zhou W, Sieniawska E, Hao X, Kai G. Current advances of endophytes as a platform for production of anti-cancer drug camptothecin. *Food and Chemical Toxicology.* 2021 May 1;151:112113.

Sabido EM, Tenebro CP, Trono DJ, Vicera CV, Leonida SF, Maybay JJ, Reyes-Salarda R, Amago DS, Aguadera AM, Octaviano MC, Saludes JP. Insights into the variation in bioactivities of closely related Streptomyces Strains from marine sediments of the Visayan Sea against ESKAPE and ovarian cancer. *Marine Drugs.* 2021 Jul 31;19(8):441.

Saia S, Corrado G, Vitaglione P, Colla G, Bonini P, Giordano M, Stasio ED, Raimondi G, Sacchi R, Rouphael Y. An endophytic fungi-based biostimulant modulates volatile and non-volatile secondary metabolites and yield of greenhouse basil (Ocimum basilicum L.) through variable mechanisms dependent on salinity stress level. *Pathogens.* 2021 Jun 23;10(7):797.

Salvatore MM, Alves A, Andolfi A. Secondary metabolites of Lasiodiplodia theobromae: Distribution, chemical diversity, bioactivity, and implications of their occurrence. *Toxins.* 2020 Jul 17;12(7):457.

Samreen T, Naveed M, Nazir MZ, Asghar HN, Khan MI, Zahir ZA, Kanwal S, Jeevan B, Sharma D, Meena VS, Meena SK. Seed associated bacterial and fungal endophytes: Diversity, life cycle, transmission, and application potential. *Applied Soil Ecology.* 2021 Dec 1;168:104191.

Šantrůček J. The why and how of sunken stomata: Does the behaviour of encrypted stomata and the leaf cuticle matter? *Annals of Botany.* 2022 Apr 22;130:285.

Sao P, Parveen R, Khattri A, Sharma S, Tiwari N, & Singh S. Understanding the metabolomics of medicinal plants under environmental pollution: Challenges and opportunities. *Environmental Pollution and Medicinal Plants.* 2022:193–205. https://doi.org/10.1201/9781003178866-11

Sawant SS, Patil SM, Gupta V, Kunda NK. Microbes as medicines: Harnessing the power of bacteria in advancing cancer treatment. *International Journal of Molecular Sciences.* 2020 Oct 14;21(20):7575.

Sharma A, Malhotra B, Kharkwal H, Kulkarni GT, Kaushik N. Therapeutic agents from endophytes harbored in Asian medicinal plants. *Phytochemistry Reviews.* 2020 Jun;19(3):691–720.

Silva LJ, Crevelin EJ, Souza DT, Lacerda-Júnior GV, de Oliveira VM, Ruiz AL, Rosa LH, Moraes LA, Melo IS. Actinobacteria from Antarctica as a source for anticancer discovery. *Scientific Reports.* 2020 Aug 17;10(1):1–5.

Singh M, Kumar A, Singh R, Pandey KD. Endophytic bacteria: a new source of bioactive compounds. *3 Biotech.* 2017 Oct;7(5):1–4.

Singh RK, Shrivastava A, Yadav A, Srivastava AK. Endophytic Bacteria as a Source of Bioactive Compounds. In *Microbial Endophytes* 2020 Jan 1 (pp. 175–88). Woodhead Publishing.

Sinha RP, editor. *Natural Bioactive Compounds: Technological Advancements.* Academic Press; 2020 Oct 6.

Srinivasa C, Pradeep S, Patil SM, Ramu R, Kollur SP, Prasad SK, Shivamallu C. Bio prospecting of Lapachol producing endophytic fungi. *International Journal of Health and Allied Sciences.* 2022;11(1):13.

Srivastava AK, Singh VK. Biological action of Piper nigrum-the king of spices. *European Journal of Biological Research*. 2017 Sep 30;7(3):223–33.

Staniek A, Woerdenbag HJ, Kayser O. Endophytes: Exploiting biodiversity for the improvement of natural product-based drug discovery. *Journal of Plant Interactions*. 2008 Jun 1;3(2):75–93.

Tanvir R, Sheikh AA, Javeed A. Endophytic actinomycetes in the biosynthesis of bioactive metabolites: Chemical diversity and the role of medicinal plants. *Studies in Natural Products Chemistry*. 2019 Jan 1;60:399–424.

Tariq F, Ahmed N, Afzal M, Khan MA, Zeshan B. Synthesis, Characterization and antimicrobial activity of Bacillus subtilis-derived silver nanoparticles against multidrug-resistant bacteria. *Jundishapur Journal of Microbiology*. 2020 May 31;13(5). https://doi.org/10.5812/jjm.91934

Techaoei S, Jirayuthcharoenkul C, Jarmkom K, Dumrongphuttidecha T, Khobjai W. Chemical evaluation and antibacterial activity of novel bioactive compounds from endophytic fungi in Nelumbo nucifera. *Saudi Journal of Biological Sciences*. 2020 Nov 1;27(11):2883–9.

Teixeira HA, de Carvalho IS, Milenna VD, TarcÃsio CD, Pereira EC, Amariz IA, Sampaio PA, Rolim-Neto PJ, Rolim LA, AraÃºjo EC. Development and validation of an analytical method for quantification of total flavonoids in Alternanthera brasiliana by ultraviolet-visible absorption spectrophotometry. *African Journal of Biotechnology*. 2019 Oct 31;18(30):920–7.

Thi Minh Le T, Thi Hong Hoang A, Thi Bich Le T, Thi Bich Vo T, Van Quyen D, Hoang Chu H. Isolation of endophytic fungi and screening of Huperzine A–producing fungus from Huperzia serrata in Vietnam. *Scientific Reports*. 2019 Nov 6;9(1):1–3.

Tiwari P, Bae H. Endophytic fungi: Key insights, emerging prospects, and challenges in natural product drug discovery. *Microorganisms*. 2022 Feb 4;10(2):360.

Todd JN, Carreón-Anguiano KG, Islas-Flores I, Canto-Canché B. Microbial effectors: Key determinants in plant health and disease. *Microorganisms*. 2022 Oct 6;10(10):1980.

Toghueo RM, Boyom FF. Endophytic Penicillium species and their agricultural, biotechnological, and pharmaceutical applications. *3 Biotech*. 2020 Mar;10(3):1–35.

Uzma F, Mohan CD, Hashem A, Konappa NM, Rangappa S, Kamath PV, Singh BP, Mudili V, Gupta VK, Siddaiah CN, Chowdappa S. Endophytic fungi—alternative sources of cytotoxic compounds: A review. *Frontiers in Pharmacology*. 2018 Apr 26;9:309.

Vallet M, Chong YM, Tourneroche A, Genta-Jouve G, Hubas C, Lami R, Gachon CM, Klochkova T, Chan KG, Prado S. Novel α-Hydroxy γ-Butenolides of kelp endophytes disrupt bacterial cell-to-cell signaling. *Frontiers in Marine Science*. 2020 Jul 30;7:601.

Venugopalan A, Srivastava S. Endophytes as in vitro production platforms of high value plant secondary metabolites. *Biotechnology Advances*. 2015 Nov 1;33(6):873–87.

Verma N, Sao P, Srivastava A, & Singh S. Physiological and molecular responses to drought, submergence and excessive watering in plants. *Harsh Environment and Plant Resilience: Molecular and Functional Aspects*. 2021:305–21.

Verma SK, Sahu PK, Kumar K, Pal G, Gond SK, Kharwar RN, White JF. Endophyte roles in nutrient acquisition, root system architecture development and oxidative stress tolerance. *Journal of Applied Microbiology*. 2021 Nov;131(5):2161–77.

Verni M, Pontonio E, Krona A, Jacob S, Pinto D, Rinaldi F, Verardo V, Díaz-de-Cerio E, Coda R, Rizzello CG. Bioprocessing of brewers' spent grain enhances its antioxidant activity: Characterization of phenolic compounds and bioactive peptides. *Frontiers in Microbiology*. 2020 Jul 31;11:1831.

Vora Z, Pandya J, Vaikuntapu PR, Singh S, Asha S, & Purama RK. Genome-Wide Association of Defense Hormone Crosstalk in Plants. In *Hormonal Cross-Talk, Plant Defense and Development* 2023 (pp. 353–71). Academic Press.

Wang T, Lu Q, Sun C, Lukianov D, Osterman IA, Sergiev PV, Dontsova OA, Hu X, You X, Liu S, Wu G. Hetiamacin E and F, new amicoumacin antibiotics from Bacillus subtilis PJS using MS/MS-based molecular networking. *Molecules*. 2020 Sep 27;25(19):4446.

Wang XJ, Min CL, Ge M, Zuo RH. An endophytic sanguinarine-producing fungus from Macleaya cordata, Fusarium proliferatum BLH51. *Current Microbiology*. 2014 Mar;68(3):336–41.

Wani ZA, Hakeem KR. Plant–Endophyte Symbiosis. In *Endophyte Biology* 2022 Oct 27 (pp. 43–63). Apple Academic Press.

Wu J, Yan D, Liu Y, Luo X, Li Y, Cao C, Li M, Han Q, Wang C, Wu R, Zhang L. Purification, structural characteristics, and biological activities of exopolysaccharide isolated from leuconostoc mesenteroides SN-8. *Frontiers in Microbiology*. 2021 Mar 26;12:644226.

Xiang YZ, Li XY, Zheng HL, Chen JY, Lin LB, Zhang QL. Purification and antibacterial properties of a novel bacteriocin against Escherichia coli from Bacillus subtilis isolated from blueberry ferments. *LWT*. 2021 Jul 1;146:111456.

Xu P, Stirling E, Xie H, Li W, Lv X, Matsumoto H, Cheng H, Xu A, Lai W, Wang Y, Zheng Z. Continental scale deciphering of microbiome networks untangles the phyllosphere homeostasis in tea plant. *Journal of Advanced Research*. 2022 Apr 16. https://doi.org/10.1016/j.jare.2022.04.002

Yadav AN, Kour D, Rana KL, Yadav N, Singh B, Chauhan VS, Rastegari AA, Hesham AE, Gupta VK. Metabolic Engineering to Synthetic Biology of Secondary Metabolites Production. In *New and Future Developments in Microbial Biotechnology and Bioengineering* 2019 Jan 1 (pp. 279–320). Elsevier.

Yadav G, Srivastva R, Gupta P. Endophytes and Their Applications as Biofertilizers. In *Microbial Technology for Sustainable Environment* 2021 (pp. 95–123). Springer.

Yan L, Zhu J, Zhao X, Shi J, Jiang C, Shao D. Beneficial effects of endophytic fungi colonization on plants. *Applied Microbiology and Biotechnology*. 2019 Apr;103(8):3327–40.

Yin H, Wang W, Fan K, Li Z. Regulatory perspective of antibiotic biosynthesis in Streptomyces. *Science China Life Sciences*. 2019 May;62(5):698–700.

Yusupov UK, Yodgorova NT, Sharipova ZO, Shakhmurov NA. Endophytic fungi of medicinal plants of uzbekistan as producers of antibiotic compounds. *European Journal of Molecular & Clinical Medicine*. 2020 Dec 14;7(3):3247–57.

Zhang S, Cao L, Wang ZR, Li Z, Ma J. Anti-cancer effect of toosendanin and its underlying mechanisms. *Journal of Asian Natural Products Research*. 2019 Mar 4;21(3):270–83.

7 Plant Development and Protection with Biocontrol Agents and Secondary Metabolites

Parul Johri, Mala Trivedi, Manish Singh Rajput, Mrunal Gokhale, and Nazleen Jahan

7.1 INTRODUCTION

The indiscriminate use of chemical pesticides and their toxic effects on humans and the environment have led to the development of alternative methods of plant protection using bio-control agents and secondary metabolites. Plant secondary metabolites provide a wide range of protection against plant pathogens, insects, and pests (Wink, 2003; Verma et al., 2021). The term "biocontrol agents" has been used for a wide range of living organisms or their derivatives that act antagonistically against plant diseases and pests/pathogens; hence, they are also referred to as natural enemies. These biocontrol agents play a significant role in improving crop quality by acting either as epiphytes or endophytes (Zubair et al., 2021). Plant protection using biocontrol agents and secondary metabolites is not a simple process. It involves a complex cascade of events that take place at a molecular level, including signaling pathways and hormonal balancing, to combat the changing conditions of the pathogen and its surroundings (Gupta et al., 2022).

In this expanding era of science and research, the scope of controlling insects and pests using biocontrol agents has now changed with the introduction of prepared extracts and products using genetic manipulations to achieve the desired the genetic changes, thereby targeting a specific group of insects and pathogens (Singh et al., 2019).

Plants have an intrinsic ability to switch to a defense mechanism during a pest attack by activating a cascade of phytohormone signaling and pausing the growth-promoting events at that time. Plants produce secondary metabolites in response to stressful environments, which helps to protect them in adverse environmental conditions. Plant secondary metabolites (PSMs) are referred to as "specialized metabolites" due to their diverse chemical nature and biosynthesis (Jabborova et al., 2022).

 DOI: 10.1201/9781003392910-7

This chapter provides current information on different types of secondary metabolites produced by plants, their biosynthesis, genetic regulation, and their role in plant protection. Additionally, it also provides information about the importance of biocontrol agent (BCA) strategies for managing plant disease, and covers the types of BCAs and the challenges faced by BCAs, which are essential for future developments in protecting plants from disease and aiding in the development of plants.

7.2 PLANT SECONDARY METABOLITES

Plant secondary metabolites constitute a major defensive role for plants to defend themselves against a wide range of phytophagous herbivores, so it is necessary to understand their biosynthesis and applications to be able to use them as eco-friendly and sustainable pest management options (Maddala et al., 2023). Recent techniques and analytical tools have enabled the role of secondary metabolites in plant defence to be explained (Wink, 2003; Patani et al., 2023).

7.3 TYPES OF PLANT SECONDARY METABOLITES

PSM are broadly classified into four major groups based on their biosynthetic pathways and chemical behaviour—terpenoids, phenolics, alkaloids (N-containing compounds), and sulfur-containing compounds—as seen in Figure 7.1.

1. Terpenoids
2. Phenolics
3. Alkaloids (N-containing compounds)
4. Sulfur-containing compounds

7.3.1 Terpenes (Complex Terpenoids)

Terpenes are the largest group among all the secondary metabolites produced by plants. Terpenes are known for their antimicrobial and insecticidal properties and therefore have a pivotal role as plant defense control agents (Gangwar et al., 2023).

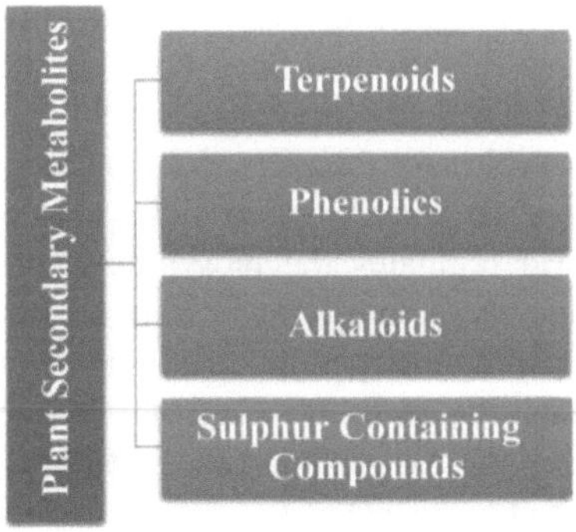

FIGURE 7.1 Types of plant secondary metabolites.

Terpenoids induce phytotoxicity against neighboring weed plants by releasing allelochemical compounds (secondary metabolites) which inhibit the germination of seeds initially and further interfere with cell division, nutrient uptake, and photosynthesis of the target weed (Mumm et al., 2008; Sao et al., 2022).

7.3.1.1 Bio-Synthesis of Terpenoids

Terpenoids are synthesized from repeating units of 5-carbon building blocks called isoprenes. Terpenes are mainly synthesized by two biochemical pathways, the mevalonic acid pathway (MAP) and the methyl erythritol phosphate (MEP) pathway in the cytosol and chloroplast, respectively (Bergman et al., 2019).

Both pathways produce isopentenyl diphosphate (IDP) and its isomer, dimethylallyl pyrophosphate (DMAPP), as their end products (Vishwakarma et al., 2024). These end products are the isoprenoid precursors, and they combine in the presence of enzyme prenyl transferases to form geranyl pyrophosphate (GPP, 10C compound, immediate precursor of monoterpenes). Further addition of a 5C isoprene (2-methyl-1,3-butadiene) unit to GPP results in farnesyl pyrophosphate (FPP, 15C), which is the precursor of sesquiterpenes (15C).

Further addition of an isoprene unit results in the formation of geranylgeranyl diphosphate (GGPP, 20C, diterpene precursor). Triterpenes (30C) and tetraterpenes (40C) are synthesized through head-to-head dimerization of FPP and GGPP precursors (Maddala et al., 2024).

7.3.1.2 Role of Terpenoids in Plant Defence

Plant protection by terpenoids occurs through two distinctive modes: direct protection and indirect protection. Direct protection involves the accumulation of phytoalexins in protective structures like trichomes, thorns, and needles. The phytoalexins are low molecular weight compounds with antimicrobial (antibiotic) properties and are released in response to biotic and/or abiotic factors (Meddya et al., 2023).

Indirect defense means that plants synthesize and release terpenoids after being invaded by pests. After being bitten by insects, some plants can release some amount of volatile terpenoids to attract natural enemies of pests (Chouhan et al., 2023).

7.3.1.3 Genetic Regulation of Terpenoid Biosynthesis

Various factors play a role in regulating the biosynthesis of terpenoids, such as developmental stages, circadian rhythm, and plant hormones, along with various biotic and abiotic stress factors.

Transcription factors such as AP2/ERF, WRKY, and bHLH are known to be involved in biosynthesis of terpenoids. In orchids, it has been shown that the transcription factor PbbHLH4 regulates the biosynthesis of different monoterpenes, imparting floral scent by interacting with PbbZIP4 and PbNAC1 (Y. C. Chuang et al., 2018). E-geraniol, a monoterpenoid, is a major contributor to the flavor of sweet oranges. In this plant, the gene CitTPS16, which is involved in the synthesis of E-geraniol, is regulated by the transcription factor CitERF71, which directly binds to its promoter region (Li et al., 2017).

The gene QH6 in *Artemisia annua*, involved in biosynthesis of the monoterpene beta-pinene, is regulated by circadian rhythms. In another plant, Liquidambar formosana, the genes LfTPS01, LfTPS02, and LfTPS03, involved in biosynthesis of sesquiterpenes, exhibit seasonal changes in their expression (L. Chuang et al., 2018). The progress of plant growth and development has a significant impact on the synthesis of terpenoids. The content and composition of essential oil in *Origanum vulgare* and *Thymus vulgaris* varied significantly at different growth stages of the same developmental stages. The expression patterns of TPS and terpene oil production also varied during different developmental stages of lemon leaves (Pandeló et al., 2012).

Plant hormones like methyl jasmonate (MeJA) are known to activate biosynthetic pathways of terpenoids in *Camellia sinensis* and *Picea abies* (Fäldt et al., 2003). Moreover, the concentration of aroma substances increased significantly after spraying grapes with MeJA, especially the content of monoterpenes (D'Onofrio et al., 2018).

Recent studies have shown that concentration variety of volatile terpenoids like farnesene increase in response to biotic stress, such as insect and microorganism attacks (Yoon et al., 2016).

Abiotic stress factors such as light, temperature, humidity, nutrient balance, PEG (polyethylene glycol), and mechanical damage have an effect on the biosynthesis of terpenoids. Certain conditions, such as high temperature exposure, have varied effects, where short-term exposure increases the concentration of terpenoids while long-term exposure decreases it (Fiedor & Burda, 2014).

7.3.2 Phenolics

Phenolics are heterogeneous secondary metabolites comprising a phenyl group on an aromatic ring. Phenolics have diverse functional roles and act as antioxidants, structural polymers (lignin), signaling molecules (flavonoids, salicylates), attractant molecules (flavonoids, carotenoids), protection from bacterial infection (tannins), natural antiseptics, and wound healing (catechols).

They can directly act as toxins or can be enzymatically catalyzed into more toxic metabolites by polyphenol oxidases, thereby inhibiting the growth and development activities in insects. In response to biotic stress, the phenylpropanoid pathway is triggered, producing excess phenolic acids. These phenolic acids start to accumulate in the cell wall and provide a second line of defense by producing cinnamic acid and benzoic acid.

7.3.2.1 Bio-Synthesis of Phenolics

Phenolics are biosynthesized via two major distinctive pathways, namely the shikimic acid pathway and the phenylpropanoid pathway (Parveen et al., 2023).

Phenolic compounds are mainly biosynthesized by the shikimic acid pathway in advanced plants, and shikimate is the key central metabolite for the synthesis of complex phenolics. Phenylalanine ammonia-lyase is a prime enzyme for the phenylpropanoid biosynthetic pathway, connecting primary metabolism to secondary metabolism.

7.3.2.2 Role of Phenolics in Plant Defence

Phenolic compounds play a major role in the induction of resistance in plants (Khare et al., 2017). Generally, phenolic compounds released from seeds, roots, or residue decomposition can act against soilborne pathogens and root-feeding insects. Various phenolic compounds, such as cajanin, medicarpin, glyceollin, rotenone, coumestrol, phaseolin, phaseolinin, isoflavonoid, and flavonoids, act as phytoalexins, phytoanticipins, and nematicides against soilborne pathogens and phytophagous insects. Accumulation of phenolic compounds at the challenge site also reinforces the cell wall, which is accompanied by localized production of reactive oxygen species driving cell wall cross-linking, antimicrobial activity, and defense signaling (Khare et al., 2024).

Phenolics are synthesized when plant pattern recognition receptors recognize potential pathogens through conserved pathogen-associated molecular patterns (PAMPs), leading to PAMP-triggered immunity. As a result, the progress of the infection is restricted long before the pathogen gains complete hold of the plant (Bhattacharya et al., 2010).

7.3.2.3 Genetic Regulation of Phenolic Biosynthesis

Genetically, the production of PSMs is regulated at the transcriptional level by a cascade of regulatory events driven by various transcription factors (TFs). TFs are sequence-specific regulatory proteins that bind to specific regions of the DNA and alter the transcription rate. These regulatory proteins are tuned by signaling molecules known as jasmonates, collectively named from jasmonic acid and its derivatives, methyl jasmonates. These molecules are conserved across gymnosperms and angiosperms in the plant kingdom.

smERF115 is an AP2/ERF transcription factor whose overexpression in hairy roots in Salvia miltiorrhiza led to increased phenolic acid production.

7.3.3 Alkaloids

Alkaloids are the largest group of secondary metabolites in plants. Basically, alkaloids contain nitrogen. The central types of alkaloids are (a) true alkaloids, (b) protoalkaloids, and (c) pseudoalkaloids.

True alkaloids and protoalkaloids are produced from amino acids, whereas, as the name indicates, pseudoalkaloids are not derived from amino acids.

True alkaloids are a major group of alkaloids. Their defining feature is that their structure contains a nitrogen-containing heterocyclic ring. Various amino acids, such as L-phenylalanine/L-tyrosine, L-ornithine, L-histidine, and L-lysine, are the main sources of true alkaloids. Cocaine, morphine, and quinine are the common true alkaloids found in nature.

Protoalkaloids form a minor group of alkaloids. The structure of this group contains a nitrogen atom, which is derived from an amino acid but is not part of the heterocyclic ring system. L-tryptophan and L-tyrosine are the main precursors of this type of alkaloid. Yohimbine, mescaline, and hordenine are the main alkaloids of this type.

The basic carbon skeleton of pseudoalkaloids is not directly derived from amino acids; instead, they are connected with amino acid pathways where they are derived from by amination or transamination reactions from forerunners or post-cursors of amino acids. Capsaicin, caffeine, and ephedrine are very common examples of pseudoalkaloids.

7.3.3.1 Bio-Synthesis of Alkaloids

Three key reactions that form the cornerstone of the biosynthesis of alkaloids are:

i) Mannich reaction of a Schiff base with a nucleophile, where an aldehyde or a ketone condenses with an amine and elimination of water occurs; the product is a Schiff base.
ii) Pictet-Spengler condensation, where the Schiff base is attacked intermolecularly by an aromatic nucleus. The product is often either a tetrahydroisoquinoline or a tetrahydro-beta-carboline.
iii) Phenolic coupling reaction, where a phenolic hydroxyl group is oxidized by the loss of a hydrogen radical. A highly reactive radical intermediate is produced, which can be trapped either internally or by another radical-containing unit, to form several different products, including those derived from carbon-oxygen bond formation and carbon-carbon bond formation at the positions ortho or para to the phenolic radical.

The alkaloids derived from ornithine are produced by a pathway where it is converted into an intermediate, which is a Schiff base, the N-methyl-Δ1-pyrrolinium cation. This compound then can lead to formation of a unit in alkaloids such as ficine, brevicolline, and cuscohygrine.

The tropane and ecgonine alkaloids, which are derived from ornithine or arginine, involve sequential addition of acetate to the N-methyl-Δ1-pyrrolinium, leading to formation of (-)-hyoscyamine and scopolamine in the tropane group and cocaine in the ecgonine group.

Alkaloids derived from phenylalanine or tyrosine amino acids involve oxidation reactions involving the NIH-shift of the 4'-proton. The oxidation product can then can undergo different types of reactions such as aromatic nucleus cleavage, reduction, and cyclization, leading to different types of alkaloids (Lichman, 2021).

7.3.3.2 Role of Alkaloids in Plant Defence

Alkaloids provide defensive elements against predators, especially animals, vertebrates, insects, and arthropods, due to their general toxicity and deterrence capability. Toxicity may arise by enzymatic alterations affecting physiological processes, or inhibition of DNA synthesis and repair mechanisms by intercalating with nucleic acids, affecting the nervous system or disrupting cell signaling.

7.3.3.3 Genetic Regulation of Alkaloid Biosynthesis

The transcription factors that are involved in the biosynthesis of several types of alkaloids include bHLH, AP2/ERF, and WRKY. In addition, regulatory factors such as MAP kinases play an important role in the genetic regulation of alkaloid biosynthesis. Most of them show jasmonate (JA) responsiveness, which suggests that the JA signaling cascade plays an important role in alkaloid biosynthesis.

7.3.4 Sulfur-Containing Compounds

Sulfur-containing compounds are unusual in plants compared to other categories of secondary metabolites. They are found in families like Brassicaceae, Alliaceae, and Asteraceae. In these families, they constitute major chemical defenses and play important role in plant-pest interactions. They play an important role in two activated plant defense systems, i.e., the glucosinolate–myrosinase system of the Brassicaceae, and the alliin–alliinase system of the Alliaceae. Apart from these systems, sulfur-containing compounds such as phytoalexins have structures ranging from elemental sulfur to the alkaloid-like camalexin, and complex polypeptides like defensins. The diversity of these sulfur-containing molecules provides plants with a versatile array of chemical defenses against a broad range of potential enemies.

7.3.4.1 Bio-Synthesis of Sulfur-Containing Compounds

In the family Brassicaceae, thio-glucosides, also known as glucosinolates, and indole-type phytoalexins are the major sulfur-containing compounds.

The glucosinolates biosynthetic pathway starts with chain-elongated methionine, which undergoes a series of biochemical reactions to form aci-nitro compounds. Glutathione donates the sulfur atoms to these aci-nitro compounds through the action of specialized glutathione S-transferases to form intermediate GSH conjugates. These GSH conjugates are then converted into aliphatic glucosinolates through a series of biochemical reactions.

The phytoalexin biosynthesis pathway starts with the amino acid tryptophan, which provides the necessary indole ring. It is converted into indole-3-acetaldoxime. There is a branching from this point, which leads to the formation of either 1-aci-nitro-2-indolyl-ethane or indole-3-acetonitrile. Both these compounds receive sulfur from glutathione through the action of glutathione S-transferases to form intermediate GSH conjugates. Through a series of biochemical reactions, the GSH-thiohydroximate conjugates derived from either 1-aci-nitro-2-indolyl-ethane can lead to formation of sulfur-containing compounds like wasalexin, brassinin, etc. On the other hand, the GSH-indole-3-acetonitrile conjugate leads to the formation of sulfur-containing compounds like camalexin (Czerniawski & Bednarek, 2018).

7.3.4.2 Role of Sulfur-Containing Compounds in Plant Defence

Various sulfur-containing secondary metabolites are termed phytoanticipins, as they are already in place before any external attack by pathogens or are synthesized immediately from inactive precursors already present in the plants with no expenditure of energy. They, therefore, anticipate the attack and are adequately equipped by nature to ward off such challenges by microorganisms and pests. These compounds are antibacterial, antifungal, and anti-insect, and prevent pests and herbivore attacks.

Allicin from garlic and Petivericin from *Petiveria alliacea* are examples of anti-bacterial and anti-fungal sulfur-containing compounds.

The sulfur compounds of glucosinolates are involved in plant defense against herbivores. The sulfur-containing compounds, thiophenes in Asteraceae, are known to be associated with defense against insects and nematodes. Additionally, they also exhibit anti-bacterial and anti-fungal activity.

Polysulfanes from diverse plants like garlic, onion, shiitake mushroom, durian fruit, and *Petiveria alliacea* exhibit antimicrobial, insecticidal, and larvicidal properties, and thus play an important role in plant defence.

7.3.4.3 Genetic Regulation of Sulfur Compound Biosynthesis

The genes involved in sulfate assimilation are light-induced and are directly correlated with glucosinolates, an important class of sulfur-containing secondary metabolites. Six genes directly associated with the GLS biosynthesis pathway—SOT16, SOT17, SOT18, MAM-L, CYP79F1, and CYP79B2—are also light-induced.

Transcription factors *MYB28*, *MYB29*, and *MYB76*, which have been shown to control biosynthesis of aliphatic glucosinolates, are also light-induced. Transcription factors MYB51, MYB34, and MYB122 are associated with the expression genes for indolic glucosinolates. They show varied responses to light induction.

The light induction of most genes in the GLS biosynthetic network is dependent upon the gene HY5. For some genes, HY5 is transcription repressor, but for others it is a transcription activator (Huseby et al., 2013).

7.4 PLANT SECONDARY METABOLITES AND STRESS MANAGEMENT

Stress response in plants is a collection of responses at molecular, cellular, and signaling levels, which are triggered upon detection of specific or combined biotic or abiotic stress factors. It results in the induction of secondary metabolites. The immune system of plants covers numerous stress detection mechanisms, including transmembrane recognition, polymorphic NB-LRR protein production by most R-genes, and production of secondary metabolites to cope with the stress situation.

Different abiotic and biotic stress factors such as high and low temperature, drought, alkalinity, salinity, UV stress, and pathogen infection lead to theproduction of different types of secondary metabolites. The concentrations of various secondary plant products are strongly dependent on the growing conditions and have an impact on the metabolic pathways responsible for the accumulation of the related natural products. Environmental stresses, such as pathogen attack, UV irradiation, high intensity light, wounding, nutrient deficiencies, temperature, and herbicide treatment, often increase the accumulation of phenylpropanoids.

Nutrient stress also has a marked effect on phenolic levels in plant tissues. Deficiencies in nitrogen and phosphate directly influence the accumulation of phenylpropanoids. Potassium, sulfur, and magnesium deficiency are also reported to increase phenolic concentrations. Low iron levels can cause increased release of phenolic acids from roots.

Salt stress usually leads to an increase in anthocyanin in salt-tolerant plants, while in salt-sensitive plants, the anthocyanin levels go down when exposed to salt stress.

Drought often causes oxidative stress and was reported to show an increase in the amounts of flavonoids and phenolic acids. Anthocyanins are reported to accumulate under drought stress and at cold temperatures. Plant tissues containing anthocyanins are usually rather resistant to drought. Flavonoids have protective functions during drought stress.

Flavonoids are implicated to provide protection to plants growing in soils that are rich in toxic metals such as aluminium. Trace metals limit anthocyanin biosynthesis by inhibiting the activity of l-phenylalanine ammonia-lyase. Co2+ and Cu2+ exhibit a stimulatory effect on the production of secondary metabolites.

Cold stress increases phenolic production and its subsequent incorporation into the cell wall either as suberin or lignin, which increases resistance to cold temperatures.

7.5 BIOCONTROL AGENTS

Biocontrol agents are living organisms or microorganisms, including parasites, predators, and disease-causing bacteria, fungi, and viruses, that control plant pathogens and also help in plant development by invading plant pathogens. They are the natural enemies of plant pathogens and can affect the life cycle of a pathogen in such a way that crop damaged is reduced. Plants are persistently exposed to various phytopathogens such as bacteria, fungi, viruses, nematodes, and oomycetes. These pathogens pose a serious threat to crop productivity worldwide. They can significantly reduce crop yields, ranging from 20% to 40%, by causing various pathogenic diseases. To provide an adequate amount of food for the growing world population, an effective management system is required to control various crop diseases (Farzand et al., 2019).

Previously, chemical pesticides and implementation of disease-resistant crops were adopted as the primary approaches, but they are not eco-friendly in nature. Nowadays, researchers are exploring the use of biocontrol agents as an eco-friendly strategy to control plant diseases and help in the development of plants (Zubair et al., 2021).

7.5.1 Types of Biocontrol Agent

Based on the interaction of plant pathogens with beneficial bacteria and fungi, biocontrol agents (Figure 7.2) are divided into two kinds: bacteria as BCA, and fungi as BCA.

7.5.1.1 Bacteria as a BCA

Many studies have investigated the role of bacteria in reducing plant disease and promoting plant development. At the present time, many bacteria from a variety of genera, including *Acinetobacter*, *Agrobacterium*, *Azospirillum*, Azotobacter, *Bacillus*,

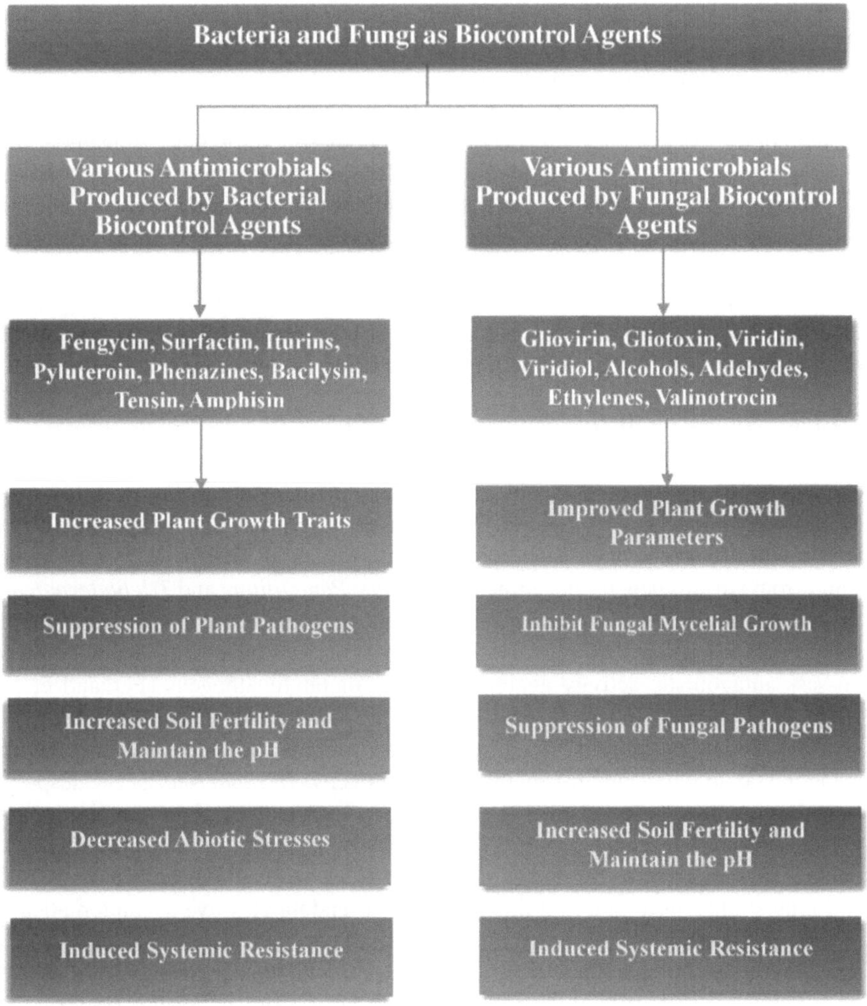

FIGURE 7.2 Important roles of different biocontrol agents. This figure shows the different antimicrobial products produced by biocontrol agents, i.e., bacteria and fungi, and their impact on plant pathogens and plant growth parameters.

Bradyrhizobium, *Paenibacillus*, *Pseudomonas*, *Rhizobium*, and *Streptomyces*, have been documented as biocontrol agents to control various plant diseases (Ayaz et al., 2021). Plant pathogens, mainly bacteria, fungi, and nematodes, lead to serious plant diseases, which is significant obstacle to sustainable crop yield. Crops are severely affected when infected by pathogenic bacteria such as *Pseudomonas savastanoi*, *Xanthomonas axonopodis*, and *Ralstonia solanacearum*. However, inoculation with Bacillus species suppresses the growth of pathogens and reduces the severity of disease in many infected crops (Massawe et al., 2018).

Bacteria that act as BCAs create biofilm and secondary metabolites, such as bacillomycin, fengycin, iturin, and surfactin, which minimize the plant pathogen population by creating plant-microbial interactions in the rhizosphere region (Ayaz et al., 2022). *Bacillus species* attach to mycelial cell walls and deform hyphae by the production of extracellular enzymes such as cellulase, chitosanase, glucanase, and protease (Zubair et al., 2021). Some lipopeptides, including fengycin, iturin, mixirin, pumilacidin, and surfactin, are antifungal peptides that function against pathogenic fungi in the rhizosphere (Farzand et al., 2019). Several bacteria, specifically *Bacillus*, *Pseudomonas*, and *Burkholderia species*, are known to suppress nematodes in several plants by affecting nematode behaviors such as reproduction and feeding. Thus, there is a large variety of bacteria that interact with plant pathogens to protect the plant from phytopathogens and help in the development of the plants.

7.5.1.2 Fungi as a BCA

Many fungi produce a large number of bioactive compounds that are used as agrochemicals for crop protection. The development of fungal strains as BCAs for crop diseases has gained a lot of attention because it has been found that many fungi inhibit the growth of various phytopathogens (Thambugala et al., 2020). At present, many fungi from various genera, including *Aspergillus*, *Penicillium*, and *Trichoderma*, are used as BCAs against both bacterial and fungal plant diseases. In addition to these, some other fungi like *Gliocladium* and *Saccharomyces* have been documented to possess antagonistic activity against a variety of phytopathogens (Farzand et al., 2019).

Many fungi, particularly *Trichoderma*, are popular for their broad-spectrum antagonistic activities against many plant pathogens. *Trichoderma* is a widespread fungal genus that is found in soil as residents, saprotrophs, plant symbionts, and mycoparasites. This genus contains filamentous fungi that have been broadly investigated and employed as BCAs against plant pathogens in agriculture.

Many studies have reported that most fungal endophytes live asymptomatically in the inner plant tissues. Fungal endophytes interact with their hosts very closely and intricately through mutualism and, in rare cases, by parasitism (Alblooshi et al., 2022). These are crucial for protecting their hosts from pathogenic bacteria and fungi because of their capacity to produce a great variety of structurally varied and physiologically active secondary metabolites (Zubair et al., 2021). Fungal endophytes are a rich source for discovering novel secondary metabolites with a wide range of potential agricultural applications. Various metabolites with numerous chemical structures, including alkaloids, terpenoids, steroids, peptides, isocoumarins, benzopyranones, and quinones, have been produced by endophytic fungi. The identification of these metabolites provides a solid chemical basis for the development of agrochemicals that may have antibacterial, antifungal, herbicidal, nematocidal, and insecticidal properties for reducing the severity of disease in many infected plants (Contreras-Cornejo et al., 2016). Thus, there is a large variety of fungi that interact with plant pathogens to protect the plant from phytopathogens and help in the plant's development.

7.6 FACTORS AFFECTING BIOCONTROL ACTIVITY OF BCAS

After decades of studies, biocontrol activity of BCAs under laboratory conditions has been documented as more effective than in open-field trials (Ayaz et al., 2021). Greater challenges are faced by the BCAs when they transition from the controlled environment of the laboratory to the harsh environment of the open field (Lahlali et al., 2022). The factors affecting the efficacy of BCAs are discussed below in detail.

7.6.1 Plant Species Influence of Biocontrol Activity of BCAs

Different plant species and genotypes have a different impact on the effectiveness of the biocontrol potential of microbes (Timper, 2014). Many studies have been conducted that show the effect of plant species on BCA activities. The degree of rhizosphere colonization, the production of antibiotics by BCA, and the stimulation of induced systemic resistance vary in different plant species (Umer et al., 2021). In the rhizosphere region, the interaction shown between BCAs and phytopathogens is mainly influenced by root exudate excretion, mineral absorption, water, and other surrounding microbes (Babalola, 2010). The expression of plant genes is reported to be involved in the rhizosphere enrichment process, which provides fitness to the plants by recruiting microbes from the surroundings (Choudhary & Johri, 2009). The complexity behind the specificity of BCAs for specific hosts must be elucidated with in-depth research to control plant disease in a well-organized manner.

7.6.2 Pathogen Influence of Biocontrol Activity of BCAs

The behavior of phytopathogens toward BCAs is one of the key factors in plant disease control. Each phytopathogen interacts with the host differently because of their genetic variety and ecological fitness diversity (Thambugala et al., 2020). Depending on the complexity of their modes of actions, plant pathogens may show a wide range of sensitivities to BCAs. According to the researchers, many phytopathogens have developed tolerance to one or more BCAs, such as Bt toxin (Lahlali et al., 2022).

The strength of the biological control of plant disease has been reported to be low (Akutse et al., 2020). Thus, weak spots in the phytopathogen life cycle must be found as opportunity windows for successful biocontrol (Bardin et al., 2015). BCAs should enter through the window of opportunity and stop the progression of pathogens. For example, the window of opportunity for some unspecialized necrotrophic phytopathogens is to block the nutrient uptake required for their growth. If BCAs are present in large quantities near the phytopathogen spores, the consumption of local nutrients from the disease spore could restrict or stop germination (Umer et al., 2021).

7.6.3 Environmental Stress Impacts on BCA Activity

The microbial makeup of the soil and its environment has a significant impact on the efficacy of biological control measures. Soil biology should investigate the distinctive qualities of different species, especially those that live in the rhizosphere of plants. Similarly, ecological study research should be able to identify all biotic and abiotic factors that significantly impact BCAs as crucial components of plant health (Lahlali et al., 2022). Many BCAs are highly susceptible to changes in the biotic and abiotic environment. It is not always possible to transfer their efficiency from the laboratory to the open field (Risoli et al., 2022). For example, many Pseudomonas and Bacillus BCAs function well in trials, but they cannot effectively manage diseases in a variety of field conditions (Mark et al., 2006). Thus, it is very important to select a BCA that maintains stable efficiency under different environmental conditions, such as soil texture, moisture, temperature extremes, or competition.

7.7 CONCLUSION

Secondary metabolites play a major role in the adaptation of plants to the environment and in overcoming biotic and abiotic stress conditions. Well-defined biosynthetic pathways and a wide array of genes supporting and controlling the production of different types of secondary metabolites indicate that the secondary metabolite-dependent mechanism of handling different types of stresses is highly evolved and is expected to keep evolving with changing climate scenarios. In addition to this, the use of biocontrol agents to control plant diseases is a safe and eco-friendly strategy that has attracted the attention of many scientists in the crop protection sector. This chapter addresses the importance of the role of BCAs in preventing diseases in crops. Moreover, for effective BCA strategies in the future, more research must be conducted on some less-developed aspects, including the types of BCAs and the challenges faced by them. Therefore, this chapter will be helpful to researchers studying plant-microbe interactions, as it provides sufficient knowledge on the current status of PSMs and BCAs for the further development and protection of plants.

REFERENCES

Akutse, K. S., Subramanian, S., Maniania, N. K., Dubois, T., & Ekesi, S. (2020). Biopesticide research and product development in Africa for Sustainable Agriculture and Food Security – Experiences From the International Centre of Insect Physiology and Ecology (ICIPE). *Frontiers in Sustainable Food Systems*, 4(September), 1–14. https://doi.org/10.3389/fsufs.2020.563016

Alblooshi, A. A., Purayil, G. P., Saeed, E. E., Ramadan, G. A., Tariq, S., Altaee, A. S., El-Tarabily, K. A., & Abuqamar, S. F. (2022). Biocontrol potential of endophytic actinobacteria against fusarium solani, the causal agent of sudden decline syndrome on date palm in the UAE. *Journal of Fungi*, 8(1). https://doi.org/10.3390/jof8010008

Ayaz, M., Ali, Q., Farzand, A., Khan, A. R., Ling, H., & Gao, X. (2021). Nematicidal volatiles from bacillus atrophaeus gbsc56 promote growth and stimulate induced systemic resistance in tomato against meloidogyne incognita. *International Journal of Molecular Sciences*, 22(9). https://doi.org/10.3390/ijms22095049

Ayaz, M., Ali, Q., Jiang, Q., Wang, R., Wang, Z., Mu, G., Khan, S. A., Khan, A. R., Manghwar, H., Wu, H., Gao, X., & Gu, Q. (2022). Salt Tolerant Bacillus Strains Improve Plant Growth Traits and Regulation of Phytohormones in Wheat under Salinity Stress. *Plants*, 11(20). https://doi.org/10.3390/plants11202769

Babalola, O. O. (2010). Beneficial bacteria of agricultural importance. *Biotechnology Letters*, 32(11), 1559–1570. https://doi.org/10.1007/s10529-010-0347-0

Bardin, M., Ajouz, S., Comby, M., Lopez-Ferber, M., Graillot, B., Siegwart, M., & Nicot, P. C. (2015). Is the efficacy of biological control against plant diseases likely to be more durable than that of chemical pesticides? *Frontiers in Plant Science*, 6(July), 1–14. https://doi.org/10.3389/fpls.2015.00566

Bergman, M. E., Davis, B., & Phillips, M. A. (2019). Medically useful plant terpenoids: Biosynthesis, occurrence, and mechanism of action. *Molecules*, 24(21), 1–23. https://doi.org/10.3390/molecules24213961

Bhattacharya, A., Sood, P., & Citovsky, V. (2010). The roles of plant phenolics in defence and communication during Agrobacterium and Rhizobium infection. *Molecular Plant Pathology*, 11(5), 705–719. https://doi.org/10.1111/j.1364-3703.2010.00625.x

Choudhary, D. K., & Johri, B. N. (2009). Interactions of Bacillus spp. and plants - With special reference to Induced Systemic Resistance (ISR). *Microbiological Research*, 164(5), 493–513. https://doi.org/10.1016/j.micres.2008.08.007

Chouhan PN, Azimov A, Jabborova D, & Singh S. (2023). Review on the role of indigenous plants and plant derivatives of Ayurveda in bone healing. *Plant Science Today*, 10(sp2), 143–151.

Chuang, L., Wen, C. H., Lee, Y. R., Lin, Y. L., Hsu, L. R., Wang, S. Y., & Chu, F. H. (2018). Identification, functional characterization, and seasonal expression patterns of five sesquiterpene synthases in Liquidambar formosana [Review-article]. *Journal of Natural Products*, 81(5), 1162–1172. https://doi.org/10.1021/acs.jnatprod.7b00773

Chuang, Y. C., Hung, Y. C., Tsai, W. C., Chen, W. H., & Chen, H. H. (2018). PbbHLH4 regulates floral monoterpene biosynthesis in Phalaenopsis orchids. *Journal of Experimental Botany*, 69(18), 4363–4377. https://doi.org/10.1093/jxb/ery246

Contreras-Cornejo, H. A., Macías-Rodríguez, L., del-Val, E., & Larsen, J. (2016). Ecological functions of Trichoderma spp. and their secondary metabolites in the rhizosphere: interactions with plants. *FEMS Microbiology Ecology*, 92(4), 1–17. https://doi.org/10.1093/femsec/fiw036

Czerniawski, P., & Bednarek, P. (2018). Glutathione S-transferases in the biosynthesis of sulfur-containing secondary metabolites in brassicaceae plants. *Frontiers in Plant Scienc*e, 871(November), 1–8. https://doi.org/10.3389/fpls.2018.01639

D'Onofrio, C., Matarese, F., & Cuzzola, A. (2018). Effect of methyl jasmonate on the aroma of Sangiovese grapes and wines. *Food Chemistry*, 242(May 2017), 352–361. https://doi.org/10.1016/j.foodchem.2017.09.084

Fäldt, J., Martin, D., Miller, B., Rawat, S., & Bohlmann, J. (2003). Traumatic resin defense in Norway spruce (Picea abies): Methyl jasmonate-induced terpene synthase gene expression, and cDNA cloning and functional characterization of (+)-3-carene synthase. *Plant Molecular Biology*, 51(1), 119–133. https://doi.org/10.1023/A:1020714403780

Farzand, A., Moosa, A., Zubair, M., Khan, A. R., Massawe, V. C., Tahir, H. A. S., Sheikh, T. M. M., Ayaz, M., & Gao, X. (2019). Suppression of sclerotinia sclerotiorum by the induction of systemic resistance and regulation of antioxidant pathways in tomato using fengycin produced by Bacillus amyloliquefaciens FZB42. *Biomolecules*, 9(10). https://doi.org/10.3390/biom9100613

Fiedor, J., & Burda, K. (2014). Potential role of carotenoids as antioxidants in human health and disease. *Nutrients*, 6(2), 466–488. https://doi.org/10.3390/nu6020466

Gangwar P, Johri P, Datta R, Singh S, & Trivedi M. (2023). Overview on 1 Phytopharmaceuticals and Biotechnology of Herbal Plants. In *Phytopharmaceuticals and Biotechnology of Herbal Plants* (pp. 1–17). CRC Press.

Gupta, G., Paul, S., Singh, S., Pietramellara, G., Pathan, S. I., Danish, S., & Jha, P. N. (2022). Exploring functional diversity and community structure of diazotrophic endophytic bacteria associated with Pennisetum glaucum growing under field in a semi-arid region. *Land*, 11(7), 991.

Huseby, S., Koprivova, A., Lee, B. R., Saha, S., Mithen, R., Wold, A. B., Bengtsson, G. B., & Kopriva, S. (2013). Diurnal and light regulation of sulphur assimilation and glucosinolate biosynthesis in Arabidopsis. *Journal of Experimental Botany*, 64(4), 1039–1048. https://doi.org/10.1093/jxb/ers378

Jabborova, D., Davranov, K., Jabbarov, Z., Bhowmik, S. N., Ercisli, S., Danish, S., & Datta, R. (2022). Dual inoculation of plant growth-promoting Bacillus endophyticus and Funneliformis mosseae improves plant growth and soil properties in ginger. *ACS Omega*, 7(39), 34779–34788.

Khare, P., Kishore, K., & Sharma, D. K. (2017). A study on the standardization parameters of Cassia angustifolia. *Asian Journal of Pharmaceutical and Clinical Research*, 10(7), 329–332.

Khare, P., Kishore, K., Sharma, D. K., & Khare, N. (2024). Investigation of memory enhancing activity of combination extracts of bauhinia variegata and madhuca longifolia leaves against colchicine: An experimental study and biochemical alterations in Mice. *Indian Journal of Pharmaceutical Education and Research*, 58(1s), s282–s288.

Lahlali, R., Ezrari, S., Radouane, N., Kenfaoui, J., Esmaeel, Q., El Hamss, H., Belabess, Z., & Barka, E. A. (2022). Biological Control of Plant Pathogens: A Global Perspective. In *Microorganisms* (Vol. 10, Issue 3). https://doi.org/10.3390/microorganisms10030596

Li, X., Xu, Y., Shen, S., Yin, X., Klee, H., Zhang, B., & Chen, K. (2017). Transcription factor CitERF71 activates the terpene synthase gene CitTPS16 involved in the synthesis of e -geraniol in sweet orange fruit. *Journal of Experimental Botany*, 68(17), 4929–4938. https://doi.org/10.1093/jxb/erx316

Lichman, B. R. (2021). The scaffold-forming steps of plant alkaloid biosynthesis. *Natural Product Reports*, 38(1), 103–129. https://doi.org/10.1039/d0np00031k

Maddala, V. K. S., & Singh, S. (2023). Essential oils from plants and their role in nanomaterial synthesis characterization and applications. In *Secondary Metabolites Based Green Synthesis of Nanomaterials and Their Applications* (pp. 191–207). Springer Nature Singapore.

Maddala, V. K. S., Singh, S., & Bandi-, J. (2024). Soil Properties Influencing Nutrient Signaling in Plant Soil Systems. In *Essential Minerals in Plant-Soil Systems* (pp. 373–393). Elsevier.

Mark, G. L., Morrissey, J. P., Higgins, P., & O'Gara, F. (2006). Molecular-based strategies to exploit Pseudomonas biocontrol strains for environmental biotechnology applications. *FEMS Microbiology Ecology*, 56(2), 167–177. https://doi.org/10.1111/j.1574-6941.2006.00056.x

Massawe, V. C., Hanif, A., Farzand, A., Mburu, D. K., Ochola, S. O., Wu, L., Tahir, H. A. S., Gu, Q., Wu, H., & Gao, X. (2018). Volatile compounds of endophytic Bacillus spp. have biocontrol activity against Sclerotinia sclerotiorum. *Phytopathology*, 108(12), 1373–1385. https://doi.org/10.1094/PHYTO-04-18-0118-R

Meddya, S., Meshram, S., Sarkar, D., Datta, R., Singh, S., Avinash, G., & Thulasinathan, T. (2023). Plant stomata: An unrealized possibility in plant defense against invading pathogens and stress tolerance. *Plants*, 12(19), 3380.

Mumm, R., Posthumus, M. A., & Dicke, M. (2008). Significance of terpenoids in induced indirect plant defence against herbivorous arthropods. *Plant, Cell and Environment*, 31(4), 575–585. https://doi.org/10.1111/j.1365-3040.2008.01783.x

Pandeló, D., Melo, T. D., Singulani, J. L., Guedes, F. A. F., Machado, M. A., Coelho, C. M., Viccini, L. F., & Santos, M. O. (2012). Oil production at different stages of leaf development in Lippia alba. *Revista Brasileira de Farmacognosia*, 22(3), 497–501. https://doi.org/10.1590/S0102-695X2012005000013

Parveen, R., Khare, N., Singh, S., & Johri, P. (2023). An In Silico Outlook for the Detection and Surveillance of Evolving and Persistent Plant Pathogens. In *Detection, Diagnosis and Management of Soil-borne Phytopathogens* (pp. 17–40). Springer Nature Singapore.

Patani A, Patel A, Prajapati D, Khare N, & Singh S. (2023). Various Agriculture Crop Plant-based Bioactive Compounds and Their Use in Nanomaterial Synthesis and Applications. In *Nanomaterials from Agricultural and Horticultural Products* (pp. 223–241). Springer Nature Singapore.

Risoli, S., Cotrozzi, L., Sarrocco, S., Nuzzaci, M., Pellegrini, E., & Vitti, A. (2022). Trichoderma-induced resistance to botrytis cinerea in solanum species: A meta-analysis. *Plants*, 11(2). https://doi.org/10.3390/plants11020180

Sao, P., Parveen, R., Khattri, A., Sharma, S., Tiwari, N., & Singh, S. (2022). Understanding the Metabolomics of Medicinal Plants Under Environmental Pollution: Challenges and Opportunities. In *Environmental Pollution and Medicinal Plants* (pp. 193–205). CRC Press.

Singh, A., Sengar, R. S., Singh, A. K., Singh, S., Kumar, A., & Ahmad, K. (2019). Impact of genetic engineering on agriculture food crops. *Biotech Today: An International Journal of Biological Sciences*, 9(1), 17–24.

Thambugala, K. M., Daranagama, D. A., Phillips, A. J. L., Kannangara, S. D., & Promputtha, I. (2020). Fungi vs. fungi in biocontrol: An overview of fungal antagonists applied against fungal plant pathogens. *Frontiers in Cellular and Infection Microbiology*, 10(November), 1–19. https://doi.org/10.3389/fcimb.2020.604923

Timper, P. (2014). Conserving and enhancing biological control of nematodes. *Journal of Nematology*, 46(2), 75–89.

Umer, M., Mubeen, M., Iftikhar, Y., Ali Shad, M., Usman, H. M., Sohail, M. A., Atiq, M. N., Abbas, A., & Ateeq, M. (2021). Role of rhizobacteria on plants growth and biological control of plant diseases: A review. *Plant Protection*, 5(1), 59–73. https://doi.org/10.33804/pp.005.01.3565

Verma, N., Sao, P., Srivastava, A., & Singh, S. (2021). Physiological and Molecular Responses to Drought, Submergence and Excessive Watering in Plants. In *Harsh Environment and Plant Resilience: Molecular and Functional Aspects* (pp. 305–321). Springer.

Vishwakarma, S. K., Ilyas, T., Shahid, M., Malviya, D., Kumar, S., Singh, S., & Singh, H. V. (2024). Bacillus spp.: Nature's Gift to Agriculture and Humankind. In *Applications of Bacillus and Bacillus Derived Genera in Agriculture, Biotechnology and Beyond* (pp. 1–36). Springer Nature Singapore.

Wink, M. (2003). Evolution of secondary metabolites from an ecological and molecular phylogenetic perspective. *Phytochemistry*, 64(1), 3–19. https://doi.org/10.1016/S0031-9422(03)00300-5

Yoon, S. J., Sukweenadhi, J., Khorolragchaa, A., Mathiyalagan, R., Subramaniyam, S., Kim, Y. J., Kim, H. Bin, Kim, M. J., Kim, Y. J., & Yang, D. C. (2016). Overexpression of Panax ginseng sesquiterpene synthase gene confers tolerance against Pseudomonas syringae pv. tomato in Arabidopsis thaliana. *Physiology and Molecular Biology of Plants*, 22(4), 485–495. https://doi.org/10.1007/s12298-016-0384-9

Zubair, M., Farzand, A., Mumtaz, F., Khan, A. R., Sheikh, T. M. M., Haider, M. S., Yu, C., Wang, Y., Ayaz, M., Gu, Q., Gao, X., & Wu, H. (2021). Novel genetic dysregulations and oxidative damage in fusarium graminearum induced by plant defense eliciting psychrophilic bacillus atrophaeus ts1. *International Journal of Molecular Sciences*, 22(22). https://doi.org/10.3390/ijms222212094

8 Synthesis of Green Nanoparticles from Plants

Applications, Production, and Characterisation

Manish Singh Rajput, Parul Johri, Mala Trivedi, Nazleen Jahan, Priyanshi Chaudhary, Vaishnavi Srivastav, and Rishika Nair*

8.1 INTRODUCTION

In the infinitesimal realm of nanotechnology, where the minuscule holds immense potential, a transformative evolution emerges, one that intertwines the power of nanoscience with an imperative for environmental sustainability. Within the area of scientific innovation lies the rapidly growing domain of green nanotechnology. The complexities surrounding sustainable materials, renewable energy, and ecological advancements have steered us towards the transformative panoramas offered by the principles and applications of green nanotechnology.

8.1.1 NANOTECHNOLOGY

Nanotechnology is a fast-expanding and wide-ranging field with many implementations in science and technology. The concept of nanotechnology first began with a lecture delivered by Richard Feynman in 1959 (Kavitha et al. 2013). This field combines key concepts from a variety of disciplines, including chemistry, engineering, physics, and biology, in order to impart novel methods for controlling and generating nanoparticles (NPs) (Hano and Abbasi 2022). Nanotechnology can be defined as the science and engineering behind the design, synthesis, characterization, manipulation, and application of utilitarian materials and devices in which at least one dimension is in the range of 1 to 100 nm. Nanoscale materials hold specific properties, including melting point, charge capacity, and tensile strength (Yadi et al. 2018). What's enthralling about these materials at the nanoscale is that they often unveil unique attributes that differ from their larger counterparts (Figure 8.1). For instance,

* Corresponding author

 DOI: 10.1201/9781003392910-8

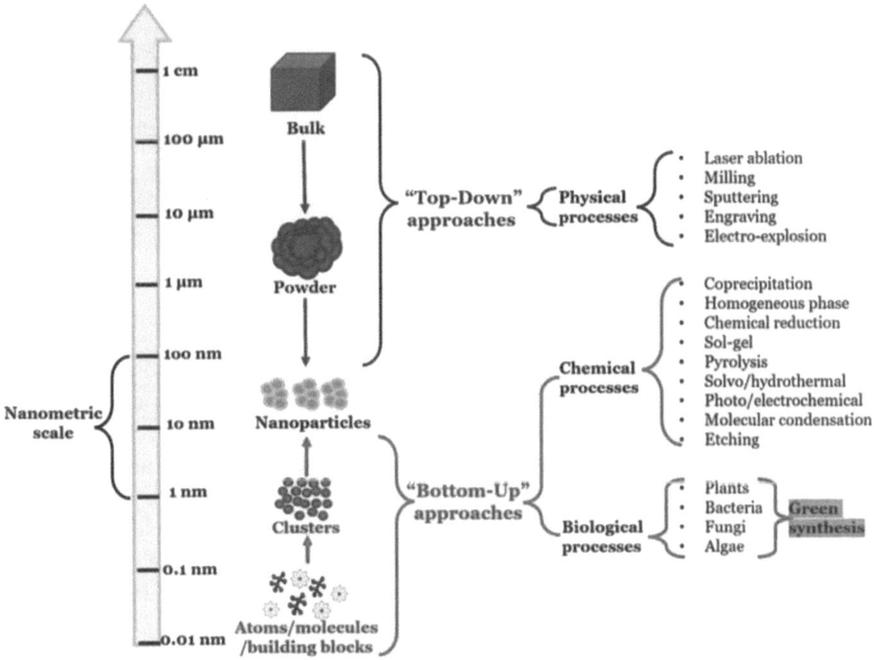

FIGURE 8.1 The various applications of green nanoparticles in multiple fields

their melting points, how much charge they can hold, or their strength can behave differently depending on their size. Sometimes, materials can become sturdier, more conductive, or have disparate optical properties at the nanoscale in contrast to their bulk form.

8.1.2 Nanoparticles

NPs are tiny flecks or particles with at least one dimension ranging from 1 to 100 nm. NPs have specific physicochemical and biological features based on their size. Subject to their morphology, size, and distribution, NPs exhibit superior properties compared to bulk materials (Yadi et al. 2018, Patani et al. 2023). These materials have gained importance and interest in recent years owing to their large number of applications. Matter at this scale confers a more compact arrangement of atoms and molecules (Figure 8.2), generating different phenomena and thereby enhancing mechanical, electrical, magnetic, optical, catalytic, and antibacterial properties that are completely different from those of their macroscopic counterparts (Álvarez-Chimal and Ángel Arenas-Alatorre 2023).

NPs can be classified based on their composition, shape, and size. The most common types of nanoparticles are metals, metal oxides, carbon-based, and quantum dots (Barot et al., 2024). Owing to their unique sizes and properties, nanoparticles have attracted significant attention in various fields, including medicine, electronics, energy, and environmental science (Chaudhry et al. 2018, Barot et al., 2024).

FIGURE 8.2 Size (a) and surface (b) of a cube.

By reducing their size, nanoparticles can have a higher surface-to-volume ratio, enabling a greater number of atoms or molecules per volume, which signifies that less material is needed to perform the same activity and exhibit certain properties (Khan, Saeed, and Khan 2019). Furthermore, the higher surface area-to-volume ratio of NPs is essential and plays a crucial role in additional efficiency in biological applications such as catalysis, biosensors, antimicrobial, photothermal therapy, and pharmaceutical production (Yadi et al. 2018, Barot et al., 2024). Noble metals, such as gold, silver, or platinum, are commonly employed to synthesise NPs by a variety of chemical and physical processes.

However, according to some studies, it has been shown that nanoparticles can harm plants, animals, and humans, but it depends on many factors such as concentration, size, and time of exposure (Yao et al. 2013). Nanoparticles can easily be released into nature in various ways, such as industrial emissions, consumer products, and medical procedures. Once they are released into the environment, it can be difficult to control and monitor them. There is potential for long-term accumulation, which raises concerns about the possibility that nanoparticles can bring prolonged harm to the environment and human health (Álvarez-Chimal and Ángel Arenas-Alatorre 2023).

8.1.3 Nanoparticle Synthesis

The process of manufacturing nanoscale particles is called nanoparticle synthesis. Techniques employed for the synthesis of nanoparticles are broadly classified under two processes, i.e., "top-down" processes and "bottom-up" processes (Figure 8.3). In the top-down approach, bulk material is broken down into particles at the nanoscale using various lithographic techniques, e.g., grinding and milling. In the bottom-up approach, atoms self-assemble to new nuclei that enlarge into particles at the nanoscale, e.g., by chemical reduction, sol-gel techniques, and biological synthesis using organisms like bacteria or plants (Kavitha et al. 2013, Barot et al., 2024, Patani et al. 2023). The choice of synthesis method often relies on the desired properties of the nanoparticles, the materials used, and the intended applications.

Many conventional methods have been used in the synthesis of nanoparticles. However, these traditional approaches come with assorted constraints such as expense and the generation of precarious toxic chemicals, which has encouraged

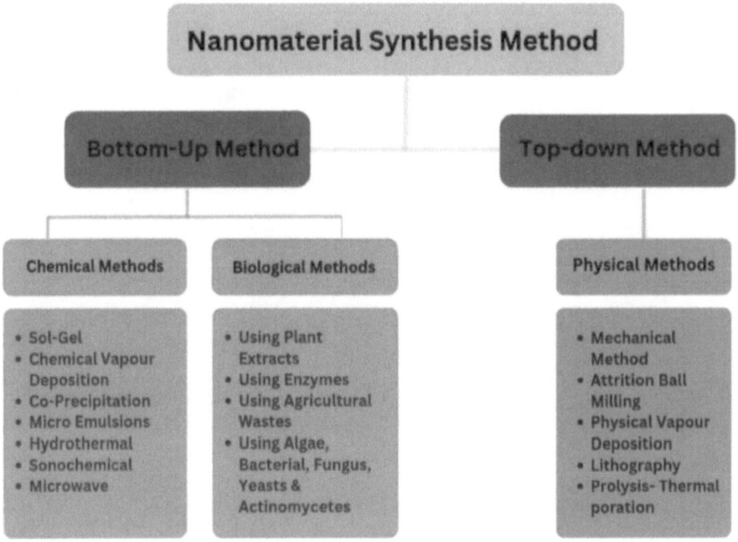

FIGURE 8.3 Synthesis of various nanoparticles.

researchers to develop safe, eco-friendly substitutes for the synthesis of nanoparticles, among which biological systems have been focused (Kavitha et al., 2013).

8.1.4 Green Nanotechnology and Nanoparticles

Green nanotechnology is a branch of green technology that employs the ideas or concepts of green chemistry and green engineering (Verma et al. 2019, Maddala & Singh 2023). It operates by minimising energy and fuel consumption by prioritising renewable resources and efficient material usage. Green nanotechnology, in phyto-formulations, significantly contributes to environmental sustainability through the fabrication of nanomaterials and nanoproducts, without causing harm to human health or the environment (Álvarez-Chimal and Ángel Arenas-Alatorre 2023, Maddala & Singh 2023). Green nanoparticles are nanosized particles created by environmentally benign processes and by utilising sustainable materials. Using plants in nanoparticle formulations leverages their ready availability and diverse range of metabolites, including vitamins, antioxidants, and nucleotides. Green nanotechnology's branches, from bioengineering to nanobiotechnology, intertwine sustainability with innovation (Figure 8.4). When it comes to environmental and climate protection, by conserving raw materials, energy, and water, as well as by reducing greenhouse gases and hazardous waste, green nanotechnological products, processes, and applications are expected to contribute significantly. The key benefits of green nanotechnology include enhanced energy efficiency, minimised waste and greenhouse gas emissions, and lower use of non-renewable resources. Green nanotechnology proposes a substantial opportunity to cease the unpropitious effects before they occur.

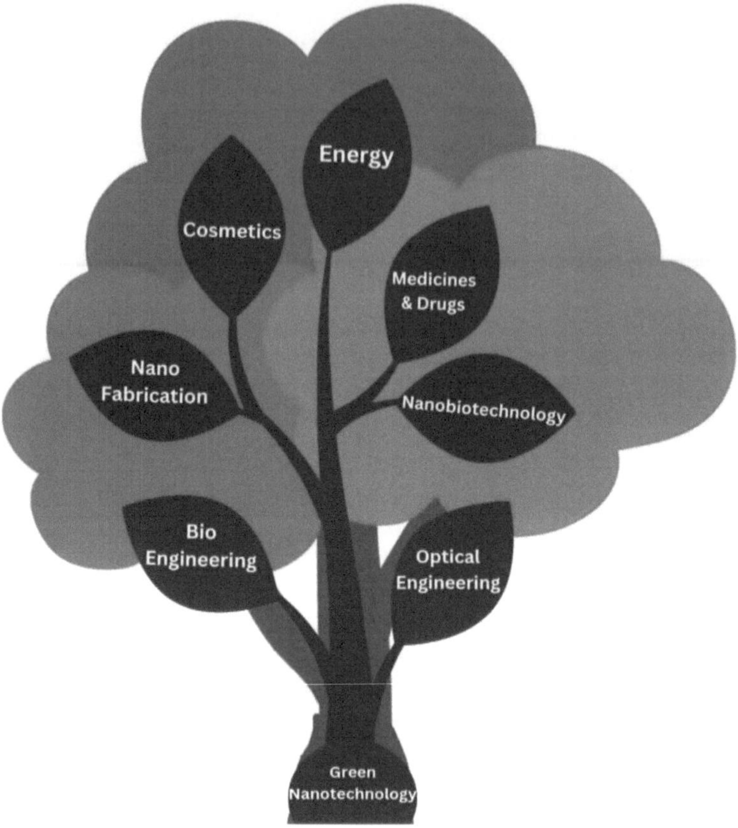

FIGURE 8.4 The seven branches of green nanotechnology science.

8.2 GREEN APPROACH

Nowadays, green approaches are gaining popularity and have triggered the synthesis of green nanoparticles using different sources (living organisms) like bacteria, fungi, algae, and plants (Ahmad et al. 2019, Patani et al. 2023, Maddala & Singh 2023). Green approaches allow large-scale production with less contamination. Green approaches can also support biocompatible, sustainable, and eco-friendly practices, minimising the harm to ecosystems generally caused by using plant extracts, yeast, or bacteria (Ahmad et al. 2019).

8.2.1 Green Synthesis of Nanoparticles Using Plants: An Overview

Green synthesis will allow the development of innovative chemical technologies that minimise or eliminate the use and production of hazardous substances in the design, manufacturing, and use of chemical products. This will lead to minimising or eliminating the synthesis processes; it will also avoid the consumption and wastage of non-renewable raw materials, which are hazardous or polluting materials in product manufacturing, and will minimise synthesis time. Paul J. Anastasia, considered widely as the father of green chemistry, defined the philosophy as preventing pollution by using alternative tools or pathways that reduce or eliminate the use of hazardous substances in the design, manufacturing, and applications of chemical products (Álvarez-Chimal and Ángel Arenas-Alatorre 2023).

8.2.1.1 Approaches

To generate nanoparticles, two approaches are generally used: top-down and bottom-up.

8.2.1.1.1 Top-Down

In the top-down approach, bulk material is broken down into particles at the nanoscale by using physical techniques (Figure 8.5), e.g., grinding, milling, and lithography (Álvarez-Chimal and Ángel Arenas-Alatorre 2023).

8.2.1.1.2 Bottom-Up

In the bottom-up approach, atoms self-assemble to new nuclei that enlarge into particles at the nanoscale (Figure 8.6), e.g., by chemical reduction, sol-gel techniques, and biological synthesis using organisms like plant species, algae, or microorganisms such as bacteria or fungi (Kavitha et al. 2013).

8.2.2 Steps Involved in Green Synthesis

In general, four phases are involved in green synthesis, which are described in the sections that follow (Álvarez-Chimal and Ángel Arenas-Alatorre 2023) (Figure 8.7).

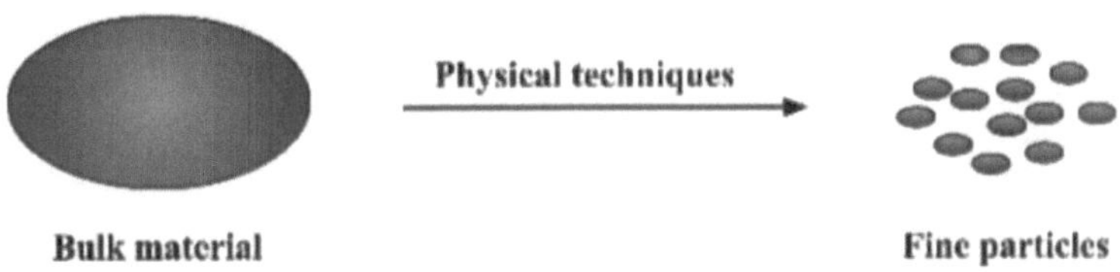

FIGURE 8.5 The top-down approach for the synthesis of green nanoparticles.

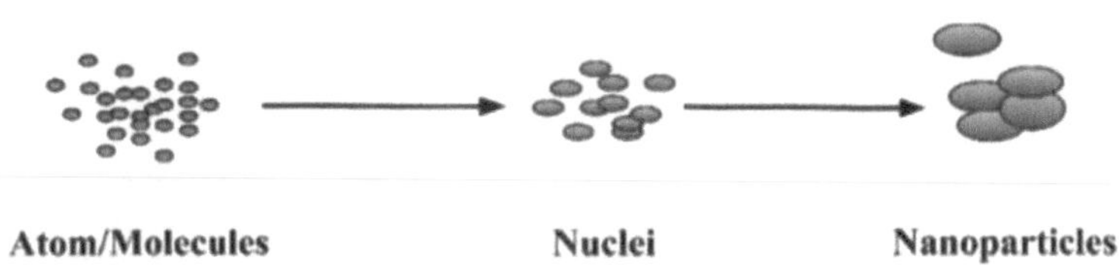

FIGURE 8.6 The bottom-up approach for the synthesis of green nanoparticles.

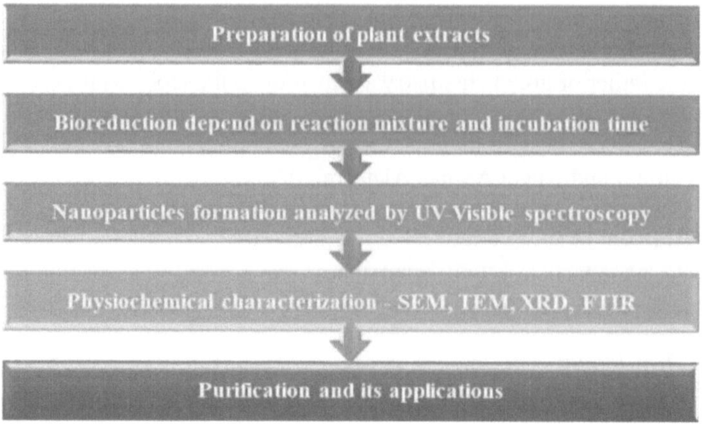

FIGURE 8.7 The various steps involved in the synthesis of the green nanoparticles.

8.2.2.1 Initial Phase

In this phase, we can create the reaction medium, in which the aqueous extract of one or several parts of the plant species is mixed with the precursor salt. In this reaction medium, plant extracts act as a culture medium for the growth of microorganisms, and precursor salt serves as the source of metal ions.

8.2.2.2 Activation Phase

In this phase, the generation of the nucleation centres is started with the help of chemical reduction of metal ions. A nucleation centre is where nanoparticles emerge and grow (Figure 8.8).

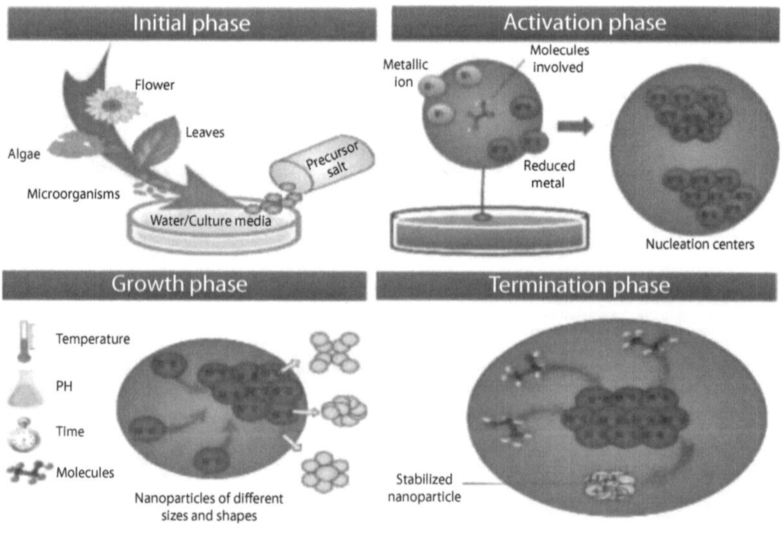

FIGURE 8.8 The various steps involved in the synthesis of the green nanoparticles.

8.2.2.3 Growth Phase

In this phase, small adjacent nanoparticles are influenced by factors like temperature, pH, and reaction time to fuse spontaneously and convert into larger nanoparticles.

8.2.2.4 Termination Phase

In this phase, the final shape and size of the nanoparticles are established.

8.2.3 Biological Resources

As studied previously, nanoparticles have attracted attention in the fields of biology, medicine, and electronics in recent years, owing to their incredible applications (Figure 8.9). Up to now, several nanoparticle synthesis techniques have been

FIGURE 8.9 The various biological resources for the synthesis of green nanoparticles.

developed, but these may require the use of toxic compounds and high-energy physical processes. To avoid these obstacles, an alternative is the use of biological methods. Bacteria, fungi, algae, and plant extracts are some of the most commonly used biological resources for the green synthesis of nanoparticles. This method is reliable, straightforward, and environmentally beneficial (Álvarez-Chimal and Ángel Arenas-Alatorre 2023, Barot et al., 2024).

8.2.4 Factors involved

Generally, in synthesis processes, reaction conditions such as temperature, pH, and reaction time play a crucial role in the shape, size, and yield of synthesised nanoparticles (Álvarez-Chimal and Ángel Arenas-Alatorre 2023).

8.2.4.1 Temperature

This parameter is one of the most notable factors because the shape, size, and synthesis depend on it. With increased temperature, the reaction rate and the formation of the nucleation centre also increase, leading to higher yields. Different temperatures stimulate different interactions between the reactants, resulting in various shapes. The higher the temperature is raised, the greater the size of the nanoparticles.

8.2.4.2 pH

This parameter plays an important role in influencing the nucleation centres to generate more centres at high pH values. Some nanoparticles can only be synthesised in acidic media or some in alkaline media. For example, metal oxide nanoparticles are generally synthesised at an acidic or neutral pH, whereas magnetic nanoparticles are synthesised at an alkaline pH.

8.2.4.3 Time

This parameter plays a crucial role in defining the size of the nanoparticles. It has been observed that as the reaction time increases, the size of nanoparticles and yields also increase because of the prolonged interaction time between reactants.

8.3 OVERVIEW OF DIFFERENT TYPES OF PLANT-DERIVED NANOPARTICLES

Using plants for nanoparticle synthesis is one of the most adapted methods because of its environmentally friendly nature. There are many advantages to using plants to extract nanoparticles; for example, the use of toxic or detrimental substances can be avoided. It is also one of the fastest and most budget-friendly methods because it includes fewer steps (Álvarez-Chimal and Ángel Arenas-Alatorre 2023). This makes it highly effective in the production of nanoparticles compared to synthesis using other resources.

Various categories of plant-derived nanoparticles are described in the sections that follow.

8.3.1 Copper and Copper Oxide Nanoparticles (CuNPs)

Copper (Cu) and copper oxide (CuO) nanoparticles are produced from various plant extracts. Cu nanoparticles from magnolia leaf extract were biologically assembled to produce stable nanoparticles ranging from 40 to 100 nm. Moreover, the effect of Cu nanoparticles shows some antibacterial activity against *Escherichia coli*. *Syzygium aromaticum* (clove) extracts have been shown to be useful for the production of Cu nanoparticles with spherical to granular conformations and a mean granule size of 40 nm. Stem latex of *Euphorbia nivulia* (common milk hedge) is helpful for the extraction of Cu nanoparticles. These nanoparticles are stabilised by peptides and terpenoids from the stem latex, and these nanoparticles are known to be toxic to human adenocarcinoma alveolar basal epithelial cells (Álvarez-Chimal and Ángel Arenas-Alatorre 2023). *Punica granatum* peel extract can be used for the extraction of Cu. The fresh peels of the plant are taken and washed many times. After that, the peel is dried and turned into powder, mixed with sterile distilled water, and boiled until the solution starts changing its colour to yellow (Ghidan, Al-Antary, and Awwad 2016).

8.3.2 Palladium Nanoparticles and Platinum Nanoparticles (PdNPs and PtNPs)

The bark extract of *Cinnamomum zeylanicum* (cinnamon) is used to extract Pd. During the production of Pd nanoparticles, the temperature, concentration, and reaction pH of the bark extract are changed, but conformation and granule size (15–20 nm) are not affected. Peel extract of *Annona squamosa* (custard apple) is used for the synthesis of Pd nanoparticles ranging in size from 75 to 85 nm. Extracts of *Camellia sinensis* (tea) and *Coffea arabica* (coffee) are used to produce Pd nanoparticles ranging from 20 to 60 nm with a cubic crystal symmetry in the centre (Verma et al. 2019). Extracts of *Filicium decipiens* characterised by antibacterial activity are used in the production of Pd NPs. Fresh leaves of *F. decipiens* are collected and washed with water; after that they are dried and chopped into small pieces, then mixed with distilled water to obtain the extract. The leaf extract is mixed with 1 mM $PdCl_2$ solution and isolated in a dark place at room temperature for a few days (Sharmila et al. 2017). Extracts of *Hippophae rhamnoides* Linn. are used to produce Pd NPs. The leaves of the plant are gathered, then turned into a powder and mixed with a hydroalcoholic solution. The extract is then added to the palladium chloride drop by drop, and after a few minutes, the yellow colour of the solution changes to dark brown, showing that the synthesis of NPs is complete. The first Pt NPs were reportedly extracted from the leaf of *Diospyros kaki* (persimmon), where nanoparticles had sizes ranging from 2 to 12 nm. Recently, particle size and shape-controlled biological synthesis of platinum nanoparticles has also been reported. Plant wood at the nanometer scale is used for this purpose. For the production of both platinum and palladium nanoparticles, it has been reported that isolated lignin from *Pinus resinosa* (red pine) was used (Verma et al. 2019).

8.3.3 Titanium Dioxide (TiNPs)

Leaf extracts of *Catharanthus roseus* were used to synthesise TiO2 nanoparticles ranging in size from 25 to 110 nm, and TiO2 were irregularly shaped (Verma et al. 2019). An extract from *Psidium guajava* contained the antioxidant and antibacterial properties of TiO2 nanoparticles. They were examined against *Pseudomonas aeruginosa*, *Aeromonas hydrophila*, *Staphylococcus aureus*, *Proteus mirabilis*, *E. coli*, and many other pathogens. Furthermore, the antioxidant and antibacterial qualities of TiO2 nanoparticles were examined at the nanometer scale and in bulk.

8.3.4 Indium Oxide (In2O3)

Aloe vera leaf extract (*Aloe barbadensis Miller*) is used for the primary synthesis of In2O3 nanoparticles. After primary synthesis, the precipitates were thermally treated at 673.15–873.15 K. The synthesised spherical nanoparticles ranged from 5 to 50 nm, and the size was dependent on the reaction temperature (Verma et al. 2019).

8.3.5 Iron Oxide (FeNPs)

Iron nanoparticles play a very significant role. Thus, various studies are centred towards green chemistry to produce iron (Fe) nanoparticles. *Sorghum bran* aqueous extracts have been used to extract iron oxide nanoparticles. Leaf extract of *Azadirachta indica* (neem) is useful in the synthesis of iron oxide nanoparticles, and the spherical iron nanoparticles have a size of 100 nm (Verma et al. 2019).

8.3.6 Gold Nanoparticles (Au NPs)

Au NPs have gained huge interest because of their easy synthesis, unique characteristics, high potential for use in medicines, low levels of toxicity, and high biocompatibility in nature (Table 8.1). Using *Ziziphus jujube* (ennab) leaf extract, then mixing it with aqueous chloric acid, the solution is seen to change its colour from pale yellow to vivid ruby-red, indicating the production of Au NPs with a diameter of 30 nm (Aljabali et al. 2018).

8.3.7 Silver Nanoparticles (Ag NPs)

Ag NPs have received attention in various fields, such as therapeutics, antimicrobial activity, biomolecular detection, optical receptors, and silver nano-coated medical devices, in recent years. Furthermore, the biological approach, in particular the usage of natural organisms, has offered a trusted, easy, nontoxic, and eco-friendly method. Leaves of Catharanthus roseus are used for the biosynthesis of Ag NPs that contain antiplasmodial activity towards P. falciparum (Ponarulselvam et al. 2012).

TABLE 8.1
Various Nanoparticles Synthesized by Different Plants

Nanoparticles	Plant	Shape
Au and Ag	Aloe vera	Spherical, triangular
Au	Eucalyptus macrocarpa	Spherical, triangular, hexagonal
Au	Psidium guajava	Spherical
Au	Pear fruit extract, Pyrus sp.	Triangular, hexagonal
Au	Terminalia catappa	Spherical
Ag	Citrullus colocynthis	Spherical
Ag	Mangifera indica	Spherical, triangular, hexagonal
Ag	Rhododendron dauricum	Spherical
Pd	Curcuma longa	Spherical
Pt	Diopyros kaki	Crystalline
In2O3	Aloe vera	Spherical
Cu	Syzygium aromaticum	Spherical
Ti	Catharanthus roseus	Irregularly shaped
Fe	Azadirachta indica	Spherical

8.4 CHARACTERIZATION

It is crucial for the nanoparticles to be accurately and extensively characterised so that consistency in their production, biological activity, and safety can be guaranteed. Thus, for the same intent, an extensive variety of physicochemical methods has been brought into use for characterising the synthesized nanoparticles precisely (Hano and Abbasi 2022). The commonly used methods for characterization are described in the sections that follow.

8.4.1 UV-Visible Spectroscopy

This is a method that is employed to examine the absorption spectra of the produced nanoparticles. The peak that emerges in the UV-visible spectrum signifies the presence of nanoparticles; it offers details and information about their dimensions and structure.

8.4.2 Fourier Transform Infrared Spectroscopy (FTIR)

This assists in recognising the functional groups that exist in the plant extract as well as on the exterior surface of the nanoparticles. It also verifies the engagement of specific biomolecules in the cohesion of the nanoparticles.

8.4.3 Dynamic Light Scattering (DLS)

This is used to assess the hydrodynamic size distribution of nanoparticles in the solution. It also provides details regarding the size and dispersion of the nanoparticles.

8.4.4 X-Ray Diffraction (XRD)

XRD is a method that is utilized to ascertain the crystalline configuration of nanoparticles. It aids in identifying the crystalline phases and provides details regarding the size of the nanoparticles.

8.4.5 Transmission Electron Microscopy (TEM)

This gives high-resolution images of an individual nanoparticle, providing valuable information about its size, shape, and distribution.

8.4.6 Scanning Electron Microscope (SEM)

SEM is employed to analyse the surface morphology of nanoparticles, and it also supplies data about their overall structure.

8.4.7 Energy Dispersive X-Ray Spectroscopy (EDS)

This is usually paired with SEM to study the elemental composition of the nanoparticles, confirming the existence of the desired metal or material.

8.4.8 Magnetic Resonance Imaging (MRI) and Nuclear Magnetic Resonance (NMR)

These are methods that can be used for the analysis of the interaction of nanoparticles with biological systems.

Thus, after integrating data from all of these characterisation methods, one can obtain a comprehensive understanding of the physical, chemical, and structural characteristics of green nanoparticles. This information is important for refining the approaches of synthesis as well as customisation of nanoparticles for specific uses.

8.5 RISK ASPECTS

In recent times, the creation and evolution of nanotechnology applications in the area of green technology have prompted us to take into consideration the impact of nanomaterials on the health and safety of workers (Figure 8.10). Therefore, there is an urgent need for action to be taken in scientific, technological, and governmental areas in order to prevent and control such risks. This includes identifying the risks

FIGURE 8.10 The three risks involved in green nanotechnologysynthesis of various nanoparticles.

generated due to the exposure of nanomaterials at the site of work (risk assessment), creating a strategy for its control (risk management), and, finally, to convey the plan (risk communication). This aims to safeguard the workers from danger and to offer the advantages of green nanotechnology to mankind (Verma et al. 2019).

8.5.1 Risk Assessment

In the assessment of risks of nanoparticles, this comprises the recognition of hazards, their characterisation, dose-response relationships, and the analysis of exposure in different cases. It includes similar steps to those used in the risk assessment of other variants of chemicals or materials. However, it has recently become clear that there is still a lack of toxicological data for various nanomaterials; this is necessary for risk assessment. Moreover, information about the implications of serious health impacts, such as genotoxicity, pulmonary toxicity, or carcinogenicity when exposed for long durations or in low dose conditions, is becoming imperative and, thus, needs urgent consideration. Thus, attempts should be made to overcome the obstacles associated with green-nanomaterial exposure scenarios, the detection and classification of nanomaterials, and assessments of the parameters required for health-related sampling. It is also necessary to conduct biological monitoring studies, which are essential to describe the possible biomarkers of nanomaterials, and that potential impacts are tested and verified for use in occupational risk evaluation (Verma et al. 2019).

8.5.2 Risk Management

In order to efficiently control the risks and hazards associated with green nanotechnology, a strategy for the management of risks consisting of an order of controls should be considered. The initial step is the determination of possible exposures to workers, which includes an assessment of the level of exposure; this may vary depending on the task being performed. This should be managed by adopting control measures that include the elimination of hazards, using green chemistry to find non-hazardous replacements, and also introducing engineering controls that include enclosed systems. Controls at the administrative level should ensure that training programs are conducted, where management interacts with workers to ensure they have the required information to understand the possible ways and nature of exposure to potential nanomaterials at the worksite, the procedures to follow, the identification of possible risks and preventive measures, and information about the policies adopted. It is essential for every worker to be provided with adequate information on safety data sheets. One final step of exposure control is personal protective equipment (PPE) kits, which consist of respiratory and eye protection, gloves, and lab coats (Verma et al. 2019).

8.5.3 Risk Communication

Risk communication is an essential step, necessary for making complex technical and health-related information accessible to the general public. It should be in a format that is easily available and understandable by the public. For this, scientists, researchers, governmental authorities, industry, and workforce representatives should be involved to discuss the potential green nanotechnological risks in order to form the appropriate perspectives and mindsets. This step is crucial for ensuring the circulation of proper and relevant information related to the advantages and issues connected to green nanotechnology, and also to prevent the public from exposure to unrealistic expectations and unwanted information in this matter from mass media (Verma et al. 2019).

8.6 APPLICATIONS

NPs are currently experiencing significant commercial demand owing to their extensive applications across various industries such as electronics, environmental science, energy, and, notably, within the biomedical domain. Some of their applications are described in the following sections.

8.6.1 Anti-Cancer Potential

Nanomedicine harnesses nanotechnology to address and combat a spectrum of diseases, including cancer. It introduces procedures and effective approaches for predicting and diagnosing cancer, preventing its onset, and administering personalised therapies, thereby enhancing the shield against this disease.

8.6.2 Anti-Leishmanial Potential

Leishmaniasis is a protozoan vector-borne illness that affects almost 350 million people worldwide. For its treatment, earlier chemotherapeutic medicines were used, but they had adverse side effects. Hence, later, a variety of nanotechnology-based techniques and products replaced with them due to their unique properties, such as bioavailability, reduced toxicity, targeted drug delivery, and biodegradability. These anti-leishmanial drugs included liposomes, lipid nanocapsules, metal and metallic oxide nanoparticles, polymeric nanoparticles, nanotubes, and nano-vaccines.

8.6.3 Antimicrobial Potential

Due to the rapid evolution of the bacterial genome, bacteria have developed resistance to antimicrobial agents. Therefore, in search of new therapy, biogenic NPs have displayed encouraging results in the treatment of multidrug-resistant bacteria and might be a potential choice in the fight against such resistant pathogenesis.

8.6.4 Antioxidant Potential

Oxidative damage is caused to various cell macromolecules like membrane lipids, proteins, and DNA due to excessive oxidative stress generated by the action of mitochondria and other internal or external sources. In order to reverse this detrimental

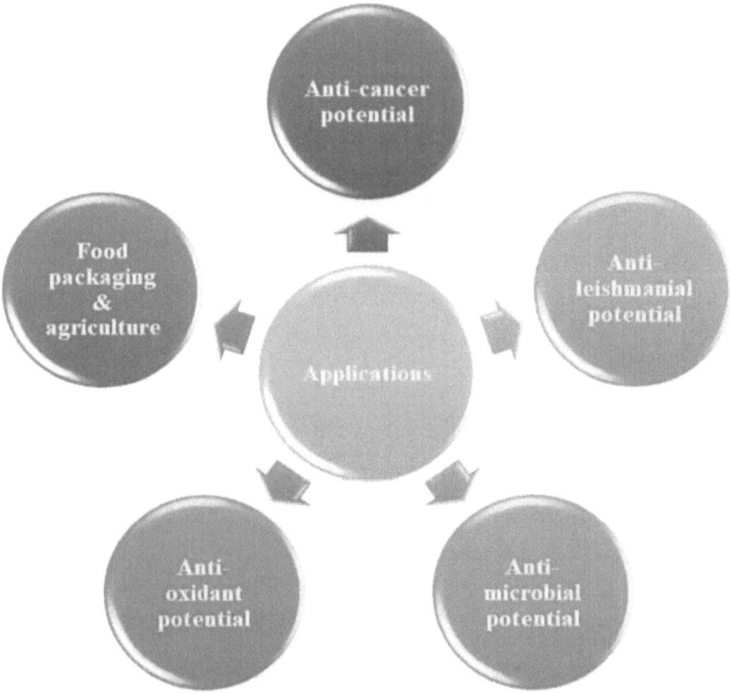

FIGURE 8.11 The various applications of green nanoparticles in multiple fields.

process, antioxidants are used (to treat ageing and age-related diseases). Some green plant-derived NPs have been described for their antioxidant potential, e.g., Ag NPs produced from a C. carandas leaf extract.

8.6.5 Food Packaging and Agriculture

New techniques for water filtration and desalination that are even more economical have been developed because of nanotechnology. Regarding the food industry, it must be mentioned that new functional materials and new instruments have been designed and developed for food preservation and biosecurity (Figure 8.11). When agricultural pathogens are targeted, the antimicrobial activity may be effective for crop protection. For example, ZnONPs have gained wide agricultural interest by showing an anti-phytopathogenic action against bacteria, as evidenced by ZnONPs derived from lemon fruit against the soft rot bacterial pathogen Dickeya dadantii.

8.7 CONCLUSION

Green nanotechnology will play a significant role in upcoming and current ecological situations. It has a pivotal objective along with multifaceted applications. Green nanotechnology has indeed amalgamated the principles of nanoscience, focusing on applying sustainable practices as its primary goal. However, despite its promising potential, it stands not without risks and challenges. There are undoubtedly potential hazards, including toxicity concerns, lifecycle impacts, and unintended environmental consequences, that need to be addressed and regulated to ensure safe implementation and disposal. The applications of green nanotechnology are indubitably wide-ranging and impactful. They have not only enhanced the efficiency of various sources of renewable energy but have also improved healthcare and environmental scenarios. In essence, green nanotechnology has enormous potential to revolutionize industries while safeguarding our planet's well-being, thus ensuring for us an advanced future aligned with sustainable development.

REFERENCES

Ahmad, Shabir, Sidra Munir, Nadia Zeb, Asad Ullah, Behramand Khan, Javed Ali, Muhammad Bilal, et al. 2019. "Green Nanotechnology: A Review on Green Synthesis of Silver Nanoparticles — An Ecofriendly Approach." *International Journal of Nanomedicine*. Dove Medical Press Ltd. https://doi.org/10.2147/IJN.S200254.

Aljabali, Alaa A. A., Yazan Akkam, Mazhar Salim Al Zoubi, Khalid M. Al-Batayneh, Bahaa Al-Trad, Osama Abo Alrob, Alaaldin M. Alkilany, Mourad Benamara, and David J. Evans. 2018. "Synthesis of Gold Nanoparticles Using Leaf Extract of Ziziphus Zizyphus and Their Antimicrobial Activity." *Nanomaterials* 8 (3). https://doi.org/10.3390/nano8030174.

Álvarez-Chimal, Rafael, and Jesús Ángel Arenas-Alatorre. 2023. "Green Synthesis of Nanoparticles: A Biological Approach." www.intechopen.com.

Barot, R. B., Gawande, N. D., Omprabha, S., Kaushal, C., Ghosh, S., Saha, J., Bhatia, D., & Sankaranarayanan, S. (2024). Novel carbon nanoparticles derived from Bougainvillea modulate vegetative growth via auxin–cytokinin signaling in Arabidopsis. Chemical Papers, 78(8), 4733–4750.https://doi.org/10.1007/s11696-024-03421-6.

Chaudhry, Niladri, Sourabh Dwivedi, Vasvi Chaudhry, Anuradha Singh, Quaiser Saquib, Ameer Azam, and Javed Musarrat. 2018. "Bio-Inspired Nanomaterials in Agriculture and Food: Current Status, Foreseen Applications and Challenges." *Microbial Pathogenesis*. Academic Press. https://doi.org/10.1016/j.micpath.2018.07.013.

Ghidan, Alaa Y., Tawfiq M. Al-Antary, and Akl M. Awwad. 2016. "Green Synthesis of Copper Oxide Nanoparticles Using Punica Granatum Peels Extract: Effect on Green Peach Aphid." *Environmental Nanotechnology, Monitoring and Management* 6 (December): 95–98. https://doi.org/10.1016/j.enmm.2016.08.002.

Hano, Christophe, and Bilal Haider Abbasi. 2022. "Plant-Based Green Synthesis of Nanoparticles: Production, Characterization and Applications." *Biomolecules*. MDPI. https://doi.org/10.3390/biom12010031.

Khan, Ibrahim, Khalid Saeed, and Idrees Khan. 2019. "Nanoparticles: Properties, Applications and Toxicities." *Arabian Journal of Chemistry*. Elsevier B.V. https://doi.org/10.1016/j.arabjc.2017.05.011.

Kavitha, K. S., Syed Baker, D. Rakshith, H. U. Kavitha, H. C. Yashwantha Rao, B. P. Harini, and S. Satish. 2013. "Plant as Green Source towards Synthesis of Nanoparticles." *International Research Journal of Biological Sciences*. 2 (6).

Maddala, V. K. S., and S. Singh. 2023. Essential Oils from Plants and Their Role in Nanomaterial Synthesis Characterization and Applications. In: Husen, A. (eds.) *Secondary Metabolites Based Green Synthesis of Nanomaterials and Their Applications*. Smart Nanomaterials Technology. https://doi.org/10.1007/978-981-99-0927-8_10.

Ponarulselvam, S., C. Panneerselvam, K. Murugan, N. Aarthi, K. Kalimuthu, and S. Thangamani. 2012. "Synthesis of Silver Nanoparticles Using Leaves of Catharanthus Roseus Linn. G. Don and Their Antiplasmodial Activities." *Asian Pacific Journal of Tropical Biomedicine* 2 (7): 574–80. https://doi.org/10.1016/S2221-1691(12)60100-2.

Patani, A., A. Patel, D. Prajapati, N. Khare, and S. Singh. 2023. Various Agriculture Crop Plant-Based Bioactive Compounds and Their Use in Nanomaterial Synthesis and Applications. In: Husen, A. (eds.) *Nanomaterials from Agricultural and Horticultural Products*. Smart Nanomaterials Technology. https://doi.org/10.1007/978-981-99-3435-5_12

Sharmila, G., M. Farzana Fathima, S. Haries, S. Geetha, N. Manoj Kumar, and C. Muthukumaran. 2017. "Green Synthesis, Characterization and Antibacterial Efficacy of Palladium Nanoparticles Synthesized Using Filicium Decipiens Leaf Extract." *Journal of Molecular Structure* 1138 (June): 35–40. https://doi.org/10.1016/j.molstruc.2017.02.097.

Verma, Ajay, Surya Gautam, Kuldeep Bansal, Neeraj Prabhakar, and Jessica Rosenholm. 2019. "Green Nanotechnology: Advancement in Phytoformulation Research." *Medicines* 6 (1): 39. https://doi.org/10.3390/medicines6010039.

Yadi, Morteza, Ebrahim Mostafavi, Bahram Saleh, Soodabeh Davaran, Immi Aliyeva, Rovshan Khalilov, Mohammad Nikzamir, et al. 2018. "Current Developments in Green Synthesis of Metallic Nanoparticles Using Plant Extracts: A Review." *Artificial Cells, Nanomedicine and Biotechnology*. Taylor and Francis Ltd. https://doi.org/10.1080/21691401.2018.1492931.

Yao, Duoxi, Zheng Chen, Kui Zhao, Qing Yang, and Wenying Zhang. 2013. "Limitation and Challenge Faced to the Researches on Environmental Risk of Nanotechnology." *Procedia Environmental Sciences* 18: 149–56. https://doi.org/10.1016/j.proenv.2013.04.020.

9 Biochar Technology for Organic Farming, Production, and Environmental Applications for Long-Term Sustainability

Hiral S. Desai, Anil Patani, Dilfuza Jabborova, Sachidanand Singh, and Rahul Dutta

9.1 INTRODUCTION

Globally, intensive agriculture has frequently resulted in the deterioration of the biological and physico-chemical characteristics of soil, causing soil degradation. This deterioration in soil quality could be brought on by nutrient and organic matter mining and erosion, which would impair the soil's ability to carry out its many tasks, including controlling the nutrient cycle, and inorganic and organic materials, storage, maintaining biological activity, filtering, and water flow. Regarding the problem of declining soil fertility, Oshunsanya, S. and Aliku, O. (2016) noted that nutrient-rich materials being accidentally or purposely deposited on farmlands have frequently resulted in an improvement in soil fertility status. When organic material is thermochemically broken down in a controlled atmosphere with little oxygen, a process known as pyrolysis produces biochar, a stable solid material rich in carbon (C) (DeLuca, T.H. and Gao, S., 2019). In order to enhance soil quality, soil fertility, and plant growth, biochar can be added as a soil amendment. It is known to reduce the requirement for chemical fertilizers and pesticides, and enhance soil water retention and crop yeids (Patani, A et al., 2023; Singh, S et al., 2022).

The idea of using charcoal in farming was first proposed thousands of years ago, even though the term "biochar" was only recently coined. According to reports, indigenous cultures first settled in the Terra Preta or "Amazon Dark Earth" soils of the Amazon River Basin thousands of years ago. These soils are still among the most productive and highly biodiverse in the region today. The production of

 DOI: 10.1201/9781003392910-9

"soil-fertilizing material" with dried woody vegetation was primarily done in the Mediterranean region up until the 1960s using an antiquated device known as a "formiguer," whose structure is similar to a charcoal kiln. Growing interest has been shown in establishing how applying biochar fits into the framework of more viable agricultural ecosystems while still providing food safety. This is especially true for organic farming systems that depend on natural soil amendments and that aim to have as little of an impact as possible on the environment (DeLuca, T.H. and Gao, S., 2019).

9.2 PRODUCTION OF BIOCHAR

Biochar is produced by thermochemical processes such as flash carbonization, pyrolysis, and gasification (Xiong et al., 2017). Pyrolysis is the process of thermally breaking down dry biomass at temperatures between 400 and 700 °C without oxygen or with a small amount of oxygen present. The pyrolysis process is further classified as slow (long vapor residence time; >1 h, low heating rate; 5–7 °C/min) or fast (short vapor residence time; <10 sec, high heating rate; more than 200 °C/min). Both pyrolysis processes produce biochar with distinct physicochemical properties because of different process parameters. Since tar-like material becomes trapped in the pores during fast pyrolysis, the yield of biochar is lower than that of slow pyrolysis, and the resulting biochar has a lower surface area (Kumar, A et al., 2020). Microwave pyrolysis, which entails effective, quick, uniform, and volumetric heating, is another method of producing biochar in addition to conventional pyrolysis. When using microwave pyrolysis, the treated biomass is heated at a rate between 0.1 and >1000 °C per second. This ensures that the microwave radiation penetrates the biomass uniformly (Li et al., 2016). In addition to pyrolysis, flash carbonization—in which the biomass bed is ignited in the presence of air—is another method for producing biochar. A brief residence time of 30 minutes is required for the process to occur, at a temperature between 300 and 600 °C and high pressure (~1–2 MPa) (Kumar, A et al., 2020). Another thermochemical process used to generate biochar is gasification. With this technique, biomass is partly oxidized in the presence of different gasifying agents, such as steam, oxygen, air, CO2, or a combination of these, at temperatures between 600 and 900 °C or higher. The primary outcome of gasification is the production of producer gas and syngas, which means that only 5–10% of the feedstock is converted into biochar (Kumar et al., 2019).

9.3 BIOCHAR'S EFFECTS ON THE CHARACTERISTICS OF SOIL

The inclusion of biochar to agricultural soils may result in changes to the soil's water holding capacity (WHC), aggregation, porosity, drainage, and bulk density.

9.3.1 Soil Surface Area

One inherent feature of soil that is influenced by the sizes of its particles is its surface area. It is possible to argue that soil surface area controls some of the key processes

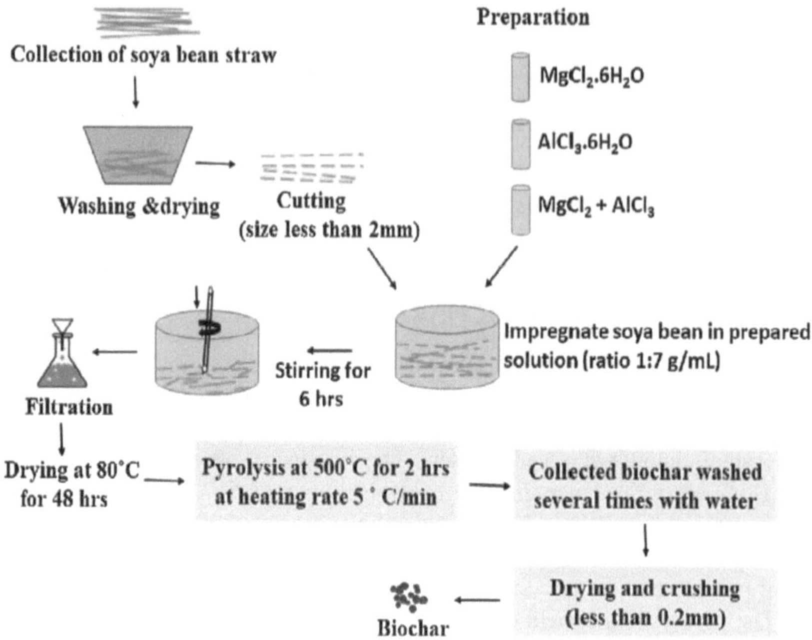

FIGURE 9.1 Production of biochar.

TABLE 9.1
Physical and Chemical Properties of Biochar

Chemical		Amount	Reference
Proximate	Fixed carbon	10.70–86.37 wt%	(Wang et al., 2018)
	Volatiles	12.3–60.6%	(Jafri et al., 2018; Wang et al., 2018)
	Ash content	0–40%	(Wang et al., 2018; S. Zhang et al., 2019)
Elemental	Inorganics	60–80%	(Hoekman et al., 2011)
	Total carbon content	60–80%	(Kantakanit et al., 2018)
Others	pH	3.5–11.3	(Liu et al., 2020; Sun et al., 2019)
Ultimate	H/C molar ratio	>1.5	(Dieguez-Alonso et al., 2018; Wang et al., 2018; Weber and Quicker, 2018)
	O/C molar ratio	>0.7	(Dieguez-Alonso et al., 2018; Wang et al., 2018; Weber and Quicker, 2018)
Physical			**Reference**
	Shape	Non-defined	(Zhang et al., 2019)
	Surface texture	Porous and rich in aromatic moieties	(Cao et al., 2011; M. M. Fu et al., 2019)
	Surface area	0.8–3320 m2/gm	(Tomczyk et al., 2020)

that contribute to soil fertility, as it is an essential physical feature that is crucial to aeration, microbial activity, and the ability of soil to hold water and nutrients (Van Zwieten, L et al., 2012). The enhanced nutrient retention capacity of soil is attributed to the high surface area of biochar, which allows for the formation of bonds and complexes with anions and cations and with metals and soil constituents (Oshunsanya, S.O. and Aliku, O., 2016). The specific surface area of a soil can be increased up to 4.8 times by incorporating biochar (Liang, B et al., 2006).

9.3.2 Porosity

One aspect that may have a considerable influence on how the traits of soils treated with biochar are altered is the overall porosity of the biochar (Oshunsanya, S.O. and Aliku, O., 2016). In contrast to the control, waste-derived biochar-enriched soil showed notable increases in mesoporosity at the expense of macropores (Jones, B.E et al., 2010).

9.3.3 Bulk Density

It has been demonstrated that bulk density—also known as the amount of soil per unit volume—has a negative correlation with surface area. When it comes to bulk

TABLE 9.2
Effects of Biochar on the Porosity and Surface Area of Amended Soils

Soil	Study Type	Biochar Type	Rate of Biochar Application	Porosity (%)	Reference
Silt loam	Field	Birch (*Betula pendula*), 400 °C	1.2	52.8	(Karhu, K., Mattila, T., Bergström, I. and Regina, K., 2011)
			0	50.9	
Sandy soil	Greenhouse	Jarrah woods (*Eucalyptus marginata*), 600 °C	2.27	62.1	(Dempster, D.N., Gleeson, D.B., Solaiman, Z.I., Jones, D.L. and Murphy, D.V., 2012)
			0.45	57.6	
			0	56.1	
Residue sand	Laboratory	Municipal green waste, 450 °C	5.2	0.51	(Jones, B.E., Haynes, R.J. and Phillips, I.R., 2010)
			2.6	0.48	
			0	0.46	

TABLE 9.3
Impact of Applying Biochar on the Bulk Density of Soil

Soil	Soil Type	Biochar Type	Rate of Biochar Application	Bulk Density g cm–3	Reference
Clarion fine loamy	Laboratory	Mixed hardwoods, 500 °C	2.0	1.08, 1.24	(Laird, D.A., Fleming, P., Davis, D.D., Horton, R., Wang, B. and Karlen, D.L., 2010)
			1.0	1.08, 1.24	
			0.5	1.10, 1.24	
			0	1.21, 1.34	
Residue sand	Laboratory	Municipal green waste, 450 °C	5.2	1.44	(Jones, B.E., Haynes, R.J. and Phillips, I.R., 2010)
			2.6	1.55	
			0	1.65	
Hydroagric stagnic anthrosol	Field	Wheat (*Triticum* spp.) straw, 350–550 °C	4.4	0.89, 0.88	(Mankasingh, U., Choi, P.C. and Ragnarsdottir, V., 2011)
			2.2	0.91, 0.86	
			1.1	0.96, 0.91	
			0	0.99, 0.94	

density, well-organized (fine texture) soils tend to have low values from 1.0 to 1.3 g cm–3, whereas weakly structured soils are known to have higher values from 1.6 to 1.8 g cm–3 (Bhupenchandra, I et al., 2019). In a study involving soil column incubation, soil columns treated with biochar exhibited a notably reduced bulk density in comparison to those without any biochar (Laird, D.A. et al., 2010; Mukherjee, A. and Lal, R., 2013). The soil column treated with biochar had a lower rate of compaction than the soil columns treated with manure or control when all of the columns were subjected to compaction caused by gravity and repeated leaching events. According to further investigation, the enhancement of soil structure or aeration and aggregation may be indicated by a decrease in bulk density of soil amended with biochar. This phenomenon may be specific to a given soil.

9.3.4 Aggregate Stability

Research has indicated that biochar reacts favorably to aggregation. The enhanced soil aggregate stability is caused by processes that result in complexes of biochar

TABLE 9.4
The Impact of Applying Biochar on Soil Aggregation

Soil Type	Study Type	Biochar Type	Rate of Biochar Application	Aggregation	Reference
Albicluvisol	Laboratory	Hydrochar, 220 °C	10	65.1	(George, C., Wagner, M., Kücke, M. and Rillig, M.C., 2012)
			5	69.0	
			0	49.8	
Norfolk loamy sand: Ap	Laboratory	Pecan shells, 700 °C	1.0	10.7, 12.3	(George, C., Wagner, M., Kücke, M. and Rillig, M.C., 2012)
			0.5	9.53, 12.7	
			0	9.95, 13.0	
Norfolk loamy sand: E	Laboratory	Pecan shells, 700 °C	2.1	12.9	(Busscher, W.J., Novak, J.M. and Ahmedna, M., 2011)
			0	14.3	

with minerals, which are formed when oxidized carboxylic acid groups interact with biochar particle surfaces (Oshunsanya, S.O. and Aliku, O., 2016). Thus, adding organic matter from biodegradation may enhance soil aggregates and pore size distribution, which in turn improves soil hydraulic properties. Some authors, however, have presented different findings. Busscher, W.J. et al. (2010) found that adding pecan-derived biochar to switchgrass enhanced aggregation; however, the effect was much less pronounced when the soil was treated with biochar alone, without mixing it with the grass. From these results, Mukherjee, A. and Lal, R. (2013) came to the conclusion that the addition of biochar and the presence of a substrate, such as switchgrass, would be necessary for a favorable impact on soil aggregate stability.

9.3.5 Soil Microorganisms

The presence, composition, and activity of soil microorganisms are crucial to almost every soil process, and as a result, they will primarily determine the agricultural land's sustainable productivity (DeLuca, T.H. and Gao, S., 2019). Biochar has the potential to impact soil microbial communities via various mechanisms :

(a) For soil microorganisms, the biochar itself may act as a surface or habitat (Quilliam et al., 2013; Jiang et al., 2016).
(b) For microbial consumption, biochar can function as a substrate or a loci where substrate accumulates (Quilliam et al., 2013).

(c) When soil toxins and chemical cues are present, biochar can adsorb them and prevent microbial development.
(d) Biochar has the ability to modify abiotic parameters such as the concentration of particular elements, pH, and moisture potentially through adsorption, which can change the number of soil microorganisms (DeLuca 2017; Yu et al., 2018).

When Dumontet et al. (2017) examined biochemical and microbial activity in soils amended with biochar, either with or without the addition of organic fertilizer, their findings indicated that both treatments had a higher potential for oxidizing carbon and a more diverse range of bacteria that break down cellulose compared to the control, indicating a positive effect of biochar in microbial heterotrophic metabolism that may be caused by inputs of carbon substrate. Recent research has examined the diversity and abundance of soil microbial populations, including fungi, bacteria, and fauna in soils amended with biochar (Teutscherova et al., 2017b; Lucheta et al., 2017; Abujabhah et al., 2017). Using the substrate-induced respiration (SIR) method, Teutscherova et al. (2017b) observed a decrease in microbial biomass in biochar-treated soils during a short-term incubation study. They attributed this decrease to the pH shift in the soil caused by the biochar, which changed the ratio of fungal to bacterial biomass. On the other hand, in a rice paddy soil, the application of 20 t biochar ha−1 resulted in a significant increase (28%) in bacterial abundance and a decrease (35%) in fungal abundance (Chen et al., 2013). The neutral pH of the soil was thought to be resistant to the addition of biochar, encouraging a varied bacterial community over acid soils, which would probably be preferred by fungi (Rousk et al., 2009). In short-term studies, the number of bacteria in soil has been shown to decline or remain unchanged but to increase in long-term studies (Imparato et al., 2016; Song et al., 2017). In a long-term study looking at the structure of microbial communities after corncob biochar was added to an agricultural soil used for soybean cultivation, researchers found that biochar-treated soils had higher bacterial activity and diversity than the control, where the bacterial communities switched from preferring to metabolize carbohydrates to xenobiotics (Sun et al., 2016). Long-term studies revealed that fungal diversity remained constant, but the structure of the fungal community was found to be substantially correlated with the amount of K, C, or N in the soil. It seems that adding biochar to soils affects the corresponding population of soil fauna, particularly earthworms (Kamau et al; 2017; Pingree et al., 2017). According to a study, the type of the biochar feedstock and the amount of charcoal in the soil were both related to the abundance of earthworms in the soil (Kamau et al., 2017).

9.4 IMPACT OF BIOCHAR ON ORGANIC AGRICULTURE CROP PRODUCTIVITY

"Organic farming is a system that relies, as much as possible, on crop residues, crop rotations, mineral grade rock additives, animal manures, off-farm organic waste, and biological systems of plant protection and nutrient mobilization. Synthetic inputs, like feed additives, hormones pesticides, etc., are avoided or largely excluded in this system." However, detractors contend that, in comparison to traditional farming methods, agriculture founded on these principles usually produces comparatively lower yields (DeLuca, T.H. and Gao, S., 2019). Therefore, even though increasing soil fertility over time is one of the goals of organic operations, one must investigate efficient crop and nutrient management techniques, such as applying biochar amendments to surface soils (DeLuca, T.H. and Gao, S., 2019). Although biochar has been the subject of numerous studies recently, few of them have concentrated on its application in organic farming systems, particularly in relation to field research, or its impact on crop yield and nutrient uptake yield (Lehmann and Joseph, 2015). In agricultural soils treated with biochar, aboveground production and yield have generally been reported to increase. The primary factor influencing a crop's response to biochar addition is the effect of biochar on the physical and biochemical properties of the soil, which is then converted to soil-plant interaction (Gao and DeLuca, 2016). It was discovered that different crop types, soil types, types of biochar, the length of time biochar stayed in the soil, and combinations of these variables affected overall responses. It was discovered that types of biochar, types of crops, soil types, and combinations of these variables affected the overall responses. Biochar's black color will increase surface albedo, which will then affect thermal dynamics related to the physical properties of the soil. This could potentially have an impact on the germination process (Genesio et al., 2012). On the other hand, it was found that the increase in crop productivity under wood- and crop-derived biochar was less responsive than the increase under manure-derived biochar; additionally, crops growing on acidic soil with a coarse texture tended to respond to biochar additions in their productivity more quickly and effectively (Liu et al., 2013).

Given the possibility of introducing contaminants from industrial and municipal organic wastes that would need to be managed without the use of synthetic chemicals, the ability of biochar to remove heavy metals and related pollutants is crucial for organic farming systems (DeLuca, T.H. and Gao, S., 2019). It has been discovered that using biochar in conjunction with organic fertilizer or compost significantly increases the availability of nutrients in the soil and its organic carbon content. This, in turn, increases crop nutrient uptake and yield in soils treated with biochar (Gao et al., 2017). An additional widely investigated agricultural advantage of biochar is its impact on root nodulation, biological N2 fixation, and legume crop growth—all of which are particularly significant in organic farming systems (DeLuca et al., 2015b). Numerous theories have suggested that this effect is directly tied to the increased availability of molybdenum (Mo) and boron (B) caused by the addition of biochar (Güereña et al., 2015). However, it has been noted that conventional farming produces yields that are 5–34% higher than those of organic farming systems (DeLuca, T.H. and Gao, S., 2019). By adding biochar to an organic management system, it may be possible to decrease nutrient loss, close the yield difference between the two farming systems, and promote soil C and fertility (Gao et al., 2017; Gao et al., 2019).

9.5 APPLICATIONS OF BIOCHAR

9.5.1 Biochar as a Substitute Fuel

Reusing renewable bioresources for energy generation through retrofitting them into current fossil fuel-based energy systems has become increasingly popular (Nizami et al., 2017). Because it is rich in carbon and has a high calorific value, biochar made from the pyrolysis process can be used to produce electricity and heat, among other things (Balat, 2007). Because it contains a high concentration of lignocellulose compounds like cellulose, hemicellulose, and lignin, biomass such as plant debris or agroforest debris has a high energy value. The utilization of raw biomass as a fuel results in decreased power generation efficiency because of its low energy density, high moisture content, hygroscopic properties, high bulk density (Tsai et al., 2012). A number of investigators turned various agricultural residues into biochar for use as fuel. In terms of heating value, biochar (with a value of 16–35 MJ/kg) is about twice as high as raw biomass and many low-grade coals for any given feedstock. Zhu et al. (2017) examined through experimentation the behavior, in terms of ignition and combustion, of single-droplet biochar slurry made from coconut shell. The water content of the fuel prepared was between 50 and 70%, and the diameter of the biochar particles was roughly 10 μm. According to their findings, a smaller droplet size improved the residual biochar's burning rate. Currently, the steel industry uses coal, which raises the industry's carbon footprint and releases CO2 emissions. In terms of the environment and metallurgy, biochar—a charcoal made from biomass that has been injected into blast furnace tubes with particles that have been ground—is a desirable substitute for coal. The effects of replacing blast furnace fuel with biochar indicate the possibility of an 18–40% reduction in CO2 emissions (Feliciano-Bruzual, 2014).

9.5.2 Biochar in the Role of Catalyst

A catalyst is used to facilitate the reaction between fat or alcohol and oil to generate biodiesel, a renewable alternative. By using concentrated H2SO4 to pyrolyze rice husk, Li et al. (2014) created a solid biochar-catalyst. The created catalyst was applied to waste cooking oil during the transesterification and esterification processes to produce biodiesel. The process demonstrated high selectivity and catalytic reactivity, suggesting the material's potential as a heterogeneous catalyst in order to produce biodiesel (Fu et al., 2012). The increased production of biodiesel was caused by the greater acidity and surface area of the catalyst. It was found that better outcomes were obtained with a molar ratio of alcohol to oil (A:O) of 18:1, a 3-hour reaction time, and a 5 wt% catalyst loading (Bazargan et al., 2015).

9.5.2.1 Adsorbent for Contaminants in Water

Biochars' ability to adsorb pollutants is influenced by its specific surface area and aromaticity-specific surface area (Sulaymon et al., 2012). In addition, biochar is utilized for the sorption of various pollutants from aqueous solutions, including organic contaminants, heavy metals, and contaminants containing phosphorus and nitrogen. With the use of feces and rice husk char, Xu et al. (2013) eliminated heavy metals like Zn, Pb, Cd, and Cu from a water solution. It was found that rice husk char is less effective at removing metal than feces char. When it comes to the sorption of heavy metal on the surface of biochar, mineral compounds and oxygen-functional groups are important factors. The feces char absorbed heavy metal on ionized hydroxyl-O-1 groups, while rice husk char adsorbed only O-groups. The biochar produced by Goswami et al. (2016) from Ipomoea fistulosa at 350–550 °C was used to extract cadmium (Cd) from aqueous solutions. A higher efficiency in the removal of Cd was achieved by activating biochar with a KOH solution to increase its surface area. Biochar was created by Liu et al. (2013) by slow pyrolysis of green macroalga. Tartaric, acetic, and citric acids were used by Sun et al. (2016) to activate the biochar they made from sawdust from eucalyptus trees. Later, methylene blue was eliminated from an aqueous suspension by utilizing activated biochar. Citric acid-activated biochar was also found to have a higher carboxyl group (COOH) content than other types using Fourier transform infrared spectroscopy (FTIR).

9.5.2.2 The Air-Purifying Adsorbent

Because of anthropogenic and natural processes, the amount of dangerous chemicals in the atmosphere is growing daily and is having a negative impact on human health. By pyrolyzing sewage sludge at 700 °C, Lee et al. (2011) created biochar. KOH (KOH/char = 0.5) was used to chemically activate the char after it had been physically activated with steam. Next, low concentration formaldehyde in indoor pollution was adsorbed using activated char. Slow pyrolysis was used by Creamer et al. (2014) to produce biochar from hickory woodand sugarcane bagasse at various temperatures, including 300, 450, and 600 °C. Analysis of CO2 adsorption was done at room temperature. CO2 adsorption was greater in biochar made from sugarcane bagasse at 600 °C (73.55 mg/g at 25 °C). The authors stated that physisorption caused carbon dioxide to firmly bind on the surface of biochar; as a result, surface area availability was a significant factor in CO2 adsorption.

9.5.2.3 Biochar in the Construction Industry

One of the major contributors to CO2 emissions has been found to be the construction industry, which accounts for an estimated 7% of all greenhouse gases released into the atmosphere (Gupta et al., 2018). This has prompted scientists all around the world to experiment with replacing cement with alternative materials (Akhtar and Sarmah 2018). Biochar is a good material for construction because of its qualities, which include high chemical stability, low thermal conductivity, and low flammability (Gupta and Kua 2017). Slow pyrolysis minimizes the surface area of the biochar, thereby decreasing carbon free radicals and resulting in low flammability. This is how the biochar gets its low flammability (Zhao et al., 2014). One characteristic of biochar is its ability to separate carbon from its structure. It can function as a system for CO2 storage in concrete structures by locking stable carbon in its structure for decades. However, at present very little data is available regarding the feasibility of using biochar in cement mortar for building purposes.

9.5.2.4 Utilizing Biochar in Sodium- and Lithium-Ion Batteries

Anodes for silicon ion batteries are commonly made of amorphous carbons, which can be obtained from biochar-based materials. In addition to being more affordable and environmentally friendly, materials composed of biochar have a higher specific ability in contrast to typical grafite. According to Zhou et al. (2017), graphitization and hydrothermal processes of wheat stalk biomass can be used to create highly graphitic carbon nanosheets (HGCNS). The HGCNS synthetization process results in an interrelated two-dimensional (2D) nanostructure with an average thickness of 1.0–2.0 nm. This type of nanostructure can store lithium ions in multiple locations and can hold between two and six single layers of graphene. Furthermore, the 2D nanostructure facilitates quick transport of electrons and Li + ions (Liu et al., 2019).

REFERENCES

Abujabhah, I. S., Doyle, R. B., Bound, S. A., & Bowman, J. P. (2017). Assessment of bacterial community composition, methanotrophic and nitrogen-cycling bacteria in three soils with different biochar application rates. *Journal of Soils and Sediments, 18*, 1–11.

Akhtar, A., & Sarmah, A. K. (2018). Novel biochar-concrete composites: Manufacturing, characterization and evaluation of the mechanical properties. *Science of the Total Environment, 616*, 408–416.

Balat, M. (2007). Influence of coal as an energy source on environmental pollution. *Energy Sources, Part A: Recovery, Utilization, and Environmental Effects, 29*(7), 581–589.

Bazargan, A., Kostić, M. D., Stamenković, O. S., Veljković, V. B., & McKay, G. (2015). A calcium oxide-based catalyst derived from palm kernel shell gasification residues for biodiesel production. *Fuel, 150*, 519–525.

Bhupenchandra, I., Basumatary, A., Saraswat, P. K., Chongtham, S. K., Singh, L. K., Kalita, P., Tamuli, B., & Sinyorita, S. (2019). Biochar: Its impact on soil and environment. *International Journal of Current Microbiology and Applied Sciences, 8*(11), 2408–2424.

Busscher, W. J., Novak, J. M., & Ahmedna, M. (2011). Physical effects of organic matter amendment of a southeastern US coastal loamy sand. *Soil Science, 176*(12), 661–667.

Cao, X., Ro, K. S., Chappell, M., Li, Y., & Mao, J. (2011). Chemical structures of swinemanure chars produced under different carbonization conditions investigated by advanced solid-state 13C nuclear magnetic resonance (NMR) spectroscopy. *Energy & Fuels, 25*, 388–397.

Chen, J., Liu, X., Zheng, J. J. J. J., Zhang, B., Lu, H., Chi, Z., et al. (2013). Biochar soil amendment increased bacterial but decreased fungal gene abundance with shifts in community structure in a slightly acid rice paddy from Southwest China. *Applied Soil Ecology, 71*, 33–44.

Creamer, A. E., Gao, B., & Zhang, M. (2014). Carbon dioxide capture using biochar produced from sugarcane bagasse and hickory wood. *Chemical Engineering Journal, 249*, 174–179.

DeLuca, T.H. and Gao, S., 2019. Use of biochar in organic farming. In S. Chandran, C. S. Thomas, & M. R. Unni (Eds.), *Organic farming: New advances towards sustainable agricultural systems* (pp. 25–49). New York: Springer.

Dieguez-Alonso, A., Funke, A., Anca-Couce, A., Rombola, A.G., Ojeda, G., Bachmann, J., & Behrendt, F. (2018). Towards biochar and hydrochar engineering-influence of process conditions on surface physical and chemical properties, thermal stability, nutrient availability, toxicity and wettability. *Energies, 11*, 496.

DeLuca, T. H., Gundale, M. J., MacKenzie, M. D., & Jones, D. L. (2015b). Biochar effects on soil nutrient transformations. In J. Lehmann & S. Joseph (Eds.), *Biochar for environmental management:Science, technology and implementation* (pp. 421–454). London: Routledge.

Dempster, D. N., Gleeson, D. B., Solaiman, Z. I., Jones, D. L., & Murphy, D. V. (2012). Decreased soil microbial biomass and nitrogen mineralisation with Eucalyptus biochar addition to a coarse textured soil. *Plant and Soil, 354*, 311–324.

Dumontet, S., Cavoski, I., Ricciuti, P., Mondelli, D., Jarrar, M., Pasquale, V., et al. (2017). Metabolic and genetic patterns of soil microbial communities in response to different amendments under organic farming system. *Geoderma, 296*, 79–85.

Feliciano-Bruzual, C. (2014). Charcoal injection in blast furnaces (Bio-PCI): CO2 reduction potential and economic prospects. *Journal of Materials Research and Technology, 3*(3), 233–243.

Fu, Z., Wan, H., Hu, X., Cui, Q., & Guan, G. (2012). Preparation and catalytic performance of a carbon-based solid acid catalyst with high specific surface area. *Reaction Kinetics, Mechanisms and Catalysis, 107*(1), 203–213.

Gao, S., & DeLuca, T. H. (2016). Influence of biochar on soil nutrient transformations, nutrient leaching, and crop yield. *Advances in Plants & Agriculture Research, 4*, 150.

Gao, S., DeLuca, T. H., & Cleveland, C. C. (2019). Biochar additions alter phosphorus and nitrogen availability in agricultural ecosystems: A meta-analysis. *Science of The Total Environment, 654*, 463–472.

Gao, S., Hoffman-Krull, K., & DeLuca, T. H. (2017). Soil biochemical properties and crop productivity following application of locally produced biochar at organic farms on Waldron Island, WA. *Biogeochemistry, 136*, 31–46.

Genesio, L., Miglietta, F., Lugato, E., Baronti, S., Pieri, M., & Vaccari, F. P. (2012). Surface albedo following biochar application in durum wheat. *Environmental Research Letters, 7*, 14025.

Güereña, D. T., Lehmann, J., Thies, J. E., Enders, A., Karanja, N., & Neufeldt, H. (2015). Partitioning the contributions of biochar properties to enhanced biological nitrogen fixation in common bean (*Phaseolus vulgaris*). *Biology and Fertility of Soils, 51*, 479–491.

Gupta, S., & Kua, H. W. (2017). Factors determining the potential of biochar as a carbon capturing and sequestering construction material: Critical review. *Journal of Materials in Civil Engineering, 29*(9), 04017086.

Gupta, S., Kua, H. W., & Koh, H. J. (2018). Application of biochar from food and wood waste as green admixture for cement mortar. *Science of the total environment, 619*, 419–435.

Goswami, R., Shim, J., Deka, S., Kumari, D., Kataki, R., & Kumar, M. (2016). Characterization of cadmium removal from aqueous solution by biochar produced from Ipomoea fistulosa at different pyrolytic temperatures. *Ecological Engineering, 97*, 444–451.

Hoekman, S. K., Broch, A., & Robbins, C. (2011). Hydrothermal carbonization (HTC) of lignocellulosic biomass. *Energy & Fuels, 25*, 1802–1810.

Imparato, V., Hansen, V., Santos, S. S., Nielsen, T. K., Giagnoni, L., Hauggaard-Nielsen, H., et al. (2016). Gasification biochar has limited effects on functional and structural diversity of soil microbial communities in a temperate agroecosystem. *Soil Biology and Biochemistry, 99*, 128–136.

Jiang, X., Denef, K., Stewart, C. E., & Cotrufo, M. F. (2016). Controls and dynamics of biochar decomposition and soil microbial abundance, composition, and carbon use efficiency during long-term biochar-amended soil incubations. *Biology and Fertility of Soils, 52*, 1–14.

Jafri, N., Wong, W. Y., Doshi, V., Yoon, L. W., & Cheah, K. H. (2018). A review on production and characterization of biochars for application in direct carbon fuel cells. *Process Safety and Environmental Protection, 118*, 152–166.

Jones, B. E., Haynes, R. J., & Phillips, I. R. (2010). Effect of amendment of bauxite processing sand with organic materials on its chemical, physical and microbial properties. *Journal of Environmental Management*, *91*(11), 2281–2288.

Kamau, S., Barrios, E., Karanja, N. K., Ayuke, F. O., & Lehmann, J. (2017). Spatial variation of soil macrofauna and nutrients in tropical agricultural systems influenced by historical charcoal production in South Nandi, Kenya. *Applied Soil Ecology, 119*, 286–293.

Kantakanit, P., Tippayawong, N., Koonaphapdeelert, S., & Pattiya, A. (2018). Hydrochar generation from hydrothermal carbonization of organic wastes. In *IOP Conference Series: Earth and Environmental Science* (Vol. 159, p. 12001). Bristol: IOP Publishing.

Karhu, K., Mattila, T., Bergström, I., & Regina, K. (2011). Biochar addition to agricultural soil increased CH4 uptake and water holding capacity–Results from a short-term pilot field study. *Agriculture, Ecosystems & Environment*, *140*(1–2), 309–313.

Kumar, A., Saini, K., & Bhaskar, T. (2020). Hydochar and biochar: Production, physicochemical properties and techno-economic analysis. *Bioresource Technology*, *310*, 123442.

Kumar, A., Saini, K., & Bhaskar, T. (2019). Advances in design strategies for preparation of biochar based catalytic system for production of high value chemicals. *Bioresource Technology*, *299*, 122564.

Kumar, A., Kumar, J., & Bhaskar, T. (2020). High surface area biochar from *Sargassum tenerrimum* as potential catalyst support for selective phenol hydrogenation. *Environmental Research*, 186, 109533.

Laird, D. A., Fleming, P., Davis, D. D., Horton, R., Wang, B. Q., & Karlen, D. L. (2010). Impact of biochar amendments on the quality of a typical midwestern agricultural soil. *Geoderma*, *158*, 443–449.

Lee, J. Y., Park, S. H., Jeon, J. K., Yoo, K. S., Kim, S. S., & Park, Y. K. (2011). The removal of low concentration formaldehyde over sewage sludge char treated using various methods. *Korean Journal of Chemical Engineering*, *28*, 1556–1560.

Lehmann, J., & Joseph, S. (2015). *Biochar for environmental management: Science, technology and implementation*. New York: Routledge.

Liang, B., Lehmann, J., Solomon, D., Kinyangi, J., Grossman, J., O'Neill, B., et al. (2006). Black carbon increases cation exchange capacity in soils. *Soil Science Society of America Journal*, *70*, 1719–1730.

Li, J., Dai, J., Liu, G., Zhang, H., Gao, Z., Fu, J., & He, Y. (2016). Biomass and bioenergy biochar from microwave pyrolysis of biomass: A review. *Biomass and Bioenergy*, *94*, 228–244.

Li, M., Zheng, Y., Chen, Y., & Zhu, X. (2014). Biodiesel production from waste cooking oil using a heterogeneous catalyst from pyrolyzed rice husk. *Bioresource technology*, *154*, 345–348.

Liu, H., Wang, L., & Lei, M. (2019). Positive impact of biochar amendment on thermal balance during swine manure composting at relatively low ambient temperature. *Bioresource technology*, *273*, 25–33.

Liu, X., Zhai, Y., Li, S., Wang, B., Wang, T., Liu, Y., Qiu, Z., & Li, C. (2020). Hydrothermal carbonization of sewage sludge: Effect of feed-water pH on hydrochar's physicochemical properties, organic component and thermal behavior. *Journal of Hazardous Materials*, 388, 122084.

Liu, X., Zhang, A., Ji, C., Joseph, S., Bian, R., Li, L., et al. (2013). Biochar's effect on crop productivity and the dependence on experimental conditions-a meta-analysis of literature data. *Plant and Soil*, *373*, 583–594.

Lucheta, A. R., Cannavan, F. d. S., Tsai, S. M., & Kuramae, E. E. (2017). Amazonian Dark Earth and its black carbon particles harbor different fungal abundance and diversity.

Marazza, D., Macrelli, S., D'Angeli, M., Righi, S., Hornung, A., & Contin, A. (2019). Greenhouse gas savings and energy balance of sewage sludge treated through an enhanced intermediate pyrolysis screw reactor combined with a reforming process. *Waste Management, 91*, 42–53.

Mukherjee, A., & Lal, R. (2013). Biochar impacts on soil physical properties and greenhouse gas emissions. *Agronomy, 3*, 313–339.

Nizami, A. S., Shahzad, K., Rehan, M., Ouda, O. K. M., Khan, M. Z., Ismail, I. M. I., & Demirbas, A. (2017). Developing waste biorefinery in Makkah: A way forward to convert urban waste into renewable energy. *Applied Energy, 186*, 189–196.

Oshunsanya, S. O., & Aliku, O. (2016). Biochar technology for sustainable organic farming. *Organic Farming-a Promising Way of Food Production, 1*(6), 111–129.

Patani, A., Prajapati, D., Singh, S., Enakiev, Y., Bozhkov, S., Jabborova, D., & Joshi, C. (2023). Beneficial effects of biochar application on mitigating the drought and salinity stress implications on plants. *Plant Science Today, 10*(sp2), 188–193.

Pingree, M. R. A., Makoto, K., & DeLuca, T. H. (2017). Interactive effects of charcoal and earthworm activity increase bioavailable phosphorus in sub-boreal forest soils. *Biology and Fertility of Soils, 53*, 1–12.

Quilliam, R. S., Glanville, H. C., Wade, S. C., & Jones, D. L. (2013). Life in the "charosphere" – Does biochar in agricultural soil provide a significant habitat for microorganisms? *Soil Biology and Biochemistry, 65*, 287–293.

Rousk, J., Brookes, P. C., & Bååth, E. (2009). Contrasting soil pH effects on fungal and bacterial growth suggest functional redundancy in carbon mineralization. *Applied and Environmental Microbiology, 75*, 1589–1596.

Singh, S., Datta, R., Johri, P., & Trivedi, M. (Eds.). (2022). *Phytopharmaceuticals and biotechnology of herbal plants*. Boca Raton: CRC Press.

Song, Y., Bian, Y., Wang, F., Herzberger, A., Yang, X., Gu, C., et al. (2017). Effects of biochar on dechlorination of hexachlorobenzene and the bacterial community in paddy soil. *Chemosphere, 186*, 116–123.

Sulaymon, A. H., Mohammed, T. J., & Al-Najar, J. (2012). Equilibrium and kinetics studies of adsorption of heavy metals onto activated carbon. *The Canadian Journal of Chemical Engineering and Technology, 3*(4), 86.

Sun, D., Meng, J., Xu, E. G., & Chen, W. (2016). Microbial community structure and predicted bacterial metabolic functions in biochar pellets aged in soil after 34 months. *Applied Soil Ecology, 100*, 135–143.

Sun, X., Li, M., & Chen, Y. (2019). Biochar facilitated bioprocessing and biorefinery for productions of biofuel and chemicals: A review. *Bioresource Technology, 295*, 122252.

Teutscherova, N., Vazquez, E., Masaguer, A., Navas, M., Scow, K. M., Schmidt, R., & Benito, M. (2017a). Comparison of lime- and biochar-mediated pH changes in nitrification and ammonia oxidizers in degraded acid soil. *Biology and Fertility of Soils, 53*, 1–11.

Teutscherova, N., Vazquez, E., Santana, D., Navas, M., Masaguer, A., & Benito, M. (2017b). Influence of pruning waste compost maturity and biochar on carbon dynamics in acid soil: Incubation study. *European Journal of Soil Biology, 78*, 66–74.

Tomczyk, A., Sokołowska, Z., & Boguta, P. (2020). Biochar physicochemical properties: Pyrolysis temperature and feedstock kind effects. *Reviews in Environmental Science and Bio/Technology, 19*, 191–215.

Tsai, W. T., Liu, S. C., Chen, H. R., Chang, Y. M., & Tsai, Y. L. (2012). Textural and chemical properties of swine-manure-derived biochar pertinent to its potential use as a soil amendment. *Chemosphere, 89*(2), 198–203.

Van Zwieten, L., Singh, B., Joseph, S., Kimber, S., Cowie, A., & Yin Chan, K. (2009). Biochar and emissions of non-CO2 greenhouse gases from soil. In J. Lehmann & S. Joseph (Eds.), *Biochar for Environmental Management*. London Sterling: Earthscan.

Wang, T., Zhai, Y., Zhu, Y., Li, C., & Zeng, G. (2018). A review of the hydrothermal carbonization of biomass waste for hydrochar formation: Process conditions, fundamentals, and physicochemical properties. *Renewable and Sustainable Energy Reviews*, *90*, 223–247.

Xiong, X., Iris, K. M., Cao, L., Tsang, D. C. W., Zhang, S., & Ok, Y. S. (2017). A review of biochar-based catalysts for chemical synthesis, biofuel production, and pollution control. *Bioresource Technology*, *246*, 254–270.

Xu, X., Cao, X., & Zhao, L. (2013). Comparison of rice husk-and dairy manure-derived biochars for simultaneously removing heavy metals from aqueous solutions: Role of mineral components in biochars. *Chemosphere*, *92*(8), 955–961.

Yu, L., Duan, L., Naidu, R., & Semple, K. T. (2018). Abiotic factors controlling bioavailability and bioaccessibility of polycyclic aromatic hydrocarbons in soil: Putting together a bigger picture. *Science of the Total Environment*, *613–614*, 1140–1153.

Zhang, Z., Zhu, Z., Shen, B., & Liu, L. (2019). Insights into biochar and hydrochar production and applications: A review. *Energy*, *171*, 581–598.

Zhao, M. Y., Enders, A., & Lehmann, J. (2014). Short-and long-term flammability of biochars. *Biomass and Bioenergy*, *69*, 183–191.

Zhu, M., Zhang, Z., Zhang, Y., Liu, P., & Zhang, D. (2017). An experimental investigation into the ignition and combustion characteristics of single droplets of biochar water slurry fuels in air. *Applied Energy*, *185*, 2160–2167.

10 Microbial Biosurfactants

Potential in Pharmaceutical Applications and Therapeutics

Shuchi Midha, Nikhil Kirtipal, Garima Gupta, Mala Trivedi, and Vibha Gajbe

10.1 BIOSURFACTANTS

Biosurfactants can be defined as surface-active compounds that are produced by microorganisms like fungi, yeast, and bacteria (Desai & Banat, 1997). They have both hydrophilic (amino acid, protein, carbohydrate) and hydrophobic (generally long-chain fatty acid) moieties that tend to accumulate at the interface between fluid phases with different degrees of polarity and hydrogen bonding, such as oil/water or air/water interfaces (Angelova & Schmauder, 1999). Such a property leads to the formation of micelles from monomers or the creation of aggregates into micellar tubes, micellar bilayers, and vesicles. Surfactants further promote the formation of emulsions from immiscible liquids such as hydrocarbons and water. These surface-active properties render surfactants capable of reducing surface and interfacial tension and promoting the formation of microemulsions where hydrocarbons can solubilize in hydrocarbons (Desai & Banat, 1997; Pattanath, 2008). Such properties make biosurfactants excellent detergents, and wetting, emulsifying, foaming and dispersing agents (Banat et al., 2010).

In the United States, biosurfactants currently create 9.4 billion dollars in sales, which is likely to increase at a rate of 3–5% annually (Subramaniam, 2020). At present, surfactants in use are as chemicals derived from petroleum. Features of microbial surfactants that give them a good potential to be exploited are their diversity, environmentally friendly nature, possibility of their production through fermentation, and their potential applications in environmental protection, crude oil recovery, healthcare, and food processing (Banat et al., 2010 & Pattanath, 2008).

Biosurfactants are superior to chemical surfactants due to their lower toxicity, rapid biodegradability, better environmental compatibility, higher foaming ability, selectivity, and specific activity at extreme temperatures, pH, and salinity (Foley et al., 2012). They can also be synthesized from renewable feedstocks.

DOI: 10.1201/9781003392910-10

10.2 CLASSIFICATION OF BIOSURFACTANTS

According to their polarity, chemical composition, and microbial origin, biosurfactants can be classified into:

(1) Glycolipids
(2) Lipopeptides or lipoproteins
(3) Phospholipids and fatty acids (mycolic acids)
(4) Polysaccharide-lipid complexes
(5) Complete cell surfaces
(6) Polymeric surfactants
(7) The particulate type

10.2.1 Biosurfactant Classification Based on Chemical Composition and Microbial Origin

Biosurfactants can be categorized based on the charge associated with the hydrophilic group as non-ionic, anionic, cationic, or amphoteric.

(1) Hydrophilic moiety, containing amino acids or polypeptides, anions, or cations (mono, di, or polysaccharides)
(2) Hydrophobic moieties, containing unsaturated and saturated fatty acids
  a. Lipopeptides
  b. Gramicidins (*B. brevis*)
  c. Fatty acids, phospholipids
  d. Polymeric

10.2.2 Biosurfactant Classification Based on Producing Source

Another classification of biosurfactants ais based on the source producing them:

(1) Microbial biosurfactants
(2) Enzymatically synthesized surfactants

10.2.3 Biosurfactant Classification by Type of Substrate Used

(1) Those producing biosurfactants with alkanes as carbon sources (*Corynebacterium* sp. and *Arthrobacterium sp*)
(2) Those producing biosurfactants with water-soluble substrates as carbon sources (*Bacillus sp*)
(3) Those producing biosurfactants with alkanes and water-soluble substrates as carbon sources (*Pseudomonas sp*)

10.3 ENZYME-SYNTHESIZED BIOSURFACTANTS

Many industrial enzymes that catalyze alcoholysis, condensation, hydrolysis, acylation, or esterification reactions have also been utilized to produce different surfactants, including phospholipids, monoglycerides, amino acids-based surfactants, and glycolipids, from cheap raw materials such as plant oils and fats (Kiss et al., 2017). Enzymatic methods of synthesis of surfactants have several advantages over conventional chemical synthesis (Johnson et al., 2021).

Enzymatic methods show minimal thermal degradation, and have low energy requirements and high biodegradability. Lipozyme from *Rhizomucor miehei* and lipase from *Candida antarctica* are the most regularly used enzymes (Sharma et al., 2001).

Some newly synthesized enzymatic surfactants are phenol glycosides, aliphatic alcohol glycosides, glycerol glycolipids, vitamin glycosides, and fructose monooleates.

The main disadvantages of enzymatic surfactants are large liquid waste volumes and high recovery costs.

10.4 MICROBIAL BIOSURFACTANTS

Some of the important classes of biosurfactants along with the producing microorganisms are listed in Table 10.1.

10.4.1 Glycolipids

Glycolipids such as rhamnolipids, trehalolipids, sophorolipids, and mannosylerythritol are the most common class of biosurfactants. They are carbohydrates with hydroxy aliphatic acids or long-chain aliphatic acids joined either by ester or ether groups. Rhamnolipids are the most common class of biosurfactants. They have two molecules of rhamnose linked to one or two molecules of β-hydroxy decanoic acid. These are produced by *Pseudomonas* and *Burkholderia* species (Ahmad et al., 2010).

Trehalolipids, another class of biosurfactants, are mostly produced by species of *Mycobacterium*, *Nocardia*, and *Corynebacterium* (Kuyukina et. al, 2015). Structurally, this group of glycolipids has a disaccharide trehalose linked at C6 and C6 and is associated with mycolic acid (Pacheco et al., 2010).

Sophorolipids, another category of glycolipids, are mainly elaborated by yeasts such as *T. petrophilum*, *Torulopsis bombicola*, and *T.apicola* (Hommel et al., 1987).

10.4.2 Lipopeptides and Lipoproteins

Lipopeptides and lipoproteins are the second largest group of biosurfactants and are of high industrial focus. Some of these in this category include surfactin, fengycin, iturin, lichenysin, and kurstakin (Das et al., 2008). The types or classifications of lipopeptide surfactants are mainly based on the amino acid sequences and various strains of *Bacillus* spp.-producing lipopeptides such as *B. subtilis*, *B. cereus*, *B.*

TABLE 10.1
Different Types of Microbial Biosurfactants and Producing Microorganism

Biosurfactant	Microorganism
Glycolipids	
Rhamnolipids	*Pseudomonas aeruginosa* *Rhodococcus erithropolis*
Trehalolipids	*Arthrobacter* sp.
Sophorolipids	Candida bombicala, C. apicola
Mannosylerythritol lipids	*Candida antartica*
Lipopeptides	
Surfactin/iturin/fengycin	*Bacillus subtilis*
Viscosin	*P. fluorescens*
Lichenysin	*B. licheniformis*
Serrawetin	*Serratia marcescens*
Phospholipids	*Acinetobacter* sp. *Corynebacterium lepus*
Surface-active antibiotics	
Gramicidin	*Brevibacterium brevis*
Polymixin	*B. polymyxa*
Antibiotic TA	*Myxococcus xanthus*
Fatty acids/neutral lipids	
Corynomycolic acids	*Corynebacterium lepus*
Polymeric surfactants	
Emulsan	*Acinetobacter calcoacetices*
Alasan	*Acinetobacter radioresistens*
Liposan	*Candida lipolytica*
Lipomanans	*C. tropicalis*
Particulate biosurfactants	
Vesicles	*Acinetobacter* sp.
Whole microbial cells	*Cyanobacteria*

thuringiensis, *B. globigii*, *B. amyloliquefaciens*, *B. megaterium*, *B. pumilus*, and *B. licheniformis* (Yu et al., 2002, Vater et al., 2002, Pueyo et al., 2009, Nihorimbere et al., 2012).

10.4.3 Fatty Acids, Phospholipids, and Neutral Lipids

Several yeasts and bacteria, such *as Acinetobacter* sp., produce phosphatidyl ethanolamine-rich vesicles during growth on n-alkane substrates. These vesicles form optically clear microemulsions of alkanes in water. Fatty acids produced from alkanes are the result of microbial oxidations that have been considered surfactant. As well as these straight-chain acids, microorganisms produce complex fatty acids containing OH groups and alkyl branches (Rahman and Gakpe, 2008).

10.4.4 Particulate Biosurfactants

Particulate biosurfactants help partition extracellular membranous vesicles, resulting in the formation of microemulsions, which play an important role in alkane uptake by microbial cells. Vesicles of *Acinetobacter* sp. have a diameter of 20–50 nm and a buoyant density of 1.158 g/cm³ and are composed of protein, phospholipids, and lipopolysaccharides (Vijaykumar & Saravanan, 2015).

10.4.5 Surface Active Biosurfactants

Surface-active agents like gramicidin, polymyxins, and antibiotic TA fall into this category and are elaborated by *Bacillus brevis*, *Bacillus polymyxa*, and *Myxococcus xanthus*, respectively. Gramicidin S is a cyclic lipopeptide (decapeptide) that has been shown to have antibacterial activity, which is attributed to its high surface activity (Lang , 2002).

Antibiotic TA, a type of macrocyclic secondary metabolite, constrains peptidoglycan synthesis by interfering with polymerization of the lipid disaccharide pentapeptide. It has shown chemotherapeutic applications (Rosenberg et al., 1973).

10.4.6 Polymeric Microbial Biosurfactants

Polymeric biosurfactants are heterosaccharide proteins. The most studied polymeric biosurfactants are liposan, lipomanans, emulsan, and alasan. These are produced by *Acinetobacter calcoaceticus*. Emulsan is an effective emulsifying agent for hydrocarbons in water (Lang, 2002). Liposan, another heteropolysaccharide, produced from *Candida lipolytica*, is an extracellular water-soluble emulsifier possessing 17% protein and 83% carbohydrate (Kaeppeli & Finnerty, 1979).

10.5 POTENTIAL APPLICATIONS IN PHARMACEUTICAL INDUSTRY AND THERAPEUTICS

10.5.1 Antimicrobial Activity

The diverse properties of biosurfactants confer them a unique advantage for industrial employability (Ajesh and Sreejith, 2009; Zhao et al., 2010). Going by structure, biosurfactants exert their toxic potential on cellular membrane permeability in a detergent-like way. Owing to this, many biosurfactants demonstrate antimicrobial action (Rahman and Ano, 2009), that is, they are antifungal, antibacterial, and antiviral.

Das et al. (2009) reported that biosurfactant extracted from marine *B. circulans* showed potential antimicrobial effects against Gram-positive and Gram-negative semi-pathogenic and pathogenic strains.

Another study by Fernandes et al. (2007) demonstrated the antimicrobial effect of lipopeptide biosurfactants from *Bacillus subtilis* R14 against 29 bacterial strains.

Rodrigues et al. (2006) stated that mannosylerythritol lipids (MELs) formed by *Candida antarctica*, rhamnolipids formed by *P. aeruginosa*, and lipopeptides formed by *B. subtilis* 31 and *B. licheniformis* show antimicrobial activity 2006).

Lipopeptides and glycolipids are the most demonstrated classes of biosurfactants having antimicrobial activity. A study by Yang et al. showed that Brevibacillin, a cationic lipopeptide formed by *Brevibacillus laterosporus*, showed significant antimicrobial activities against Gram-positive bacteria.

A lipopeptide-based biosurfactant showed an outstanding inhibitory effect against *Enterobacter* sp., *Salmonella typhimurium, Klebsiella pneumoniae, Salmonella enterica, Micrococcus luteus, E. faecalis, and Listeria monocytogenes* (Ghribi et al., 2012). In another study, lipopeptide biosurfactants produced by three strains of *Bacillus cereus* isolated from garden soil showed significant antifungal and antibacterial activity against *Escherichia coli, K. pneumoniae, P. aeruginosa, Candida albicans*, and *Aspergillus niger*, with MIC values between 0.52 to 7.6 mg/mL (Basit et al., 2018). The rhamnolipids and surfactin mixtures formed by *P. aeruginosa* ST5 and *B. amyloliquefaciens* ST34, respectively, showed significant antimicrobial activity against a wide range of pathogenic microorganisms (Ndlovu et al., 2017).

Moreover, the antimicrobial properties of microbial biosurfactants have been demonstrated on cell membranes due to the amphiphilic nature of these compounds. However, evidence is also showing the role of biosurfactants in quorum-sensing signalling (Chrzanowski et al., 2012, Khan et al., 2019, Yan et al., 2019).

Lactic acid bacteria (LAB) are believed to have a significant effect on human health and immune systems. The biosurfactants formed by *Lactobacillus gasseri* P65 and Lactobacillus *jensenii* P6A have shown their remarkable antimicrobial activity against *E. coli* (MIC = 16 μg/mL) from the genitourinary tract, and in clinical isolates of *Enterobacter aerogenes, Staphylococcus saprophyticus*, and *K. pneumoniae* (MIC = 128 μg/mL) (Morais et al., 2017).

10.5.2 Anti-adhesive and Antibiofilm Activity

Biosurfactants effectively prohibit the adhesion of pathogenic microbes to solid surfaces or infection sites (Das et al., 2009). Colonization of pathogenic microorganisms may be inhibited by prior adhesion of biosurfactants to solid surfaces (Rivardo et al., 2009). Furthermore, Rodrigues et al. (2004) stated that biosurfactants effectively reduce microbial numbers on prostheses and induced a decrease in the resistance that occurs on voice prostheses after biofilm formation.

Cordeiro et al. (2018) studied the co-incubation of a biosurfactant (a mixture of surfactin, iturin, and fengicin, named TIM96) from *B. subtilis* and found that TIM96 prevented biofilm formation in *Trichosporon* spp. by inhibiting cell adhesion (up to 96.89%) in a dose-dependent manner and reduced metabolic activity (up to 99.2%), decreasing cell width and viability. Liu et al. (2019) demonstrated that, in co-incubation conditions, surfactin formed by a *B. subtilis* strain strongly affected *S. aureus* adhesion on various materials (polystyrene, glass, and stainless steel) and resulted in biofilm dislodging.

Furthermore, another study demonstrated the antibiofilm activity of rhamnolipids, glycolipids, and sophorolipids against different oral bacterial pathogens, such *as S. sanguinis, A. naeslundii, S. oralis, and N. mucosa* (Elshikh et al., 2017). Biosurfactants significantly prevented biofilm establishment in these bacterial pathogens by 60–90%.

Another study conducted by Satpute et al. (2018; 2019) found the effectiveness of biosurfactants formed by *Lactobacillus acidophilus* as biofilm inhibitors. This work focused on the use of cell-free biosurfactant (CFBS) as a therapeutic coating agent for polydimethylsiloxane (PDMS) surfaces, PDMS-based microfluidics (MF) channels, and silicone catheters. It significantly demonstrated the inhibition of biofilm formation by various bacterial strains. In all these methods, CFBS-coated samples showed no biofilm creation up to 48 hours, using the crystal violet staining and scanning electron microscopy techniques, when compared to control surfaces. Additionally, the pre-treatment of polystyrene surfaces with cell-associated biosurfactants (CABS) efficiently prohibited the adhesion of all verified strains. Approximately 80% inhibition was reported for *B. subtilis* and *S. aureus*, and 59–65% for *P. vulgaris*, *P. aeruginosa, P. putida, and E. coli.*

Giordani et al. (2019) assessed the capability of a biosurfactant isolated from *Lactobacillus gasseri* BC9 and biosurfactant-enriched liposomes (BS-LP) against biofilms of methicillin-resistant *S. aureus* strains. They stated that the free biosurfactant prohibited biofilm formation and encouraged biofilm eradication for five clinical isolates of MRSA strains, a gentamicin-resistant clinical isolate, and a penetrating reference strain, in a dose-independent manner in the *S. aureus* strain (Giordani et al., 2019).

10.5.3 Antiviral Activity

Biosurfactants have also been shown to have antiviral properties against different enveloped viruses. The inhibitory effects of biosurfactants were a result of their ability to form ion channels in viral capsids and lipid envelopes, the loss of proteins involved in viral adsorption/penetration processes, and the inhibition of viral membrane fusion (Jung & Kim, 2000, Seydlová & Svobodová, 2008, Yuan et al., 2018). Naruse et al. (1990) were the first researchers to observe the antiviral efficacy of pumilacidins against herpes simplex virus type 1. Additionally, the mixture of surfactin and fengycin from *B. subtilis* fmbJ was found to be inhibitory to pseudorabies virus, porcine parvovirus, Newcastle disease, and infectious bursal disease in vitro, and inhibited the infection and replication processes of Newcastle disease and infectious bursal disease in porcine kidney and chicken embryo fibroblast cell lines (Huang et al., 2006).

Parallel studies on other classes of biosurfactants have also reported interesting findings regarding their antiviral potential. It was confirmed that the dose of trehalose 6,6-dimycolate in mice boosted their potential to resist influenza virus infection by encouraging the propagation and accretion of lung gamma delta TCR+ lymphocytes (Hoq et al., 1997). Sophorolipids showed antiviral and sperm-immobilizing activities against human immunodeficiency virus and herpes virus (Shah et al., 2005, Gross & Shah, 2007). Remarkably, the diacetate ethyl ester by-products presented the maximum virucidal potential against the human immunodeficiency virus type 1 and established a total reduction in the viral titer of more than 5.2 log units within 2 min (Shah et al., 2005).

In a study of Jin et al. (2021), the efficacy of rhamnolipid 222B to neutralize herpes simplex virus 1 (HSV-1) and bovine coronavirus was confirmed. In addition, the possibility of using rhamnolipids as coating agents on plastic and fabric surfaces for antiviral shields and masks was evaluated.

Moreover, the exclusive properties of biosurfactants to form micellar structures at their critical micellar concentration may be crucial for their application as liposomes for drug delivery to the infection site, thereby emphasizing their function in harsh environments in the human body (Nakanishi et al., 2009). Their significant applications in pharmaceuticals for gene delivery and the design of molecules to interact with immune system mechanisms have also been considered (Çelik et al., 2021).

Importantly, as natural products, biosurfactants are sustainable with low cytotoxicity, which allows them to be used in handwashes to avoid the spread of viruses. Recognizing their potential use in such key areas, they were thought be of growing importance in dealing with the COVID-19 pandemic (Smith et al., 2020). The difficulty with such possible applications is the need for cost-effective mass manufacture. More likely applications are as sanitizers for the surfaces of materials and the skin, especially as chemical substitutes can be harsher for some people's skin.

10.5.4 Immuno-Modulatory Activity

Interestingly, several biosurfactants have been demonstrated to modulate immune responses by affecting cellular and humoral level (Anuradha, 2010). Interestingly, it has been demonstrated that microorganisms establish host infection by exploiting the immunomodulatory activity of many biosurfactants. For example, rhamnolipids produced by *P. aeruginosa* aid in subverting the host immune responses by inhibiting the production of antimicrobial peptides (e.g., human beta defensin-2), impairing innate immune function by inhibiting phagocytic activity and inducing the lysis of neutrophils and macrophages (Yakimov et al., 1995). However, such an immunosuppressive effect of biosurfactants can also be exploited for the treatment of different immune-related diseases.

Moreover, the lipopeptide surfactin produced by *Bacillus* sp. significantly reduced inflammatory response through various mechanisms. It was observed that surfactin specifically inhibits phospholipase A2, which releases arachidonic acid from the cell membrane and subsequently produces inflammatory mediators, such as leukotrienes, platelet-activating factor, and prostaglandins (Ainsworth et al., 1947).

Trehalose dimycolate (TDM), the cord factor of *Mycobacterium tuberculosis*, has the potential to elicit the activation of both the innate and adaptive limbs of immunity, despite its high toxic potential (McInerney et al., 1990).

10.5.5 Anticancer Activity

According to the World Health Organization (WHO), cancer is the second highest cause of death worldwide (GBD 2019 Cancer Risk Factors Collaborators). Approximately 8.8 million deaths in 2015 were on account of different forms of cancer, and the number has risen over the years since then. Chemotherapy remains the cornerstone of anticancer treatment despite the development of new promising anticancer therapeutics. Most drugs used to treat cancer cells are developed from molecules that are extracted from natural sources (e.g., microorganisms, plants, invertebrates, and vertebrates) (Singh & Cameotra, 2004), showing potential cytotoxic activity for cancer cells. In the development of chemotherapeutic drugs, molecules having low specific activity to tumor cells become the principal obstacle (Schneider et al., 1999). Therefore, many efforts are being made to identify new anticancer mediators that selectively target and deactivate cancer cells more effectively than the currently used chemotherapeutics.

10.6 BIOSURFACTANTS FOR COSMETICS

The majority of biosurfactants show specific properties, such as foaming, wetting, emulsification and de-emulsification, water-binding capacity, dispersal, and product stability and viscosity, that can be exploited by the cosmetic industry.

Surfactants are already being utilized as foaming agents, emulsifiers, wetting agents, solubilizers, antimicrobial agents, cleansers, and intermediaries of enzymatic action in different forms such as lotions, creams, pastes, powders, liquids, gels, and sprays. These could be substituted by biosurfactants (Tugrul and Cansunar, 2005).

10.7 CONCLUSION AND FUTURE PERSPECTIVES

Biosurfactants, as surface-active molecules, show high therapeutic and pharmaceutical potential. They possess significant anti-adhesive, antimicrobial (antifungal, antibacterial, and antiviral), and biofilm-disrupting properties. These molecules can be used either alone or in combination with other chemotherapeutic drugs and show future potential for being effective strategies in the inhibition and prevention of microbial infections, proliferation, and biofilm formation. Moreover, biosurfactants have received attention from the scientific community as a new generation of pharmaceutics involved in anticancer treatments, wound healing, immunomodulatory applications, drug delivery agents, and in cosmetics. The advantages of biosurfactants surely lay the foundation for the development of environmentally friendly pharmaceuticals and therapeutics. Challenges associated with biosurfactants' cost-effectiveness, potential applications, and availability, however, remain to be resolved.

REFERENCES

Ainsworth, G.C., Brown, A.M., Brownlee, G. (1947) Aerosporin, an antibiotic produced by *Bacillus aerosporus* Greer. *Nature*, 160:263.

Ajesh, K., Sreejith, K. (2009) Peptide antibiotics: An alternative and effective antimicrobial strategy to circumvent fungal infections. *Peptide*, 30:999–1006.

Angelova, B., Schmauder, H-P. (1999) Lipophilic compounds in biotechnology—interactions with cells and technological problems. *Biotechnol.*, 67:13–32.

Anuradha, S.N. (2010) Structural and molecular characteristics of lichenysin and its relationship with surface activity. In *Biosurfactants*; Advances in Experimental Medicine and Biology Book Series; Spinger Nature: Cham, Switzerland, 672:304–315.

Banat, I.M., Franzetti, A., Gandolfi, I., Bestetti, G., Martinotti, M.G., Fracchia, L., Smyth, T.J., Marchant, R. (2010) Microbial biosurfactants production, applications. *Appl. Microbiol. Biotechnol.*, 87:427–444. doi:10.1007/s00253-010-2589-0.

Ahmad, M. S. A., Ashraf, M., Ali, Q. (2010) Soil salinity as a selection pressure is a key determinant for the evolution of salt tolerance in Blue Panicgrass (*Panicum antidotale* Retz.). Flora (Jena), 205(1):37–45.

Basit, M., Rasool, M.H., Naqvi, S.A.R., Waseem, M., Aslam, B. (2018) Biosurfactants production potential of native strains of *Bacillus cereus* and their antimicrobial, cytotoxic and antioxidant activities. *Pak. J. Pharm. Sci.*, 31:251–256.

Çelik, P.A., Manga, E.B., Çabuk, A., Banat, I.M. (2021) Biosurfactants' potential role in combating COVID-19 and similar future microbial threats. *Appl. Sci.*, 11:334.

Chrzanowski, L., Ławniczak, L., Czaczyk, K. (2012) Why do microorganisms produce rhamnolipids? *World J. Microbiol. Biotechnol.*, 28:401–419.

Cordeiro, R.A., Weslley Caracas Cedro, E., Andrade, R.C., et al. (2018) Inhibitory effect of a lipopeptide biosurfactant produced by Bacillus subtilis on planktonic and sessile cells of Trichosporon spp. *Biofouling.*, 34(3):309–319.

Das, P., Mukherjee, S., Sen, R. (2008) Antimicrobial potential of a lipopeptide biosurfactant derived from a marine *Bacillus circulan. J. Appl. Microbiol.*, 104(6):1675–1684.

Das, P., Mukherjee, A.K., Sen, R. (2009) Antiadhesive action of a marine microbial surfactant. *Colloids Surf B: Biointerfaces*, 71:183–186.

Desai, J.D., Banat, I.M. (1997). Microbial production of surfactants and their commercial potential. *Microbiol and Mol Biol Rev.*, 61:47–64.

Elshikh, M., Moya-Ramírez, I., Moens, H., Roelants, S., Soetaert, W., Marchant, R., Banat, I.M. (2017). Rhamnolipids and lactonic sophorolipids: Natural antimicrobial surfactants for oral hygiene. *J. Appl. Microbiol.*, 123:1111–1123.

Fernandes, P.A.V., Arruda, I.R., Santos, A.F.B, Araujo, A.A., Maior, A.M.S., Ximenes, E.A. (2007). Antimicrobial activity of surfactants produced by *Bacillus subtilis* R14 against multidrug-resistant bacteria. Braz *J. Microbiol.*, 38:704–709.

Foley, P., Kermanshahi, P.A., Beach, E.S., Zimmerman, J.B. (2012) Derivation and synthesis of renewable surfactants. *Chem. Soc. Rev.*, 41:1499–1518.

Ghribi, D., Abdelkefi-Mesrati, L., Mnif, I., Kammoun, R., Ayadi, I., Saadaoui, I., Maktouf, S., Chaabouni-Ellouze, S. (2012) Investigation of antimicrobial activity and statistical optimization of Bacillus subtilis SPB1 biosurfactant production in solid-state fermentation. *J. Biomed. Biotechnol.*, 2012:373682.

Giordani, B., Costantini, P. B., Fedi, S., et al. (2019) Liposomes containing biosurfactants isolated from Lactobacillus gasseri exert antibiofilm activity against methicillin resistant Staphylococcus aureus strains. *Eur. J. Pharm. Biopharm.*, 139: 246–252.

Gross, R.A., Shah, V. (2007) Anti-herpes virus properties of various forms of sophorolipids. WO2007130738 A1. *Patent.* Nov 15.

Hommel, R., Stiiwer, O., Stuber, W., et al. (1987) Production of water-soluble surface-active exolipids by *Torulopsis apicola. Appl. Microbiol. Biotechnol.*, 26:199–205.

Hoq, M.M., Suzutani, T., Toyoda, T., Horiike G., Yoshida, I., Azuma, M. (1997) Role of gamma delta TCR+ lymphocytes in the augmented resistance of trehalose 6,6′-dimycolate-treated mice to influenza virus infection. *J. Gen. Virol.*, 78:1597–1603.

Huang, X., Lu, Z., Zhao, H., Yang, S. (2006) Antiviral activity of antimicrobial lipopeptide from *Bacillus subtilis* fmbj against pseudorabies virus, porcine parvovirus, newcastle disease virus and infectious bursal disease virus in vitro. *Int. J. Pept. Res. Ther.*, 12:373–377.

Jin, L., Black, W., Sawyer, T. (2021) Application of environment-friendly rhamnolipids against transmission of enveloped viruses like SARS-CoV2. *Viruses.*, 13:322.

Johnson, P., Trybala, A., Starov, V., Pinfield, V.J. (2021) Effect of synthetic surfactants on the environment and the potential for substitution by biosurfactants. *Adv. Colloid Interface Sci.*, 288:102340.

Jung, M., Lee, S., Kim, H. (2000) Recent studies on natural products as anti-HIV agents. *Curr. Med. Chem.*, 7:649–661.

Kaeppeli, O., Finnerty, W.R. (1979) Partition of alkane by an extracellular vesicle derived from hexadecane-grown Acinetobacter. *J. Bacteriol.*, 140:707–712.

Khan, F., Oloketuyi, S.F., Kim, Y.-M. (2019) Diversity of bacteria and bacterial products as antibiofilm and antiquorum sensing drugs against pathogenic bacteria. *J. Hazard. Mater.*, 364:441–448.

Kiss, K., Ng, W.T., Li, Q. (2017) Production of rhamnolipids-producing enzymes of Pseudomonas in *E. coli* and structural characterization. *Front. Chem. Sci. Eng.*, 11:133–138..

Kuyukina, M.S., Ivshina, I.B., Baeva, T, A., Kochina, O.A., Gein, S.V., Chereshnev, V.A. (2015) Trehalolipid biosurfactants from nonpathogenic Rhodococcus actinobacteria with diverse immunomodulatory activities. *N. Biotechnol.*, 32(6):559–568.

Lang, S. (2002) Biological amphiphiles (microbial biosurfactants) *Curr. Opin. Coll. Interface Sci.*, 7:12–20.

Liu, J., Li, W., Zhu, X., Zhao, H., Lu, Y., Zhang, C., Lu, Z. (2019) Surfactin effectively inhibits *Staphylococcus aureus* adhesion and biofilm formation on surfaces. *Appl. Microbiol. Biotechnol.*, 103:4565–4574.

McInerney, M.J., Javaheri, M., Nagle, D.P. (1990) Properties of the biosurfactant produced by *Bacillus licheniformis* strain JF-2. *J. Ind. Microbiol.*, *5*, 95–101.

Morais, I.M.C., Cordeiro, A.L., Teixeira, G.S., Domingues, V.S., Nardi, R.M.D., Monteiro A.S., Alves, R.J., Siqueira, E.P., Santos, V.L. (2017) Biological and physicochemical properties of biosurfactants produced by *Lactobacillus jensenii* P6A and *Lactobacillus gasseri* P65. *Microb. Cell Fact.*, 16:155.

Nakanishi, M., Inoh, Y., Kitamoto, D., Furuno, T. (2009) Nano vectors with a biosurfactant for gene transfection and drug delivery. *J. Drug Deliv. Sci. Technol.*, 19:165–169.

Naruse, N., Tenmyo, O., Kobaru, S., Kamei, H., Miyaki, T., Konishi M., Oki, T. (1990) Pumilacidin, a complex of new antiviral antibiotics. Production, isolation, chemical properties, structure and biological activity. *J. Antibiot.*, 43:267–280.

Ndlovu, T., Rautenbach, M., Vosloo, J.A., Khan, S., Khan, W. (2017) Characterisation and antimicrobial activity of biosurfactant extracts produced by *Bacillus amyloliquefaciens* and *Pseudomonas aeruginosa* isolated from a wastewater treatment plant. AMB Express, 7:108.

Nihorimbere, V., Cawoy, H., Seyer, A., Brunelle, A., Thonart, P., and Ongena, M. (2012). Impact of rhizosphere factors on cyclic lipopeptide signature from the plant beneficial strain *Bacillus amyloliquefaciens* S499. *FEMS Microbiol. Ecol.*, 79(1):176–191.

Pacheco, G. J., de Barros Gomes, E.M.P.C., Pereira, J.R.N. (2010) Biosurfactant production by *Rhodococcus Erythropolis* and its application to oil removal. *Braz. J. Microbiol.*, 41(3):685–693.

Pattanath, K.M., Rahman, K.S., Gakpe, E. (2008) Production, characterization and applications of biosurfactants—Review. *Biotechnology*; 7:360–370.

Pueyo, M.T., Bloch, C., Carmona-Ribeiro, A.M., Di Mascio, P. (2009) Lipopeptides produced by a soil Bacillus megaterium strain. *Microbial Ecol.*, 57(2):367–378.

Rahman, M.S., Ano, T. (2009) Production characteristics of lipopeptide antibiotics in biofilm fermentation of *Bacillus subtilis*. *J. Environ. Sci.*, 21:s36–s39.

Rahman, P., & Gapke, E., (2008) Production, characterisation and applications of biosurfactants – Review. *Biotechnology*, 7: 360–370.

Rivardo, F., Turner, R.J., Allegrone, G., et al. (2009) Anti-adhesion activity of two biosurfactants produced by *Bacillus* spp. prevents biofilm formation of human bacterial pathogens. *Appl. Microbiol. Biotechnol.*, 83:541–553.

Rodrigues, L., Bnat, I., M., Teixeira, R., et al. (2006) Biosurfactants: Potential applications in medicine. *J Antimicrobial and Chemotherapy*, 57:609–618.

Rodrigues, L., Mei, H.C.V., Teixeira, J., Oliveira, R. (2004) Influence of biosurfactants from probiotic bacteria on formation of biofilms on voice prostheses. *Appl. Environ. Microbiol.*, 70:4408–4410.

Rosenberg, E., Vaks, B., Zuckerberg, A. (1973) Bactericidal action of an antibiotic produced by *Myxococcus xanthus. Antimicrob. Agents Chemother.*, 4:507–513.

Satpute, S.K., Mone N.S., Das, P., Banat, I.M., Banpurkar, A.G. (2019) Inhibition of pathogenic bacterial biofilms on PDMS based implants by *L. acidophilus* derived biosurfactant. *BMC Microbiol.*, 19:39.

Satpute, S.K., Mone, N.S., Das, P., Banpurkar, A.G., Banat, I.M. (2018) *Lactobacillus acidophilus* derived biosurfactant as a biofilm inhibitor: A promising investigation using microfluidic approach. *Appl. Sci.*, 8:1555.

Schneider, J., Taraz, K., Budzikiewicz, H., Deleu, M., Thonart, P., Jacques, P. (1999) The structure of two fengycins from *Bacillus subtilis* S499. *Zeitschrift für Naturforschung*, 54:859–866.

Seydlová, G., Svobodová, J. (2008) Review of surfactin chemical properties and the potential biomedical applications. *Cent. Eur. J. Med.*, 3:123–133.

Shah, V., Doncel, G.F., Seyoum, T., Eaton K.M., Zalenskaya, I., Hagver, R., Azim, A., Gross, R. (2005) Sophorolipids, microbial glycolipids with anti-human immunodeficiency virus and sperm-immobilizing activities. *Antimicrob. Agents Chemother.*, 49:4093–4100.

Sharma, R., Chisti, Y., Banerjee U.C., (2001). Production, purification, characterization, and applications of lipases. *Biotechnol Adv.*, 19(8):627–662.

Singh, P., Cameotra, S.S. (2004) Potential applications of microbial surfactants in biomedical sciences. *Trends Biotechnol.*, 22:142–146.

Smith, M.L., Gandolfi S., Coshall P.M., Rahman P.K.S.M. (2020) Biosurfactants: A Covid-19 perspective. *Front. Microbiol.*, 11:1341.

Subramaniam, M.D., Venkatesan, D., Iyer, M., et al. (2020) Biosurfactants and anti-inflammatory activity: A potential new approach towards COVID-19. *Curr. Opin. Environ. Sci. Health.*, 17:72–81.

Tugrul, T., Cansunar, E. (2005) Detecting surfactant-producing microorganisms by the drop-collapse test. *World J. Microbiol. Biotechnol.*, 21:851–853.

Vater, J., Kablitz, B.,Wilde, C., Franke, P., Mehta, N., Cameotra, S.S. (2002) Matrix-assisted laser desorption ionization-time of flight mass spectrometry of lipopeptide biosurfactants in whole cells and culture filtrates of Bacillus subtilis C-1 isolated from petroleum sludge. *Appli. Environ. Microbiol.*, 68(12):6210–6219.

Vijayakumar, S., Saravanan, V. (2015) Biosurfactants-types, sources and applications. *Res. J. Microbiol.*, 10:181–192.

Yakimov, M.M., Timmis, K.N., Wray, V., Fredrickson, H.L. (1995) Characterization of a new lipopeptide surfactant produced by thermotolerant and halotolerant subsurface *Bacillus licheniformis* BAS50. *Appl. Environ. Microbiol.*, 61:1706–1713.

Yan, X., Gu, S., Cui X., Shi Y., Wen S., Chen H., Ge J. (2019) Antimicrobial, Anti-Adhesive and Anti-Biofilm potential of Biosurfactants Isolated from *Pediococcus acidilactici* and *Lactobacillus plantarum* against *Staphylococcus aureus* CMCC26003. *Microb. Pathogen.*, 127:12–20.

Yuan, L., Zhang, S., Wang Y., Li, Y., Wang, X., Yang, Q. (2018) Surfactin inhibits membrane fusion during invasion of epithelial cells by enveloped viruses. *J. Virol.*, 92(21):e00809-18. doi:10.1128/JVI.00809-18. PMID: 30068648; PMCID: PMC6189506.

Zhao, Z., Wang, Q., Wang, K., Brian, K., Liu, C., Gu, Y. (2010) Study of the antifungal activity of *Bacillus vallismortis* ZZ185 *in vitro* and identification of its antifungal components. *Bioresour. Technol.*, 101:292–297.

11 Novel Tools in Phytoremediation Technology
Proteomics and In Silico Tools

Lovely Sinha, Jitendra Kumar Sharma, Amit Lath, Jyoti Gulia, Anita Rani Santal, and Nater Pal Singh**

11.1 INTRODUCTION

Environmental contamination has increased significantly over the years due to rapid industrialization, urbanization, and human activities. Pesticides, agrochemicals, petroleum hydrocarbons, and heavy metals are continuously introduced into the environment, posing a serious threat to both the environment and food security (Dhir 2017).

The Industrial Revolution marked the beginning of a substantial rise in heavy metal contamination in the biosphere. Soil pollution with hazardous metals such as Cd, Pb, Cr, Zn, Ni, and Cu has risen significantly due to global industrialization, with factors such as continuous industrial growth and the production of numerous new chemicals each year contributing to this problem. More than 450 million kilograms of toxins have polluted air and water worldwide, according to Third World Network reports. The persistent nature of these pollutants is a major contributor to ecological pollution in a variety of ecosystems, including the hydrosphere, lithosphere, and biosphere (Bilal et al. 2018). Toxic and non-toxic metal ions, as well as other components, are present in waste from dye manufacturers, power plants, refineries, mines, and pharmaceutical businesses. These contaminants are immediately released into bodies of water, creating water pollution and a variety of health problems in both people and plants. Human health problems may include infectious or non-infectious diseases like cancer, neurodegeneration, cholera, vomiting, gastroenteritis, diarrhea, typhoid, skin issues, and kidney problems (Haseena et al. 2017; Khan and Ghouri 2011). In plants, these contaminants affect seed germination, increase the production of reactive oxygen species (ROS), impact cellular components, and alter various

* Corresponding authors

DOI: 10.1201/9781003392910-11

metabolic processes, which, in turn, affect plant growth, crop productivity, and biomass yield (Stambulska et al. 2018; Sao et al., 2021).

The extensive cultivation of agricultural lands has been driven by the increasing demand for agricultural products. To maintain the quality and quantity of these products, agrochemicals such as fertilizers, pesticides, and herbicides are widely employed. However, excessive use of these chemicals gives rise to the accumulation of pollutants in the soil and their uptake by plants, leading to further environmental problems (Trivedi et. al., 2021). Furthermore, pollution from various sources, including industrial activities, rural activities, and domestic activities, continuously releases heavy metals into terrestrial and aquatic ecosystems, exacerbating the issue. In recent decades, annual releases of heavy metals into the environment have reached alarming levels, with significant quantities of cadmium (22,000 t), copper (939,000 t), zinc (1,350,000 t), and lead (738,000 t) being introduced (Singh et al., 2003).

Because they are not biodegradable, heavy metals may spread to other environmental media. Usually found in the sky in the form of aerosols, they may be deposited onto soil and water by precipitation from rainfall. Heavy metals in the soil have the potential to seep into the groundwater. According to Jiang et al. (2018), irrigation is another way that these heavy metals in water might get into the soil. The build-up and migration of heavy metals in environmental media increases the possibility that they may enter the food chain, posing a serious threat to human health and ecological stability (Ali et al., 2019; Shen et al., 2022). Similarly, excessive pesticide use in agriculture and soil de-acidification waste are other factors contributing to soil pollution (Szczygłowska et al., 2011). As a result, global environmental pollution caused by organic xenobiotics such as pesticides, pharmaceuticals, petroleum compounds, PAHs, PCBs, and similar substances has become a widespread issue. The development of innovative remediation technologies for decontaminating affected sites is crucial. Various approaches, including physical, chemical, and biological methods, can be employed to remediate contaminated sites. Nevertheless, phytoremediation has been acknowledged as a cost-effective technique for purifying soil and water resources.

Bioremediation has developed as a safe, dependable, efficient, cost-effective, and eco-friendly alternative technology for the sustainable remediation of hazardous contaminants and pollutants. This method allows for on-site treatment of pollutants in a low-cost and environmentally friendly way, with no by-products or intricate setups and operations.

Bioremediation technology eliminates, mitigates, or converts both organic and inorganic contaminants into less harmful substances such as carbon dioxide and water by utilizing biological processes and the inherent catabolic activity of plants and microbes (Abruscia et al. 2007; Pandey and Fulekar 2012). Biological agents such as fungi, yeast, bacteria, and plants are actively engaged in removing contaminants through biodegradation and biotransformation mechanisms. The physiological and metabolic activities of these organisms play a crucial role in the degradation of contaminants, transforming them into products that are both environmentally safe and non-toxic. In this approach, the targeted compound functions as a carbon source, and the thorough mineralization of contaminants leads to the generation of water and carbon dioxide (Strong and Burgess 2008; Sharma and Fulekar 2009).

Phytoremediation, a type of bioremediation, has recently garnered significant attention (Singh et al., 2019). It relies on the utilization of plants and encompasses various procedures such as translocation of pollutants, accumulation of pollutants, transportation, pollutant transformation, and volatilization (Shah and Daverey, 2020). Moreover, nearly 400 plant species have shown phytoremediation potential (Ali et al., 2013, Shen et al., 2022). With few drawbacks, this technique is efficient and promising (Steliga and Kluk, 2020). Phytoremediation has long been recognized as a solution for environmental cleanup, particularly in the elimination of dangerous heavy metals from polluted areas. Scientific discoveries along with interdisciplinary research have transformed bioremediation into a promising, low-cost, and eco-friendly technology (Sao et al., 2022).

Proteomics and bioinformatics-based in silico methods are considered innovative phytoremediation tools. This technique offers important insights into the complex behaviors of proteins, metabolites, and enzymes, as well as the related biochemical pathways that these entities use to break down waste. This approach paves the way for sustainable solutions to environmental pollution. In response to the growing challenges of environmental cleanup, researchers have turned to innovative tools like proteomics and in silico methods (Patani et al., 2023).

11.2 PHYTOREMEDIATION

Phyto, derived from "plant," and remediates, meaning "to treat," combine to form phytotechnologies, which encompass a range of approaches utilizing plants for the remediation or containment of contaminants in soil, sediments, surface water, and groundwater. Phytoremediation, also known as "green remediation" or "botanical bioremediation," involves the deployment of plants to eliminate, transfer, or stabilize pollutants. This eco-friendly technique has gained popularity as a sustainable alternative to conventional cleanup methods due to its aesthetically pleasing qualities and cost-effectiveness (Suresh and Ravishankar 2004, Prasad 2007).

The United States Environmental Protection Agency (USEPA) described phytoremediation as a suite of technologies that employ various plants for the containment, destruction, or extraction of contaminants from media such as air, water, and soil. This emerging technology leverages different plants to break down, extract, contain, or immobilize contaminants present in soil and water. Van Ginneken et al. (2007) emphasized that phytoremediation, in general, utilizes plants (in conjunction with their associated microorganisms) for the removal, degradation, or stabilization of contaminants. Some plant species can accumulate or extract pollutants from the environment. The key features of phytoremediation are the reliance on solar energy and cost-effectiveness as a cleanup method; the applicability to shallow contamination sites; the effectiveness in treating a wide range of environmental contaminants; and the compatibility with mechanical cleanup methods, or in some cases, as a substitute for them (Gangwar et al., 2023).

The main functions of the phytoremediation process include the absorption and accumulation of hazardous substances; the degradation and detoxification of contaminants; the stabilization of contaminants around the roots; and the activation of microbes around the roots to facilitate degradation and detoxification (Sao et al., 2022).

11.2.1 Phytoremediation Methods

Various phytoremediation methods have previously been described by a number of researchers (Sharma et al., 2023; Schwitzguébel et al., 2022; Singh and Santal, 2015) (Figure 11.1).

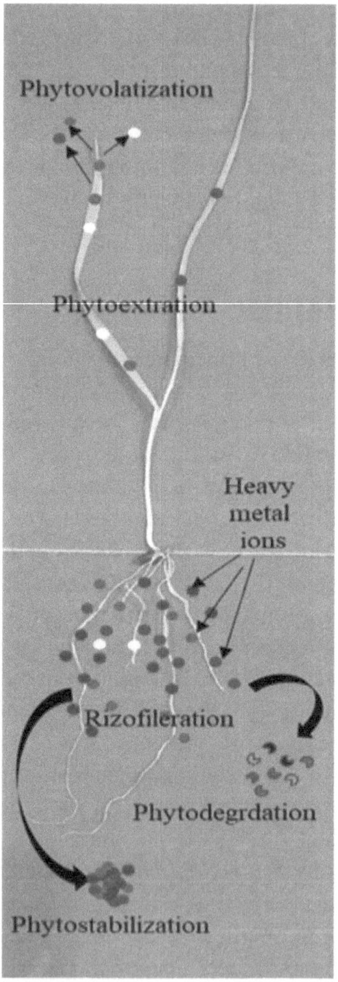

FIGURE 11.1 The phytoremediation strategies and general response of plants to harmful ions.

i. **Phytoextraction**: This method involves the removal of pollutants such as metals or organics from the soil using plants by concentrating them in plant parts. There are two different types of phytoextraction.
   a) Chelate-assisted phytoextraction: Artificial chelates are introduced to boost the absorption rate of pollutants, especially metals, thereby increasing their mobility and uptake.
   b) Continuous or natural phytoextraction: In this method, the removal of metals relies on the natural remediation ability of the plant. Some species, like those from the Brassicaceae, Euphorbiaceae, Asteraceae, and Lamiaceae families, have been identified as hyperaccumulator species.

ii. **Phytotransformation or phytodegradation**: This process involves the metabolic transformation of pollutants into inactive forms. The plant's metabolic system utilizes enzyme activities in the rhizosphere with the help of rhizosphere bacteria to reduce the toxicity of metal elements. Phytotransformation is labor-intensive and often requires soil amendments. It is considered less reliable, especially when dealing with inorganic contaminants.

iii. **Phytostimulation**: This method involves plant-assisted bioremediation, microbial stimulation, and fungal degradation through the release of enzymes/exudates into the rhizosphere.

iv. **Phytovolatilization**: Phytovolatilization converts metal contaminants into a gaseous form, which is then released into the atmosphere. However, this process only transfers pollutants to other parts of the environment and may lead to redeposition into the soil following precipitation, making it less popular compared to other methods like phytoextraction and phytofiltration.

v. **Rhizofiltration**: Rhizofiltration, also known as phytofiltration, involves the adsorption of pollutants by plant roots. It is commonly observed in aquatic plants. In rhizofiltration, the roots of plants take up and trap pollutants, primarily metals, from polluted soils and liquid waste streams. It plays a crucial role in preventing the transmission of toxic elements to different environmental components, including underground water. Terrestrial species also demonstrate phytofiltration, with microbial bio-filters aiding in the rhizosphere region.

vi. **Dendroremediation**: This method uses trees to enhance water evaporation, extracting contaminants from the soil.

vii. **Phytostabilization**: Plants in phytostabilization reduce pollutant toxicity by converting them into different forms or altering their bioavailability. This process reduces the bioavailability of pollutants in the environment using plant systems. Plants stabilize pollutants in the soil, making them inert and decreasing the likelihood of additional environmental damage through leaching into groundwater or airborne dispersion. This method helps prevent pollutants from entering the food chain.

As highlighted, plants can be employed in various ways to clean up or remediate contaminated sites by removing, degrading, or containing pollutants in soil, sediment, water, and air. They act as effective traps or filters for a wide range of metals, and inorganic and organic contaminants, providing valuable alternatives for environmental remediation.

11.2.2 Why Phytoremediation?

In the past century, there has been a significant increase in the emission of heavy metals and a substantial release of these heavy metals into our surroundings, such as soil, water, and air (Parveen et al., 2023). This has been attributed to various sources, such as industrial activities, the processing of sewage water, polluted sludge, phosphorus-based fertilizers, and a range of agricultural chemicals. Heavy metals are defined as those with a specific gravity greater than 5 g cm-3 and an atomic number exceeding 20. Some heavy metals, including Cu, Fe, Zn, Mo, and Mn, are essential micronutrients for plant physiological processes, metalloproteins, and ion-dependent enzymes like cytochrome peroxidase, superoxide dismutase, catalase, ascorbate oxidase, and nitrate reductase, among others. However, these metals can turn toxic when their concentrations surpass the levels required for the metabolism of plants. Conversely, metals such as Pb, Hg, and Cd are not essential for plant metabolism and therefore become harmful at a lower concentration. Additionally, non-physiological ions can substitute essential ions for a certain duration without causing harm to a plant; for example, during heat stress in moss, strontium ions can substitute calcium ions. The pollution of soil and water with hazardous heavy metals poses a substantial risk to environmental and human health (Cvjetko et al., 2014).

Conventional physical and chemical methods for environmental pollutant cleanup have several limitations, including high costs, alterations to native plants, changes in the physiochemical properties of the soil, and the need for intensive labor. These pollutants are virtually resistant to chemical or physical processes, except for nuclear fission and fusion, making their remediation a unique scientific and technical challenge (Vora et al., 2017). Therefore, new approaches for more effective treatment of polluted environments are imperative. In this context, adopting biological treatment strategies is a cost-effective and environmentally friendly option. Phytoremediation is a promising method that has garnered attention in recent years as the most feasible alternative. Phytoremediation leverages plants' ability to tolerate, accumulate, and translocate potentially toxic elements across their aerial tissues (Alsafran et al., 2022; Raza et al., 2020). Phytoremediation is generally acknowledged as an eco-friendly, non-invasive, and low-cost technology that is useful for remediating metal-contaminated areas.

11.2.3 Uptake and Translocation of Toxic Elements in Plants

Plants, as sessile organisms, unavoidably encounter heavy metals of geological or anthropogenic origin (Jung et al., 1996; Liao et al., 2005). Heavy metals in the environment are absorbed by the shoot parts of the plant from the soil and water through the root xylem, facilitated by apoplast and symplast pathways of transport mechanisms. Following this, heavy metal ions can be transported from the roots to the aerial parts of the plant, such as stems and leaves, via the xylem (Clemens, 2006). In response to exposure to heavy metal stress, plants exhibit the capability of accumulating elevated concentrations of heavy metals, typically following gradient patterns of accumulation in various plant organs, with the hierarchy being roots > stems > leaves > seeds (Rogers et al., 2008).

The harmful effects of some heavy metals depend on several factors, including the species of the plant and its stage of development, the concentration, and the chemical properties of the metal. All plants inherently exhibit tolerance to baseline levels of heavy metals, but elevated concentrations can lead to detrimental effects and disruptions, including inhibition of transpiration and photosynthesis, interference with carbohydrate metabolism, and nutrient and oxidative stress. These effects collectively have an impact on plant development and growth (Kramer and Clemens, 2005). The absorption of nonessential and potentially highly toxic heavy metal ions, such as Cd^{2+}, Hg^{2+}, and Pb^{2+}, as well as metalloids like arsenic (As) in ion forms such as arsenate and arsenite, and selenium (Se) in ion forms selenate and selenite, is a matter of significant concern (Clemens, 2006). Heavy metal stress causes symptoms that resemble symptoms of essential nutrient deficiency. These symptoms include reduced growth, decreased biomass, the appearance of necrotic spots on leaves, and damage primarily at the root level (Bona et al., 2007; Kieffer et al., 2008; Kieffer et al., 2009). The accumulation of higher concentrations can lead to decreased crop yields (Giordano et al., 1975) and pose a potential threat to human health, as heavy metals can enter the food chain (Islam et al., 2007; Fu et al., 2008).

Plants, to alleviate stress and restore cellular homeostasis and antioxidant capacity, have evolved highly effective techniques to control the uptake, storage, distribution, and detoxification of heavy metal ions (Hossain and Komatsu, 2013). Key components of these processes include metal transporters that facilitate metal uptake and vacuolar transport, chelators that contribute to detoxification and tolerance of heavy metals, and chaperones that play a pivotal role in the delivery and transport of heavy metal ions (Clemens, 2001; Sharma et al., 2008). All these components consist of protein molecules. Therefore, the identification of protein markers serves as a valuable starting point for uncovering new insights into the response of plants to heavy metal stress (Bona et al., 2007). Numerous proteomic studies have been conducted to investigate how plants respond to heavy metal stress, as highlighted in Table 11.1.

11.3 NOVEL APPROACHES TO STUDYING PHYTOREMEDIATION

Plant biologists have long grappled with the ongoing challenge of unraveling the roles and functions of individual genes and their products. This ongoing effort aims to enhance our understanding of the interaction of biological macromolecules in shaping a fully functional organism. Over time, various high-throughput analyses have provided a wealth of useful experimental data, due to which new terminology has surfaced. To comprehensively describe datasets of specific biomolecules, two related suffixes have been introduced: "-ome," which denotes a specific population of molecules within the cell (e.g., genome, proteome, translatome, and metabolome),

TABLE 11.1
Heavy Metal Phytoremediation Involving the Use of Comparative Proteomics

Heavy Metals	Plant Species	Plant Part Studied	Technology Used	References
Ni	*Alyssum lesbiacum* (Ni treated vs control plants)	Root, shoot	2D-GE, LC MS/MS	Ingle et al., 2005
	Noccaeacaerulescens	Shoot	2D-LC MALDI TOF/MS	Visioli et al., 2010
	Noccaea caerulescens Monte Prinzera sub-populations	Shoots	2D-LC MALDI TOF/MS	Visioli et al., 2012
As	*Artemisia annua* L.	Shoot, root	2-DE PAGE, MALDI-TOF-MS	Kumari and Pandey-Rai, 2018
	Brassica napus	Leaves	LC-MS/MS, SEM, TOF-MS, qRT-PCR	Farooq et al., 2021
	Oryza sativa L.	Leaves, root	2-DE, MALDI-TOF, TOF	Dixit et al., 2015
	Populus	Leaves, root	MALDI-TOF/TOF, MS, 2-DE, RT-PCR	Liu et al., 2017
Cd	*Arabidopsis thaliana* L.	Leaves, root	2D-GE, MALDI-TOF/TOF-MS	Klimenko et al., 2019
	Brassica campestris	Root	2D-GE, MALDI-TOF/TOF-MS	Su et al., 2019
	Brassica napus	Xylem sap	LC-MS/MS	Luo and Zhang, 2019
	Medicago sativa	Stem	2D-GE, MALDI-TOF/TOF-MS	Gutsch et al., 2019a
	Microsorumpteropus	Leaves, root	2D-GE, MALDI-TOF/TOF-MS	Lan et al., 2018
	Sorghum bicolor	Shoot	2D-GE, MALDI-TOF/TOF-MS	Roy et al., 2016
Cr	*Brassica napus* L.	Leaves	2D-GE, MALDI-TOF/TOF-MS	Yildiz and Terzi, 2016
	Callitriche cophocarpa	Shoot	SDS-PAGE, 2DE, MS/MS	Kaszycki et al., 2018
	Nicotiana tabacum	Shoot	2D-GE, MALDI-TOF/TOF-MS	Bukhari et al., 2016
	Pteris alba	Leaves, root	2D-GE, Nano HPLC MS/MS	Szuba and Lorenc-Plucinska, 2018
Cu	*Agrostis capillaris* L.	Shoot	2D-GE, LC-MS/MS	Hego et al., 2016
	Hyoscyamus albus L.	Root	MALDI-QIT-TOF-MS	Sako et al., 2016
	Triticum aestivum L.	Root, leaves	2D-GE, HPLC-Chip	Li et al., 2013

(*Continued*)

TABLE 11.1 (CONTINUED)
Heavy Metal Phytoremediation Involving the Use of Comparative Proteomics

Heavy Metals	Plant Species	Plant Part Studied	Technology Used	References
Hg	*Paspalum distichum* L.	Root	LC-MS/MS	Ding et al., 2019
	Triticum aestivum L.	Root, shoot	2D-GE, LC-MS/MS	Kang et al., 2015
Pb	*Cannabis sativa* L.	Leaves	LC-ESI-MS/MS. SWATH-MS	Xia et al., 2019
	Chrysopogonzizanioides	Root, shoot	LC-MS/MS	Pidatala et al., 2018
	Raphanus sativus L.	Root	GC-MS	Pang et al., 2015
	Glycine max L.	Nodules	2D-GE, MALDI-TOF, MS/MS	Baig et al., 2018
	Zea mays	Root	Nano-LC-MS/MS	Li et al., 2016

and "-omics," which emphasizes a research focus (e.g., transcriptomics, proteomics, translatomics, and metabolomics).

The utilization of modern "-omics" techniques has significantly enriched our understanding of the molecular pathways associated with tolerance to various environmental stresses. It is important to note that the vast volumes of omics data generated daily serve as a valuable resource for a wide range of biological applications. However, interpreting this data relies heavily on in silico databases and tools specifically designed to decode the information in alignment with distinct research objectives.

Additionally, numerous omics datasets are readily accessible in various databases, providing opportunities for innovative discoveries through analysis (Khare et al., 2017). This section delves into the intricacies of several omics datasets and the in-silico methodologies used for their analysis. As we strive for a deeper comprehension of plant stress responses to both biotic and abiotic stresses, it becomes increasingly evident that traditional approaches, either biochemical or physiological, are no longer adequate. Novel methodologies are imperative for advancing the field of phytoremediation.

11.3.1 Proteomics and Phytoremediation

Proteomics, a branch of biology, focuses on the study of encoded proteins in living organisms playing pivotal roles across different species at molecular and physiological levels. This biological process serves as a crucial pathway for conveying information regarding plant growth and adaptation to metal toxicity (Khan et al., 2023). In plant circulatory networks, nutrients, water, and potentially harmful ions are transported from roots to leaves through xylem sap. This process significantly detoxifies harmful ions within the xylem sap through proteomics (Jain et al., 2022). The detoxification of heavy metals involves various defense mechanisms, including exclusion, compartmentalization, and the formation of complexes with different proteins, such as phytochelatins (PCs) and metallothioneins (MTs), in different plant species.

Proteomics allows us to investigate every protein within a cell, characterizing its location, interactions, and modifications at specific time points. It is essential because intricate regulatory mechanisms are not always accessible through transcriptomics analysis. Protein levels are often regulated at the post-transcriptional modification level, leading to instances where mRNA abundance does not correspond to protein levels. Therefore, proteomics is necessary to determine the translation extent of mRNA, its accumulation locations, and post-translational modifications (Coleman 2005). Comparative studies of protein expression profiles and protein-protein interactions induced by various contaminants enable the assessment of the phytoremediation potential of different plant species, each with its unique capacity to remove contaminants from the environment. Proteomics research enhances our understanding of plant physiological responses to various contaminants by analyzing protein expression patterns (Mohanty, 2021; Yadav et al., 2023).

A traditional approach in proteomics involves the use of two-dimensional gel electrophoresis (2DE), which separates proteins based on two independent properties: isoelectric focusing in the initial dimension, and apparent molecular weight determination in the second dimension. This method is often combined with subsequent protein identification using mass spectrometry, such as MALDI-TOF (Matrix-Assisted Laser Desorption Ionization–Time of Flight) spectrometry, which provides masses of tryptic peptides obtained from protein spots and allows protein identification from data banks. This review highlights the initial results of proof-of-concept studies relevant to phytoremediation using "omic" technologies. In recent times, advancements in high-throughput methods for expressing and purifying recombinant proteins have contributed to the emergence of protein array technology specifically designed for plants. This technology connects recombinant proteins to characterized clones, creating a direct link between the gene catalog and a functional catalog.

Key participants in phytoremediation may be identified with the use of a comparative proteomics investigation of plant-microbe interactions in a polluted environment. Although a number of papers have analyzed proteome investigations of plant responses to heavy metals, there are presently (as of this writing) no thorough reviews of proteomic research in plants that hyperaccumulate heavy metals. Distinctly expressed proteins (DEPs) may be compared across different plant sections thanks to proteomic analyses of detrimental ion accumulators (Table 11.1) (Kryvych 2011).

11.3.2 In Silico Tools

In silico phytoremediation technology refers to the use of computational and modeling techniques to design and optimize phytoremediation strategies for the cleanup of contaminated environments (Khare et al., 2023). In silico (Latin for "in silicon") means that these processes are simulated and analyzed using computer-based methods before actual implementation. An overview of in silico phytoremediation technology is as follows:

1. Plant selection: In silico modeling allows researchers to analyze various plant species' genetic, physiological, and biochemical characteristics. This helps in selecting the most appropriate plants for a given contaminated site based on their ability to absorb, accumulate, or detoxify specific contaminants.
2. Contaminant fate and transport: Computational models simulate how contaminants move through the environment, tracking their behavior in soil, water, and plant tissues. This information is crucial for understanding how pollutants are taken up by plants and predicting their distribution and transformation.
3. Optimization: Through modeling, researchers can explore and optimize various parameters of phytoremediation, such as planting density, growth conditions, and the use of soil amendments. This process ensures that the remediation strategy is as efficient and cost-effective as possible.
4. Genetic engineering: In silico phytoremediation can assist in designing and evaluating genetic modifications of plants to enhance their pollutant-removing capabilities. By simulating the effects of genetic changes, researchers can identify the most promising modifications.

5. Risk assessment: Computational models are used to assess the potential risks and benefits associated with different phytoremediation strategies. This helps in making informed decisions and complying with regulatory requirements.
6. Monitoring and control: In silico approaches can be integrated with real-time monitoring systems, allowing for continuous assessment and control of phytoremediation processes. Adjustments can be made based on the data collected during the remediation project.
7. Data integration: To effectively apply in silico phytoremediation, various types of data are integrated, including environmental data, plant genetics, and contaminant properties. Advanced data analytics and machine learning techniques can help make sense of this information.

In silico phytoremediation technology has the potential to reduce the time and resources required for successful remediation projects. It offers a systematic and scientific basis for decision-making, enabling the development of more effective and sustainable solutions for environmental cleanup. While this field is still evolving, researchers and environmental scientists are increasingly turning to computational methods to enhance the outcomes of phytoremediation projects. By optimizing and customizing the approach to specific contaminants and environments, in silico phytoremediation technology has the potential to significantly contribute to environmental sustainability.

In this method, various proteins and phytochemicals of plants are docked with different pollutants like heavy metals (as a ligand) to identify phytochemicals that can be identified where the pollutants are binding. Primary sequences of the plant proteins can be downloaded from different databases, such as UniprotKB, NCBI, EMBL, and DDBJ. Different software can be used for the docking process, such as AutoDock, GOLD, Glide, and iGEMDOCK (Coleman, 2005; Verma et al., 2022). The steps of phytoremediation analysis by proteomics and in silico tools are shown in Figure 11.2.

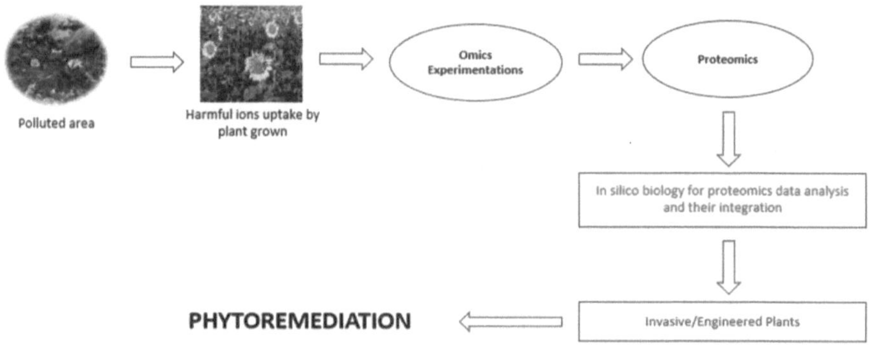

FIGURE 11.2 Phytoremediation analysis by proteomics and in silico tools.

11.4 CONCLUSION

Ecological imbalance in the ecosystem has led to the exploration of inventive techniques, methods, and devices for degrading environmental contaminants. In plant system biology, omics approaches have displayed remarkable adaptability, addressing gaps in our comprehension of the mechanisms involved in various bioremediation processes. The omic branches associated with genes, protein, transcript, metabolism, and others have notably succeeded in pinpointing new genes in microbial populations and plants proficient at efficiently breaking down pollutants. Their cost-effectiveness and overall effectiveness surpass their limitations, as evidenced by numerous ongoing research projects incorporating in silico support for this purpose. In contemporary times, bioremediation stands out as the preferred choice over conventional methods, playing a crucial role in preserving two of our most vital resources: soil and water. Bioremediation has great potential for reinstating ecosystem health and serves as a superior option for management to protect life and the environment from harmful pollutants.

REFERENCES

Abrusci, C., Marquina, D., Del Amo, A., & Catalina, F. (2007). Biodegradation of cinematographic gelatin emulsion by bacteria and filamentous fungi using indirect impedance technique. *International Biodeterioration & Biodegradation*, 60(3), 137–143.

Ali, H., Khan, E., & Ilahi, I. (2019). Environmental chemistry and ecotoxicology of hazardous heavy metals: Environmental persistence, toxicity, and bioaccumulation. *Journal of Chemistry*. https://doi.org/10.1155/2019/6730305

Ali, H., Khan, E., & Sajad, M. A. (2013). Phytoremediation of heavy metals—concepts and applications. *Chemosphere*, 91(7), 869–881.

Alsafran, M., Usman, K., Ahmed, B., Rizwan, M., Saleem, M. H., & Al Jabri, H. (2022). Understanding the phytoremediation mechanisms of potentially toxic elements: A proteomic overview of recent advances. *Frontiers in Plant Science*, 13, 881242.

Baig, M. A., Ahmad, J., Bagheri, R., Ali, A. A., Al-Huqail, A. A., Ibrahim, M. M., et al. (2018). Proteomic and ecophysiological responses of soybean (Glycine max L.) root nodules to Pb and hg stress. *BMC Plant Biology*, 18, 283–221.

Bilal, M., Rasheed, T., Sosa-Hernández, J. E., Raza, A., Nabeel, F., & Iqbal, H. M. (2018). Biosorption: An interplay between marine algae and potentially toxic elements—a review. *Marine Drugs*, 16(2), 65.

Bona, E., Marsano, F., Cavaletto, M., & Berta, G. (2007). Proteomic characterization of copper stress response in Cannabis sativa roots. *Proteomics*, 7, 1121–1130.

Bukhari, S. A. H., Zheng, W., Xie, L., Zhang, G., Shang, S., & Wu, F. (2016). Cr-induced changes in leaf protein profile, ultrastructure and photosynthetic traits in the two contrasting tobacco genotypes. *Plant Growth Regulation*, 79, 147–156.

Clemens, S. (2001). Molecular mechanisms of plant metal tolerance and homeostasis. *Planta*, 212, 475–486.

Clemens, S. (2006). Toxic metal accumulation, responses to exposure and mechanisms of tolerance in plants. *Biochimie*, 88, 1707–1719.

Coleman, J. O., Haslam, R. P., & Downie, A. L. (2005). Transcriptomics and proteomics: Tools for optimising phytoremediation activities. *Zeitschrift für Naturforschung C*, 60(7–8), 544–548.

Cvjetko, P., Zovko, M., & Balen, B. (2014). Proteomics of heavy metal toxicity in plants. *Arhiv za higijenu rada i toksikologiju*, 65(1), 1–17.

Dhir, B. (2017). Bioremediation technologies for the removal of pollutants. In *Advances in Environmental Biotechnology* (pp. 69–91). Springer.

Dixit, G., Singh, A. P., Kumar, A., Dwivedi, S., Deeba, F., Kumar, S., et al. (2015). Sulfur alleviates arsenic toxicity by reducing its accumulation and modulating proteome, amino acids and thiol metabolism in rice leaves. *Scientific Reports*, 5, 1–16.

Farooq, M. A., Hong, Z., Islam, F., Noor, Y., Hannan, F., Zhang, Y., et al. (2021). Comprehensive proteomic analysis of arsenic induced toxicity reveals the mechanism of multilevel coordination of efficient defense and energy metabolism in two Brassica napus cultivars. *Ecotoxicology and Environmental Safety*, 208, 111744.

Fu, J., Zhou, Q., Liu, J., Liu, W., Wang, T., Zhang, Q., & Jiang, G. (2008). High levels of heavy metals in rice (Oryza sativa L.) from a typical E-waste recycling area in southeast China and its potential risk to human health. *Chemosphere*, 71, 1269–1275.

Gangwar, P., Johri, P., Datta, R., Singh, S., & Trivedi, M. (2023). Overview on 1 phytopharmaceuticals and biotechnology of herbal plants. In *Phytopharmaceuticals and Biotechnology of Herbal Plants* (pp. 1–17). CRC Press.

Giordano, P. M., Mortvedt, J., & Mays, A. (1975). Effect of municipal wastes on crop yields and uptake of heavy metals. *Journal of Environmental Quality*, 4, 394–399.

Gutsch, A., Sergeant, K., Keunen, E., Prinsen, E., Guerriero, G., Renaut, J., et al. (2019a). Does long-term cadmium exposure influence the composition of pectic polysaccharides in the cell wall of Medicago sativa stems? *BMC Plant Biology*, 19, 1–17.

Haseena, M., Malik, M. F., Javed, A., Arshad, S., Asif, N., Zulfiqar, S., & Hanif, J. (2017). Water pollution and human health. *Environmental Risk Assessment and Remediation*, 1(3), 16–19.

Hego, E., Vilain, S., Barré, A., Claverol, S., Dupuy, J., Lalanne, C., et al. (2016). Copper stress-induced changes in leaf soluble proteome of cu-sensitive and tolerant Agrostis capillaris L. populations. *Proteomics*, 16, 1386–1397.

Hossain, Z., & Komatsu, S. (2013). Contribution of proteomic studies towards understanding plant heavy metal stress response. *Frontiers in Plant Science*, 3, 310.

Ingle, R. A., Smith, J. A., & Sweetlove, L. J. (2005). Responses to nickel in the proteome of the hyperaccumulator plant Alyssum lesbiacum. *Bio Metals*, 18, 627–641.

Islam, E., Yang, X., & He, Z. (2007). Mahmood Q. Assessing potential dietary toxicity of heavy metals in selected vegetables and food crops. *Journal of Zhejiang University Science B*, 8, 1–13.

Jain, T., Srivastava, K., Singh, R., Ansari, M. J., Singh, S., & Dutta, P. K. (2022). Role of nanotechnology 18 in the field of phytopharmaceuticals for the delivery of herbal drugs. In: Singh S, Dutta R, Johri P, Trivedi M, editors. *Phytopharmaceuticals and biotechnology of herbal plants*. Boca Raton: CRC Press, 14.

Jiang, D., Zeng, G., Huang, D., Chen, M., Zhang, C., Huang, C., & Wan, J. (2018). Remediation of contaminated soils by enhanced nanoscale zero valent iron. *Environmental Research*, 163, 217–227.

Jung, M. C., & Thorton, I. (1996). Heavy metal contamination of soils and plants in the vicinity of a lead-zinc mine, Korea. *Applied Geochemistry*, 11, 53–59.

Kang, G., Li, G., Wang, L., Wei, L., Yang, Y., Wang, P., et al. (2015). Hg-responsive proteins identified in wheat seedlings using iTRAQ analysis and the role of ABA in hg stress. *Journal of Proteome Research*, 14, 249–267.

Kaszycki, P., Dubicka-Lisowska, A., Augustynowicz, J., Piwowarczyk, B., & Wesołowski, W. (2018). Callitriche cophocarpa (water starwort) proteome under chromate stress: Evidence for induction of a quinone reductase. *Environmental Science and Pollution Research*, 25, 8928–8942.

Khan, I. U., Qi, S. S., Gul, F., Manan, S., Rono, J. K., Naz, M., & Du, D. L. (2023). A green approach used for heavy metals 'phytoremediation' via invasive plant species to mitigate environmental pollution: A review. *Plants*, 12(4), 725.

Khan, M. A., & Ghouri, A. M. (2011). Environmental pollution: Its effects on life and its remedies. *Researcher World: Journal of Arts, Science & Commerce*, 2(2), 276–285.

Khare, N., Bhardwaj, M., Chawla, S., Praveen, R., & Singh, S. (2023). Redox responsive smart nanomaterials to tackle hypoxia associated oxidative damage and inflammatory mediators using phytocompounds. In *Smart Nanomaterials Targeting Pathological Hypoxia* (pp. 313–325). Springer Nature Singapore.

Khare, P., Kishore, K., & Sharma, D. K. (2017). A study on the standardization parameters of Cassia angustifolia. *Asian Journal of Pharmaceutical and Clinical Research*, 10(7), 329–332.

Kieffer, P., Dommes, J., Hoffmann, L., Hausman, J. F., & Renaut, J. (2008). Quantitative changes in protein expression of cadmium-exposed poplar plants. *Proteomics*, 8(12), 2514–2530.

Kieffer, P., Planchon, S., Oufir, M., Ziebel, J., Dommes, J., Hoffmann, L., & Renaut, J. (2009). Combining proteomics and metabolite analyses to unravel cadmium stress-response in poplar leaves. *Journal of Proteome Research*, 8(1), 400–417.

Klimenko, O., Pernis, M., Danchenko, M., Skultéty, L., Klubicová, K., & Shevchenko, G. (2019). Natural ecotype of Arabidopsis thaliana (L.) Heynh (Chernobyl-07) respond to cadmium stress more intensively than the sensitive ecotypes oasis and Columbia. *Ecotoxicology and Environmental Safety*, 173, 86–95.

Krämer, U., & Clemens, S. (2005). Molecular biology of metal homeostasis and detoxification. In: Tamäs M, Martinoia E, editors. *Topics in current genetics*. New York, NY: Springer.

Kryvych, S., Kleessen, S., Ebert, B., Kersten, B., & Fisahn, J. (2011). Proteomics–the key to understanding systems biology of Arabidopsis trichomes. *Phytochemistry*, 72(10), 1061–1070.

Kumari, A., & Pandey-Rai, S. (2018). Enhanced arsenic tolerance and secondary metabolism by modulation of gene expression and proteome profile in Artemisia annua L. after application of exogenous salicylic acid. *Plant Physiology and Biochemistry*, 132, 590–602.

Lan, X.-Y., Yan, Y.-Y., Yang, B., Li, X.-Y., & Xu, F.-L. (2018). Differential expression of proteins in the leaves and roots of cadmium-stressed Microsorum pteropus, a novel potential aquatic cadmium hyperaccumulator. *Science of the Total Environment*, 642, 1369–1377.

Li, G. K., Gao, J., Peng, H., Shen, Y. O., Ding, H. P., Zhang, Z. M., et al. (2016). Proteomic changes in maize as a response to heavy metal (lead) stress revealed by iTRAQ quantitative proteomics. *Genetics and Molecular Research*, 15, 1–14.

Li, G., Peng, X., Xuan, H., Wei, L., Yang, Y., Guo, T., et al. (2013). Proteomic analysis of leaves and roots of common wheat (Triticum aestivum L.) under copper-stress conditions. *Journal of Proteome Research*, 12, 4846–4861.

Liao, X. Y., Chen, T. B., Xie, H., & Liu, Y. R. (2005). Soil As contamination and its risk assessment in areas near the industrial districts of Chenzhou City, Southern China. *Environment International*, 31, 791–798.

Liu, Y., Damaris, R. N., & Yang, P. (2017). Proteomics analysis identified a DRT protein involved in arsenic resistance in populus. *Plant Cell Reports*, 36, 1855–1869.

Luo, J.-S., & Zhang, Z. (2019). Proteomic changes in the xylem sap of Brassica napus under cadmium stress and functional validation. *BMC Plant Biology*, 19, 1–14.

Mohanty, M. (2021). Proteomics and bioinformatics as novel tools in phytoremediation technology-an overview. *Journal of Botanical Research*, 3(3), 49–54.

Pandey, B., & Fulekar, M. H. (2012). Bioremediation technology: a new horizon for environmental clean-up. *Biology and Medicine*, 4(1), 51.

Pang, X.-Q., Jia, C.-Z., & Wang, W.-Y. (2015). Petroleum geology features and research developments of hydrocarbon accumulation in deep petroliferous basins. *Pet Science*, 12, 1–53.

Parveen, R., Khare, N., Singh, S., & Johri, P. (2023). An in silico outlook for the detection and surveillance of evolving and persistent plant pathogens. In *Detection, diagnosis and management of soil-borne phytopathogens* (pp. 17–40). Springer Nature Singapore.

Patani, A., Patel, A., Prajapati, D., Khare, N., & Singh, S. (2023). Various agriculture crop plant-based bioactive compounds and their use in nanomaterial synthesis and applications. In *Nanomaterials from agricultural and horticultural products* (pp. 223–241). Springer Nature Singapore.

Pidatala, V. R., Li, K., Sarkar, D., Wusirika, R., & Datta, R. (2018). Comparative metabolic profiling of vetiver (Chrysopogon zizanioides) and maize (Zea mays) under lead stress. *Chemosphere*, 193, 903–911.

Prasad, M. N. V. (2007). Phytoremediation in India. In S. Khan (Ed.), *Phytoremediation: Methods and reviews* (pp. 435–454). University of the West of England.

Raza, A., Habib, M., Kakavand, S. N., Zahid, Z., Zahra, N., Sharif, R., & Hasanuzzaman, M. (2020). Phytoremediation of cadmium: physiological, biochemical, and molecular mechanisms. *Biology*, 9(7), 177.

Rogers, S., Girolami, M., Kolch, W., Waters, K. M., Liu, T., Thrall, B., & Wiley, H. S. (2008). Investigating the correspondence between transcriptomic and proteomic expression profiles using coupled cluster models. *Bioinformatics*, 24(24), 2894–2900.

Roy, S. K., Cho, S.-W., Kwon, S. J., Kamal, A. H. M., Kim, S.-W., Oh, M.-W., et al. (2016). Morpho-physiological and proteome level responses to cadmium stress in sorghum. *PLoS One*, 11, e0150431.

Sako, A., Kandakar, J., Tamari, N., Higa, A., Yamaguchi, K., & Kitamura, Y. (2016). Copper excess promotes propagation and induces proteomic change in root cultures of Hyoscyamus albus L. *Plant Physiology and Biochemistry*, 103, 1–9.

Sao, P., Chand, Y., Kumar, A., & Singh, S. (2021). Potential drug target identification in porphyromonas gingivalis using in-silico subtractive metabolic pathway analysis. *Bangladesh Journal of Medical Science*, 20(4), 887–896.

Sao, P., Parveen, R., Khattri, A., Sharma, S., Tiwari, N., & Singh, S. (2022). Understanding the metabolomics of medicinal plants under environmental pollution: Challenges and opportunities. *Environmental Pollution and Medicinal Plants*, 2022, 193–205.

Sao, P., Vats, S., & Singh, S. (2022). Porphyromonas gingivalis resistance and virulence: An integrated functional network analysis. *Gene*, 839, 146734.

Schwitzguébel, J. P., van der Lelie, D., Baker, A., Glass, D. J., & Vangronsveld, J. (2002). Phytoremediation: European and American trends successes, obstacles and needs. *Journal of Soils and Sediments*, 2, 91–99.

Shah, V., & Daverey, A. (2020). Phytoremediation: A multidisciplinary approach to clean up heavy metal contaminated soil. *Environmental Technology & Innovation*, 18, 100774.

Sharma, S. K., Goloubinoff, P., & Christen, P. (2008). Heavy metal ions are potent inhibitors of protein folding. *Biochemical and Biophysical Research Communications*, 372, 341–345.

Sharma, J., & Fulekar, M. H. (2009). Potential of Citrobacter freundii for bioaccumulation of heavy metal–copper. *Biology and Medicine*, 1(3), 7–14.

Sharma, J. K., Kumar, N., Singh, N. P., & Santal, A. R. (2023). Phytoremediation technologies and their mechanism for removal of heavy metal from contaminated soil: An approach for a sustainable environment. *Frontiers in Plant Science*, 14, 1076876.

Shen, X., Dai, M., Yang, J., Sun, L., Tan, X., Peng, C., & Naz, I. (2022). A critical review on the phytoremediation of heavy metals from environment: Performance and challenges. *Chemosphere*, 291, 132979.

Singh, A., Sengar, R. S., Singh, A. K., Singh, S., Kumar, A., & Ahmad, K. (2019). Impact of genetic engineering on agriculture food crops. *Biotech Today: An International Journal of Biological Sciences*, 9(1), 17–24.

Singh, N. P., & Santal, A. R. (2015). Phytoremediation of heavy metals: The use of green approaches to clean the environment. *Phytoremediation: Management of Environmental Contaminants*, 2, 115–129.

Singh, O. V., Labana, S., Pandey, G., Budhiraja, R., & Jain, R. K. (2003). Phytoremediation: an overview of metallic ion decontamination from soil. *Applied Microbiology and Biotechnology*, 61, 405–412.

Stambulska, U. Y., Bayliak, M. M., & Lushchak, V. I. (2018). Chromium (VI) toxicity in legume plants: modulation effects of rhizobial symbiosis. *BioMed Research International*, 2018. https://doi.org/10.1155/2018/8031213

Steliga, T., & Kluk, D. (2020). Application of Festuca arundinacea in phytoremediation of soils contaminated with Pb, Ni, Cd and petroleum hydrocarbons. *Ecotoxicology and Environmental Safety*, 194, 110409.

Strong, P. J., & Burgess, J. E. (2008). Treatment methods for wine-related and distillery wastewaters: a review. *Bioremediation Journal*, 12(2), 70–87.

Su, N., Wu, Q., Chen, H., Huang, Y., Zhu, Z., Chen, Y., & Cui, J. (2019). Hydrogen gas alleviates toxic effects of cadmium in Brassica campestris seedlings through up-regulation of the antioxidant capacities: Possible involvement of nitric oxide. *Environmental Pollution*, 251, 45–55.

Suresh, B., & Ravishankar, G. A. (2004). Phytoremediation—a novel and promising approach for environmental clean-up. *Critical Reviews in Biotechnology*, 24(2–3), 97–124.

Szczygłowska, M., Piekarska, A., Konieczka, P., & Namieśnik, J. (2011). Use of brassica plants in the phytoremediation and biofumigation processes. *International Journal of Molecular Sciences*, 12(11), 7760–7771.

Szuba, A., & Lorenc-Plucińska, G. (2018). Field proteomics of Populus alba grown in a heavily modified environment–An example of a tannery waste landfill. *Science of The Total Environment*, 610, 1557–1571.

Trivedi, M., Mathur, M., & Tiwari, R. K. (2021). Emerging pollutants: A tacticturn killer. In N. Goyal & V. Shukla (Eds.), *Persistent* organic pollutants *in the* environment (pp. 181–202). Taylor & Francis, CRC Press. doi:10.1201/9781003053170, ISBN: 978-0-367-51288-0.

Van Ginneken, L., Meers, E., Guisson, R., Ruttens, A., Elst, K., Tack, F. M., & Dejonghe, W. (2007). Phytoremediation for heavy metal-contaminated soils combined with bioenergy production. *Journal of Environmental Engineering and Landscape Management*, 15(4), 227–236.

Verma, S., Kour, S., & Pathak, R. K. (2022). In silico approaches in bioremediation research and advancements. *Bioremediation of Environmental Pollutants: Emerging Trends and Strategies*, 221–238.

Visioli, G., Pirondini, A., Malchevshi, A., & Marmiroli, N. (2010). Comparison of proteomic variation in Thlaspi caerulescens populations from metalliferous and non-metalliferous soils. *International Journal of Phytoremediation*, 12, 805–819.

Visioli, G., Vincenzi, S., Marmiroli, M., & Marmiroli, N. (2012). Correlation between phenotype and proteome in the Ni hyperaccumulator Noccaea caerulescens subsp. caerulescens. *Environmental and Experimental Botany*, 77, 156–164.

Vora, Z., Pandya, J., Vaikuntapu, P. R., Singh, S., Asha, S., & Purama, R. K. (2017). Genome-wide association of defense hormone crosstalk in plants. In *Hormonal cross-talk, plant defense and development* (pp. 353–371). Academic Press.

Xia, C., Hong, L., Yang, Y., Yanping, X., Xing, H., & Gang, D. (2019). Protein changes in response to lead stress of lead-tolerant and lead-sensitive industrial hemp using SWATH technology. *Genes*, 10(5), 396.

Yadav, R., Singh, G., Santal, A. R., & Singh, N. P. (2023). Omics approaches in effective selection and generation of potential plants for phytoremediation of heavy metal from contaminated resources. *Journal of Environmental Management*, 336, 117730.

Yıldız, M., & Terzi, H. (2016). Proteomic analysis of chromium stress and sulfur deficiency responses in leaves of two canola (Brassica napus L.) cultivars differing in Cr (VI) tolerance. *Ecotoxicology and Environmental Safety*, 124, 255–266.

12 Caryophyllene Oxide

A Multifaceted Terpenoid with Potent Biological Activities

Barkha Kumari, Satish Dubey, Ashwini Kumar Dixit, Shraddha Manish Gupta, Durgesh Nandini Chauhan, Kamal Shah, and Nagendra Singh Chauhan

12.1 INTRODUCTION

Caryophyllene oxide is a two-cycle structure known to have an epoxide group, which is a group of oxygen atoms attached to two neighbouring carbon atoms to form a three-membered ring, as well as a sesquiterpene, and is considered to be one of the most significant chemicals in the terpenoid class (Fidyt et al., 2016). The molecular formula for it is C15H24O, and its molecular weight is 220.35 g/mol. These are the characteristics that define it (Morales-arce, 2022). It smells slightly sweet, spicy, and woodsy all at the same time. According to Gyrdymova and Rubtsova (2022), sesquiterpenoids are able to accomplish the synthesis of a wide variety of compounds that exhibit a wide range of biological activities because of their distinctive structure (Figure 12.1).

The presence of caryophyllene oxide is typically observed in plants that are accompanied by β-caryophyllene, α-humulene, and isocaryophyllene, all of which are classified as sesquiterpenes. They serve as an indirect defence mechanism against infections and play an important part in the interactions that take place between biotic organisms.

Through the process of isolating caryophyllene oxide from the essential oils of various plant parts, including leaves, seeds, stems, flowers, and roots, we create this compound. Camellia sinensis, Melissa officinalis, Syzygium aromaticum, and Artemisia thuscula are some of the plants that contain this naturally occurring phytochemical. It is particularly prevalent in these plants (Terbot et al., 2023).

Caryophyllene oxide is a flavouring ingredient that is generally regarded as safe (GRAS) according to the Code of Federal Regulations. It is permissible to use it in a

DOI: 10.1201/9781003392910-12

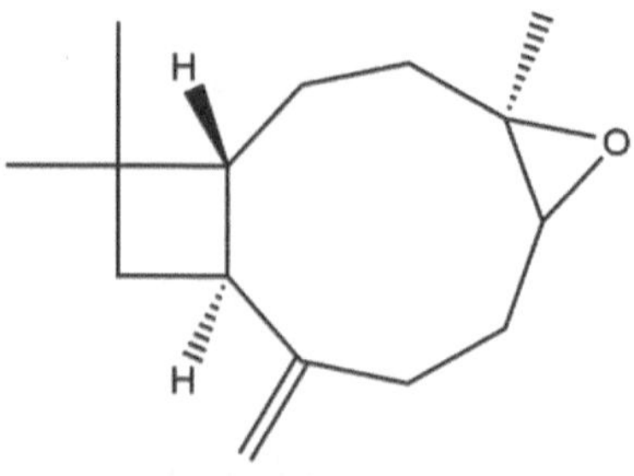

FIGURE 12.1 Chemical structure of caryophyllene oxide.

variety of common foods, including fruit juices, soft beverages, baked products, ice cream, and pudding.

According to Weimer et al. (2022), it possesses a wide range of pharmacological activities that are effective against significant human disorders. These activities include antibacterial, anti-inflammatory, anticancerous, neuroprotective, lymphoma, analgesic, and genologic effects.

12.2 BIOSYNTHETIC PATHWAYS

The mevalonate (MVA) route produces caryophyllene oxide through a series of significant steps that each contribute to the final product (Table 12.1). These phases start with the fundamental components that make up isoprene units (Tholl, 2015).

TABLE 12.1
Some Natural Sources of Caryophyllene Oxide

Plant Source	Part Used	Activity	Type of Study	Reference
Psidium cattleianum Sabine	Leaves	Anticancerous	In vitro	Jun et al. 2011
Psidium guajava	Leaves	Antifungal	In vitro	Jassal et al. 2021
Copaiferamultijuga	Oleoresin	Antiedematogenic activity	In vivo	Weimer et al. 2022
Origanum vulgare	Whole plant	Antinociceptive, anti-inflammatory	In vivo	Moghrovyan et al. 2022
Chromolaenaodorata	Leaves	Sedative activity	In vivo	Dougnon and Ito 2021
Annona squamosa	Bark	Analgesic and anti-inflammatory activity	In vivo	Chavan et al. 2010
Salvia verticillata	Aerial part	Anticholinesterase and anticancerous	In vitro	Karakaya et al. 2020

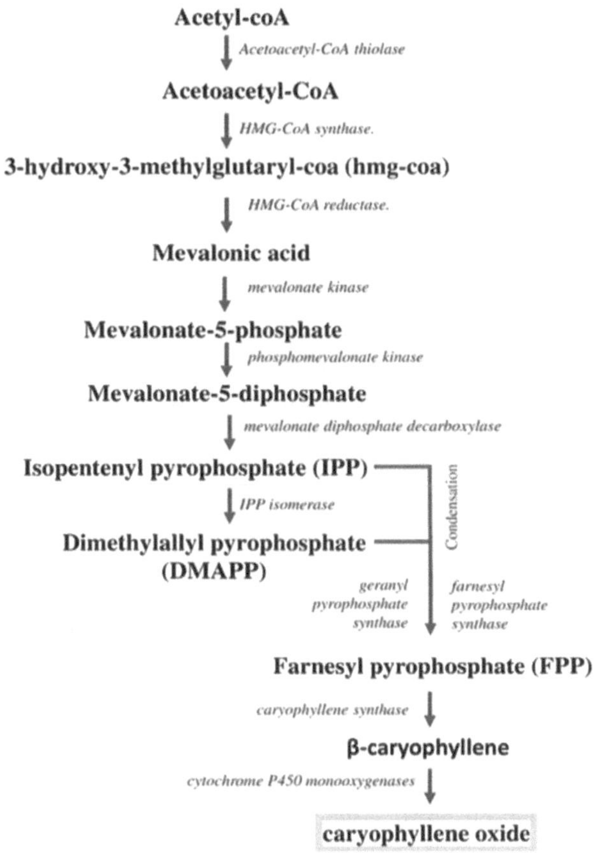

FIGURE 12.2 Biosynthetic pathway of caryophyllene oxide.

An outline of the biosynthetic pathway is described below, as well as presented in Figure 12.2.

In the first step, two molecules of acetyl-CoA come together, and acetoacetyl-CoA is thereby produced. Later, this compound joins another molecule called acetyl-CoA to produce 3-hydroxy-3-methylglutaryl-CoA, or HMG-CoA.

The second step of the biosynthetic process involves the enzyme HMG-CoA reductase, which is responsible for facilitating the reduction of HMG-CoA to mevalonate (Mcgrath et al., 2014).

In the third step, mevalonate undergoes a series of phosphorylations and decarboxylations to produce isopentenyl pyrophosphate (IPP). IPP molecules undergo condensation, which results in the formation of dimethylallyl pyrophosphate (DMAPP). There is a condensation between IPP and DMAPP, which aids in the creation of geranyl pyrophosphate (GPP). This GPP subsequently combines with another IPP to produce FPP (Cheng et al., 2007).

In the final step, the enzyme caryophyllene synthase is responsible for catalysing the cyclization of FPP into caryophyllene. The subsequent oxidation of caryophyllene to caryophyllene oxide is carried out by a particular cytochrome P450 enzyme known as caryophyllene synthase. This enzyme provides caryophyllene with an oxygen molecule, which results in the transformation of caryophyllene into its oxide form.

12.3 BIOLOGICAL ACTIVITIES

Caryophyllene oxide is a sesquiterpene found in essential oils. It is an interesting biomedical chemical because it has many biological roles. It helps with pain and inflammation by changing important pathways like NF-κB and MAPK signaling, which are needed for inflammatory reactions. Caryophyllene oxide stops breast, liver, and colon cancer cell types from growing, and causes cells to die. It has been shown to kill many types of bacteria and fungi because it damages their cell walls. Caryophyllene oxide and other antioxidants remove free radicals and lower reactive stress, which protects cells. Because it works in many different ways, it may be able to help with inflammatory, infectious, cancerous, and oxidative stress-related conditions. In this chapter, we discuss selected activities of this compound.

12.4 ANTIMICROBIAL ACTIVITY

Caryophyllene oxide is known for having strong antimicrobial qualities that work against a wide range of microorganisms. This chapter looks at a number of studies that show how useful the compound is in different situations. In a study, the main parts of the essential oil of *Satureja coerulea* were found to be β-caryophyllene (10.6%) and caryophyllene oxide (8.0%). This oil showed strong antimicrobial action against several types of microbes, highlighting the importance of these compounds for antimicrobial use (Azaz et al., 2002). The main sesquiterpene found in the essential oil of *Eugenia chlorophylla* stem (17.2%) was caryophyllene oxide. There was strong antimicrobial action in this essential oil only against Gram-positive bacteria and yeasts. It did not work against Gram-negative bacteria. The uniqueness of caryophyllene oxide's antimicrobial action provides clues about how it might be used (Stefanello et al., 2008).

Significant components of *Artemisia phaeolepis* essential oil were identified to include caryophyllene oxide, eucalyptol (11.30%), camphor (8.21%), terpine-4-ol (7.32%), germacrene D (6.39%), and caryophyllene (5.37%). A study by Hsouna et al. (2013) demonstrated how using these chemicals together increased their antibacterial activity, pointing to the possible advantages of synergistic formulations in antimicrobial therapy. In a similar vein, *Streptococcus mitis*, *S. sanguinis*, *S. sobrinus*, *S. salivarius*, and *S. mutans* were among the oral pathogens tested against the essential oil extracted from fresh *Psidium myrtoides* leaves. GC-FID analysis revealed that the main active component in the essential oil, caryophyllene oxide, showed encouraging antibacterial activity. The potential of caryophyllene oxide in treating oral microbial infections is highlighted by this study (Dias et al., 2019).

In the present scenario, a study specifically done on several caryophyllane sesquiterpenoids reported the anti-influenza activity of caryophyllane oxide. The antiviral characteristics of improved caryophyllane compounds were investigated. Scientists synthesised a range of thiosesquiterpenoids from caryophyllene oxide, including thiols, sulphides, and disulfides. Tests of these compounds against strains of the influenza virus revealed that, in comparison to the original caryophyllene oxide, the addition of sulphur-containing groups often increased antiviral effectiveness while decreasing cytotoxicity. The results suggest that these altered chemicals may have use in the creation of novel antiviral treatments (Gyrdymova et al., 2019).

Chagas syndrome is caused by *Trypanosoma cruzi*, and the chronic phase of Chagas disease is characterised by difficulties in the digestive and cardiovascular systems. Benznidazole and nifurtimox are the only two medications that are currently available for commercial use to treat this condition, notwithstanding the fact that they can cause serious problems like those described above. In order to find innovative possibilities for controlling *Trypanosoma cruzi*, the diterpene copalic acid and the sesquiterpene β-caryophyllene oxide, both of which were extracted from copaiba plants, demonstrated trypanocidal efficacy. This was accomplished by modifying β-caryophyllene oxide through multiple chemical changes, which resulted in the possession of six variants. By examining the possibility of β-caryophyllene oxide variants as possibilities for controlling *T. cruzi* (trypomastigote strain) and inhibiting heme enzymes, it is evident that both this sesquiterpene and its derivatives possess intriguing adaptability, as demonstrated by researchers (Portapilla et al., 2020).

12.5 ANTI-INFLAMMATORY ACTIVITY

The potential of plant-produced chemicals to mitigate inflammatory reactions has been a topic of great interest in the field of alternative therapies. Together with α-humulene, (–)-trans-caryophyllene was taken from *Cordia verbenacea* essential oil and thorough studied to see how well it could reduce inflammation. The outcomes were remarkable; both substances showed significant inhibitory effects in a range of rodent inflammatory models. Platelet activating factor, bradykinin, and ovoalbumin all caused significant swelling in mice paws, which the compounds effectively reduced. Carrageenan also caused swelling in the paws of mice and rats, which the compounds effectively reduced. Also, it was shown that α-humulene was particularly good at lowering the production of TNFα and IL-1β, while (–)-trans-caryophyllene was mostly effective at lowering TNFα release. Additionally, the substances demonstrated a reduction in prostaglandin E2 (PGE2) production, cyclooxygenase (COX-2) activity, and induced nitric oxide synthase, indicating their potential use as anti-inflammatory medicines. Remarkably, their actions paralleled those of the traditional anti-inflammatory medication dexamethasone (Fernandes et al., 2007).

Simultaneously, researchers investigated the characteristics of caryophyllene oxide extracted from *Annona squamosa* bark. Their results demonstrated strong analgesic and anti-inflammatory properties comparable to those of commonly used benchmark medications. The levels used showed analgesic effects in both the central and peripheral areas, suggesting a variety of uses for caryophyllene oxide in pain treatment (Chavan et al., 2010).

The noteworthy inflammation-reducing action of the essential oil of *Origanum vulgare* was emphasised by virtue of its high content of β-caryophyllene oxide as well as β-caryophyllene. Prior research has explained the mechanisms underlying their activity. β-caryophyllene works on central pain receptors to stop the production of inflammatory mediators. On the other hand, caryophyllene oxide acts on CB2 receptors to trigger the release of endorphins (Moghrovyan et al., 2022).

Meanwhile, Li et al. (2021) examined the impact of *Liquidambar fructus* on rheumatoid arthritis, given its long-standing use in Chinese medicine for treating joint pain. The study revealed myrtenal and β-caryophyllene oxide as essential components, exhibiting noteworthy anti-inflammatory properties. It's interesting that these substances stopped fibroblast-like synovial cells from activating the NLRP3 inflammasome in people with rheumatoid arthritis (Snijesh et al., 2018). This suggests their potential as medicinal agents (Li et al., 2021).

12.6 NEUROPROTECTIVE EFFECT

Many plants contain caryophyllene, a natural bicyclic sesquiterpene that is used in many studies to protect neurones because it is not harmful to them. At this dose, an intraperitoneal injection of β-caryophyllene, a neuroprotective substance, at 10 mg/kg greatly decreased the volume of a brain infarction (67%) just prior to the blockage of the main cerebral artery. Additionally, β-caryophyllene greatly improved the cognitive problems caused by a blocked middle cerebral artery (Chang et al., 2013). It was shown that β-caryophyllene reduces the rate of cell death caused by oxidative stress, raises the level of expression as well as potential of NAD(P)H quinone oxidoreductase (NQO1), enhances motor function, and safeguards dopaminergic neurones from detrimental MPTP damage. Therefore, NQO1 activity plays a part in β-caryophyllene's neuroprotective effects, making it a potential candidate for slowing the progression of Parkinson's disease (Flores-Soto et al., 2021). Furthermore, adding β-caryophyllene to cells greatly improved their ability to live and keep their shape while stopping the overexpression and hyperphosphorylation of the JAK2, STAT3, and BACE1 proteins. JAK2 is the most stable target protein for β-caryophyllene, according to molecular docking. Blocking the "JAK2-STAT3-BACE1" signalling system may help β-caryophyllene guard nerves and counteract the neurotoxicity of Aβ (Zhang et al., 2022).

12.7 CONCLUSION AND FUTURE OUTLOOK

Caryophyllene oxide is a renowned sesquiterpene and epoxide substance. Because of its unique structure, it has many different biological effects. Its native presence and importance are shown by the fact that it is found in many plants, including *Camellia sinensis* and *Artemisia thuscula*. Regulatory bodies have recognised caryophyllene oxide as a safe flavouring agent. It also has a lot of promise in the pharmaceutical field.

Researchers have found that it can kill microbes, reduce inflammation, protect neurons, and fight cancer. This shows how versatile it is as a healing agent. The mevalonate pathway and particular enzyme activities are used to make caryophyllene oxide, which demonstrates its importance in the biochemical world.

Future studies on caryophyllene oxide look promising, and there are many possible directions to explore. More in-depth pharmacological investigations are required to fully comprehend the molecular mechanisms that make it effective in living organisms, which will help scientists develop new medicines. Moving from in vitro and in vivo research to clinical research is very important for determining if caryophyllene oxide is effective and safe for humans. It might work better as a medicine if it is combined with other natural compounds that work well together, especially in antimicrobial and anti-inflammatory therapies. Further research into the biosynthetic pathways might lead to bioengineered plants or bacteria that are better at producing large amounts of the compound. Additionally, synthesising and trying different forms of caryophyllene oxide could lead to compounds with enhanced pharmacological properties, resulting in better and have fewer side effects. Caryophyllene oxide could become a valuable natural product with many applications in medicine and industry with further study and development.

REFERENCES

Azaz, D., Demirci, F., Satıl, F., Kürkçüoğlu, M., Hüsnü, K. and Başerb, C., 2002. Antimicrob ial activity of some Satureja essential oils. ZeitschriftfürNaturforschung C, 57(9–10), pp.817–821.

Chang, H.J., Kim, J.M., Lee, J.C., Kim, W.K. and Chun, H.S., 2013. Protective effect of β-caryophyllene, a natural bicyclic sesquiterpene, against cerebral ischemic injury. *Journal of Medicinal Food*, 16(6), pp.471–480.

Chavan, M.J., Wakte, P.S. and Shinde, D.B., 2010. Analgesic and anti-inflammatory activity of Caryophyllene oxide from Annona squamosa L. bark. *Phytomedicine*, 17(2), pp.149–151.

Cheng AiXia, C.A., Lou Yong Gen, L.Y., Mao Ying Bo, M.Y., Lu Shan, L.S., Wang Ling Jian, W.L. and Chen XiaoYa, C.X., 2007. Plant terpenoids: biosynthesis and ecological functions. *Journal of Integrative Plant Biology*, 49(2), 179–186. https://doi.org/10.1111/j.1744-7909.2007.00395.x

Dias, A.L., Batista, H.R., Estevam, E.B., Alves, C.C., Forim, M.R., Nicolella, H.D., Furtado, R.A., Tavares, D.C., Silva, T.S., Martins, C.H. and Miranda, M.L., 2019. Chemical composition and in vitro antibacterial and antiproliferative activities of the essential oil from the leaves of Psidium myrtoides O. Berg (Myrtaceae). *Natural Product Research*, 33(17), pp.2566–2570.

Dougnon, G. and Ito, M., 2021. Essential oil from the leaves of Chromolaenaodorata, and sesquiterpene caryophyllene oxide induce sedative activity in mice. *Pharmaceuticals*, 14(7), p.651.

Fernandes, E.S., Passos, G.F., Medeiros, R., da Cunha, F.M., Ferreira, J., Campos, M.M., Pianowski, L.F. and Calixto, J.B., 2007. Anti-inflammatory effects of compounds alpha-humulene and (−)-trans-caryophyllene isolated from the essential oil of Cordia verbenacea. *European Journal of Pharmacology*, 569(3), pp.228–236.

Fidyt, K., Fiedorowicz, A., Strządała, L. and Szumny, A., 2016. β -caryophyllene and β -caryophyllene oxide—natural compounds of anticancer and analgesic properties. *Cancer -Medicine*, 5(10), pp.3007–3017.

Flores-Soto, M.E., Corona-Angeles, J.A., Tejeda-Martinez, A.R., Flores-Guzman, P.A., Luna-Mujica, I., Chaparro-Huerta, V. and Viveros-Paredes, J.M., 2021. β-Caryophyllene exerts protective antioxidant effects through the activation of NQO1 in the MPTP model of Parkinson's disease. *Neuroscience Letters*, 742, p.135534.

Gyrdymova, Y.V. and Rubtsova, S.A., 2022. Caryophyllene and caryophyllene oxide: A variety of chemical transformations and biological activities. *Chemical Papers*, 76, pp.1–39.

Gyrdymova, Y.V., Sinegubova, E.O., Muryleva, A.S., Zarubaev, V.V. and Rubtsova, S.A., 2019. Anti-influenza activity of several caryophyllanehiosesquiterpenoids. *Chemistry of Natural Compounds*, 55, pp.1179–1181.

Hsouna, A.B., Halima, N.B., Abdelkafi, S. and Hamdi, N., 2013. Essential oil from Artemisia phaeolepis: chemical composition and antimicrobial activities. *Journal of Oleo Science*, 62(12), pp.973–980.

Jassal, K., Kaushal, S. and Rani, R., 2021. Antifungal potential of guava (Psidium guajava) leaves essential oil, major compounds: Beta-caryophyllene and caryophyllene oxide. *Archives of Phytopathology and Plant Protection*, 54(19–20), pp.2034–2050.

Jun, N.J., Mosaddik, A., Moon, J.Y., Ki-Chang, J., Dong-Sun, L., Ahn, K.S. and Cho, S.K., 2011. Cytotoxic activity of [beta]-Caryophyllene oxide isolated from jeju guava (Psidium cattleianumsabine) leaf. *Records of Natural Products*, 5(3), p.242.

Karakaya, S., Yilmaz, S.V., Özdemir, Ö., Koca, M., Pınar, N.M., Demirci, B., Yıldırım, K., Sytar, O., Turkez, H. and Baser, K.H.C., 2020. A caryophyllene oxide and other potential anticholinesterase and anticancer agent in Salvia verticillata subsp. amasiaca (Freyn&Bornm.) Bornm.(Lamiaceae). *Journal of Essential Oil Research*, 32(6), pp.512–525.

Li, W.X., Qian, P., Guo, Y.T., Gu, L., Jurat, J., Bai, Y. and Zhang, D.F., 2021. Myrtenal and β-caryophyllene oxide screened from Liquidambaris Fructus suppress NLRP3 inflammasome components in rheumatoid arthritis. *BMC Complementary Medicine and Therapies*, 21, pp.1–16.

Mcgrath, C.L., Gout, J., Johri, P., Doak, T.G. and Lynch, M., 2014. *Differential retention and divergent resolution of duplicate genes following whole-genome duplication* (pp. 1665–1675). https://doi.org/10.1101/gr.173740.114.24

Moghrovyan, A., Parseghyan, L., Sevoyan, G., Darbinyan, A., Sahakyan, N., Gaboyan, M., Karabekian, Z. and Voskanyan, A., 2022. Antinociceptive, anti-inflammatory, and cytotoxic properties of Origanum vulgare essential oil, rich with β-caryophyllene and β-caryophyllene oxide. *The Korean Journal of Pain*, 35(2), p.140.

Morales-arce, A.Y., 2022. *Inferring the distribution of fi tness effects in patient-sampled and experimental virus populations: Two case studies. December 2021.* https://doi.org/10.1038/s41437-021-00493-y

Portapilla, G.B., Pereira, L.M., Soldi, R.A., Filho, P.G.A., Gallo, I.F.L., Clososki, G.C. and de Albuquerque, S., 2020. Activity of β-caryophyllene oxide derivatives against Trypanosoma cruzi, mammalian cells, and horseradish peroxidase. *RevistaBrasileira de Farmacognosia*, 30, pp.824–831.

Snijesh, V.P., Matchado, M.S. and Singh, S., 2018. Classifying Rheumatoid Arthritis gene network signatures for identifying key regulatory molecules and their altered pathways by adopting network biology approach. *Gene Reports*, 13, pp.199–211. https://doi.org/10.1016/j.genrep.2018.10.013

Stefanello, M.É.A., Cervi, A.C., Ito, I.Y., Salvador, M.J., Wisniewski Jr., A. and Simionatto, E.L., 2008. Chemical composition and antimicrobial activity of essential oils of Eugenia chlorophylla (Myrtaceae). *Journal of Essential Oil Research*, 20(1), pp.75–78.

Terbot, J.W., Johri, P., Liphardt, S.W., Soni, V., Pfeifer, S.P., Cooper, B.S., Good, J.M. and Jensen, J.D., 2023. Plos pathogens developing an appropriate evolutionary baseline model for the study of SARS-CoV-2 patient samples. *PLOS Pathogens, 1*, pp.1–19. https://doi.org/10.1371/journal.ppat.1011265

Tholl, D., 2015. Biosynthesis and biological functions of terpenoids in plants. *Biotechnology of Isoprenoids*, 148, pp.63–106.

Weimer, P., Kreutz, T., Limberger, R.P., Rossi, R.C., de Lima, Á.A., Veiga Jr, V.F., de Araújo, B.V. and Koester, L.S., 2022. Correlation between the skin permeation profile of the synthetic sesquiterpene compounds, beta-caryophyllene and caryophyllene oxide, and the antiedematogenic activity by topical application of nanoemulgels. *Biomolecules*, 12(8), p.1102.

Zhang, Y., Huang, Q., Wang, S., Liao, Z., Jin, H., Huang, S., Hong, X., Liu, Y., Pang, J., Shen, Q. and Wang, Q., 2022. The Food Additive β-Caryophyllene Exerts Its Neuroprotective Effects Through the JAK2-STAT3-BACE1 Pathway. *Frontiers in Aging Neuroscience*, 14, p.814432.

13 Strategies to Evaluate the Quality of Herbal Products with Special Emphasis on Toxicity, Chemical Profiling, and Pharmacokinetics

*R. Thirumurugan, A. Umamaheswari, A. Puratchikody, Dinesh Kumar Chellappan, and S. Lakshmana Prabu**

13.1 INTRODUCTION

Since human beings have been on this planet, herbal drugs have been used. Therefore, one could say they are virtually as old as life itself. The use of herbs as medicine is the oldest form of healthcare and has been practiced in all cultures throughout history. The reasons behind the popularity of herbal medicine are its effective and curative nature for different chronic diseases and lower levels of toxicity. For the last two centuries, plant-based medicines largely dominated in medical practice. However, between the years 1950 and 1970, various synthetic drugs and antibiotics were discovered, leading to a decline in the use of herbal medicine (Aminov, 2010, Jain et al., 2022).

Unfortunately, access to modern pharmaceutical products remains difficult for the general population, and it seems that it will continue this way. In spite of the great advances in modern medicine, observed adverse effects have raised alarm and have now led to the recognition of natural medicine as providing health benefits (Calixto, 2000). The goal of human beings at large is "health for all" (Mosihuzzaman and Choudhary, 2008, Trivedi et al., 2024). Therefore, researchers have started searching for natural sources based on their knowledge: in simple terms, we could say "returning to nature." The World Health Organization (WHO) reported that nearly 75–80% of the world population still uses herbs and traditional medicines for their

* Corresponding author

DOI: 10.1201/9781003392910-13

primary healthcare. Indigenous traditional medicinal systems, existing from ancient times, are still practiced in many countries, such as Ayurvedic medicine in India, Traditional Chinese Medicine in China, Kampo medicine in Japan, and Unani medicine in the Middle East and South Asia (Kunle et al., 2012).

Nowadays, plant-derived products are utilized as home remedies, health foods, and as over the counter medications (Pravin et al., 2012, Singh et al., 2022), and more pharmaceutical companies are entering the herbal market. About 13 new drugs from natural products were approved during 2005–2007 (Katiyar et al., 2012). Herbal medicine has therefore led to the discovery of new drugs (Jain et al., 2022, Singh et al., 2022). For example, in modern medicine, about 25% of drugs are derived either directly or circuitously from herbal plants. By 2050, the global herbal market is predicted to grow to US $5 trillion (Dhiman et al., 2016, Gupta et al., 2024).

13.2 IMPORTANCE OF HERBAL STANDARDIZATION

Since plants grow in different environmental stress conditions and geographical locations, they can develop a huge complex nature of phytoconstituents for their survival (Gangwar et al., 2021). Herbal drugs are converted into herbal products by various processes. During the processes, phytoconstituents can vary; further, during storage these variations can multiply (Kunle et al., 2012). In order to get the desired therapeutic benefit, the required quantity of phytoconstituents must be present in the herbal product.

Safety and health issues are two factors that are increasingly recognized among consumers. Some herbal products have potential, whereas some remain untested (Shukla et al., 2018). A single herbal drug may contain hundreds of phytoconstituents, while mixed herbal products contain many more phytoconstituents. Several factors, such as selection of the wrong species, adulteration, hazardous excess dosages, and usage with other medicines, can cause adverse effects (Khor, 2014; Zhou et al., 2013; Zhang et al., 2012). In addition, lack of documentation about the taxonomic botany of herbal drugs poses challenges in the collection and identification of the right medicinal plants used for herbal product preparation (Farah et al., 2000, Singh et al., 2022).

The main impediment for herbal products is the lack of protocol in standardization and ensuring quality with respect to regulatory standards. These challenges have made the herbal market globally unsafe (Kunle et al., 2012). The WHO provided guidelines for herbal standardization and encouraged governments to utilize local traditional knowledge for health promotion and disease prevention (Mosihuzzaman and Choudhary, 2008). Hence, regulatory authorities have established fundamental standards of quality and efficacy in herbal products through good manufacturing practices (Kunle et al., 2012).

Standardization is the process of developing a set of technical standards by compiling comprehensive data on herbs to provide an assurance of quality, efficacy, safety, and reproducibility of the main active ingredients in each dose (Sagar et al., 2005; Trivedi et al., 2024). However, standardization of herbal medicines is not an easy task, because factors like the absence of reference standards, variation of phytoconstituents, and therapeutic efficacy of the herbs can influence standardization (Jaiswal et al., 2018).

In order to maintain the quality of herbal drugs and herbal products, a coordinated effort must be made in developing specific, reliable, and sensitive analytical methods as quality control to assess the quality of the process.

Problems associated with herbal drugs not applicable to synthetic drugs are:

1. Herbal drugs are of a hugely complex nature, having many phytoconstituents.
2. Main phytoconstituents may be unknown.
3. Non-availability of reference standards and selective analytical methods.
4. Variations of source, quality, and phytoconstituents with respect to the origin of raw material.
5. Chemo-varieties and chemo-cultivars exist (Kunle et al., 2012).

Systematic qualitative and quantitative evaluation of herbal products is a fundamental requirement of an organization to ensure the quality of the herbal drugs and herbal products from cultivation to clinical application (Bijauliya et al., 2017). Regulatory authorities classify products based on the quantity of active ingredients and their concentration on the product label.

Ensuring the quality of products is essential in the healthcare industry for successful survival in this competitive global market. Different parameters towards the standardization of herbal drugs (Bijauliya et al., 2017) are shown in Figure 13.1.

13.3 KEY FACTORS IN HERBAL DRUG STANDARDIZATION

The key factors in herbal drug standardization are:

1. Authentication of genuine species: Inter or intra-species variation, botanical identity, environmental factors, regional status, stage of collection and harvesting, plant parts used, post-harvesting factors, presence of unknown active ingredients, non-availability of selective analytical methods, natural and chemical variability of plant materials.
2. Morphological or organoleptic evaluation of herbal drugs: Size, shape, taste, and color.
3. Macroscopic evaluation: Size, shape, color, surface and fracture character, texture, and appearance of the cut surface.
4. Microscopic evaluation: Identification of powdered materials, determination of vein islet and palisade ratio (Bijauliya et al., 2017).

13.4 OTHER FACTORS IN HERBAL PRODUCT STANDARDIZATION

Continuous increases of environmental pollution in water, air, soil, etc., can affect the quality and safety of herbs. Various forensic investigations have highlighted that herbal products from different parts of the world have heavy metal toxicity.

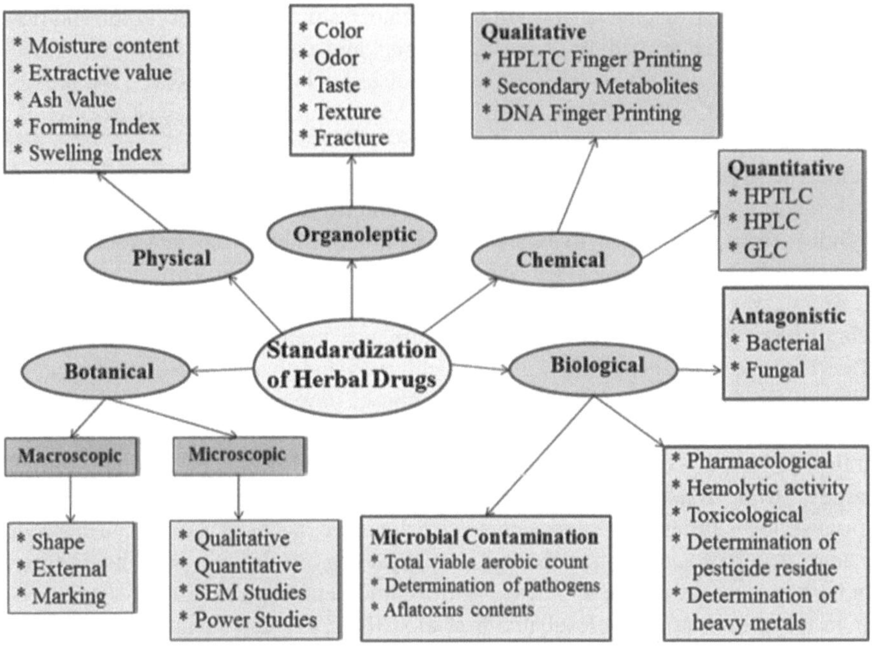

FIGURE 13.1 Schematic representation for the standardization and quality evaluation of herbal drugs.

13.4.1 Toxicity Assessment

A toxicity assessment is also required to identify levels of toxicity.

Approaches in the determination of toxicity are:

1. *In vitro* techniques.
2. *In vivo* techniques.
3. Cell line techniques.
4. Micro-array and other modern techniques (Mosihuzzaman and Choudhary, 2008).

13.4.1.1 Toxicity Testing in Zebrafish Model

Genetically, each plant species can synthesize various toxic substances over a period of time as a defense mechanism for its survival against a highly polluted environment, diseases, insects, and other organisms (Yumnamcha et al., 2015).

Consumers can also be exposed to potential health risks triggered by the specific phytoconstituents of herbal products (Gebrelibanos et al., 2019). Some of the phytoconstituents of medicinal plants, such as phorbol esters, cardiac glycosides, cyanogenic glycosides, alkaloids, and lectins have toxic effects (Yumnamcha et al., 2014). Hence, evaluation of herbal product toxicity is imperative to ensure its safety (Jayasinghe and Jayawardena 2019). Initially, *in vitro* toxicity studies are performed.

In vivo studies can also be performed on laboratory animals like mice, rats, and rabbits; however, they are tedious, time-consuming, and expensive (Rizzo et al., 2013). The obtained *in vivo* studies are being extrapolated to humans; however, the prediction of chemical risks remains based on vertebrate toxicity studies (Koziol et al., 2019). In addition, ethical considerations and recent restrictions on the use of higher animals have become more contentious (Modarresi et al., 2020). These restrictions have made it imperative to identify a gold standard for the predictive analysis of chemical risks. Alternative organisms like daphnia, brine shrimp, and zebrafish are utilized in toxicity assessment (Jayasinghe and Jayawardena, 2019).

Among these, the zebrafish embryo toxicity model has been considered. The zebrafish is a lower vertebrate and is predominantly used in the toxicity assessment of herbal drugs because of the nature of embryos and that it is low cost, requires a short time for analysis, allows easy detection of developmental phases, endpoint evaluation, and visualization of organogenesis through fluorescence and transgenic strains, and has a genetic similarity to humans and high fecundity (Sant and Timme-Laragy, 2018). In addition, more than 70% of the genes associated with disease in zebrafish are similar to human genes. The digestive, cardiovascular, and nervous systems also have similar morphology and physiology to those in humans. Further, signaling mechanisms, metabolites, cognitive and neurological structures are similar to those in mammals (Jesubatham et al., 2018). The full genome sequence of zebrafish is available as a ready reference for any verification. Recently, the National Health Institute (NIH), USA, and Food and Drug Administration (FDA) recognized and encouraged the use of the zebrafish model for evaluating different diseases occurring genetically (Chakraborty et al., 2016). The zebrafish model has been used to study the toxicity of anticancer drugs, metastasis, angiogenesis, de novo mutations, and to assess teratogenicity, carcinogenicity, behavioral toxicity, reproductive and endocrine disorders, acute toxicity, cardiotoxicity, ocular toxicity, neurotoxicity, and vascular toxicity (Caballero and Candiracci 2018).

13.4.2 Safety

Three categories of safety are considered to classify the levels of potential harm to users (A. V. Johri et al., 2023). They are:

Category 1: Established safety by use over a long time.
Category 2: Safe under specific conditions of use.
Category 3: Uncertain safety of herbal medicines (Dhiman et al., 2016; WHO Guidelines for the Regulation of Herbal Medicines in the South-East Asia Region; June, 2003).

13.4.3 Risk Assessment

There are three types of risk assessment.

Risk assessments are:

1. Identification of hazard.
2. Assessment of dose.
3. Exposure assessment (Ricci, 2006).

13.5 ANALYTICAL METHODS IN HERBAL STANDARDIZATION

Generally, the basic requirements in the preparation of medicines, either synthetic or of plant origin, are that the medicine should be safe and effective. The term "herbal drugs" indicates that plants or plant parts have been converted into herbal products, which may have variations in the active constituents (Long et al., n.d.). Plants consist of active phytoconstituents along with several inert substances; the active phytoconstituents must be isolated and purified from the plants for their selective pharmacological activity (P. Johri et al., 2021). Due to the complex nature and inherent variability of the constituents, scientists face a major challenge in establishing specific quality control parameters. The development of an authentic analytical method using modern analytical techniques that can reliably profile the phytochemical composition, including quantitative analysis of bioactive phytoconstituents, is expected to help in circumventing the above-mentioned problems (Patra et al., 2010).

Different analytical techniques employed in the standardization of herbal drugs are shown in Table 13.1.

TABLE 13.1
Techniques for Fingerprint Analysis in Herbal Standardization

Technique	Instrument
Chromatography	TLC
	HPTLC
	HPLC
	HILIC
	Gas chromatography
	Paper chromatography
	Column chromatography
	Super critical chromatography
Spectroscopy	UV
	IR & NIR
	NMR
	Mass spectrophotometer
	Atomic absorption spectrometer
Hyphenated technique	GC-MS
	HPLC-MS
	LC-NMR
	LC-MS

13.5.1 Recent Techniques

13.5.1.1 Chemometric Method

With the naked eye, it is very difficult to identify slight spectral differences between samples, and data cannot be evaluated easily. Therefore, a data processing tool is needed to produce a comprehensive analytical report from the complicated spectra. Chemometrics is a data processing tool that provides an accurate analysis report from the data by application of mathematical and statistical techniques (Singh et al., 2013). The International Chemometrics Society defines chemometrics as the science of using mathematical and statistical methods to analyze chemical data (Gemperline, 2006). Along with the computer technology, chemometric methods have become a leading and essential tool in scientific communities for the collection of multivariate data, protocol analysis, process modeling, calibration, pattern classification and recognition, signal compression, and correction in a shorter product development time. This technique can be used for solving both descriptive and predictive issues (Singh et al., 2013).

There are three kinds of chemometric method:

1. Mathematical pretreatments.
2. Classification methods.
3. Regression methods.

Chemometric models are classified into:

1. Bilinear models
    i. Principal component analysis (PCA).
    ii. Partial least squares (PLS).
2. Multiway models
    i. Parallel factor analysis (PARAFAC).
    ii. Parallel factor analysis-2 (PARAFAC-2).
    iii. Tucker-3.
    iv. N-partial least squares (N-PLS) (Smilde et al., 2004).

Practical aspects in qualitative methods based on the chemometric technique are:

1. Sampling and data pre-processing.
2. Library management (calibration set management).
3. Calibration transfer and model update (Roggo et al., 2007).

13.5.1.2 Chemometric Methods in Herbal Drug Standardization

1. Principal component analysis.
2. Linear discriminant analysis.
3. Spectral correlative chromatography.
4. Information theory.
5. Local least square.
6. Heuristic evolving latent projections.
7. Orthogonal projection analysis (Sanchez et al., 1996; Sanchez et al., 1997; Li et al., 2004b).

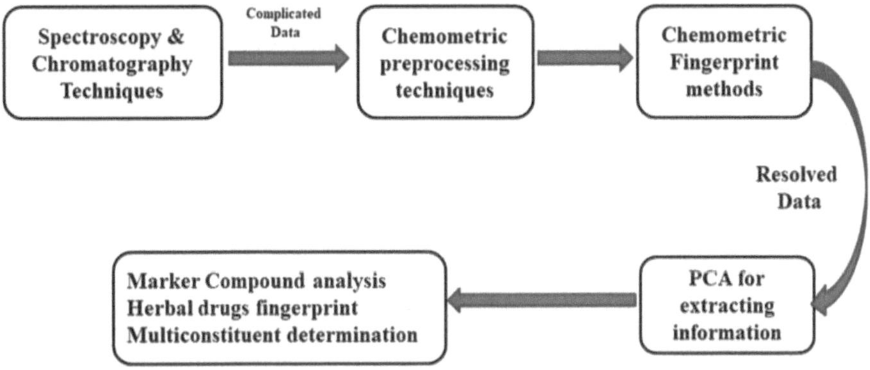

FIGURE 13.2 Steps in data handling of herbal products.

Of all the techniques, PCA is most often used in herbal drugs/herbal product standardization, because it can analyze multivariate data without prior information. A schematic diagram of the data-handling steps for herbal products is shown in Figure 13.2.

13.5.1.3 Omics Techniques

Omics technique is based on omic science. The term "omics" is derived from the Latin suffix "ome" meaning many or mass (Bajpai et al., 2023). The omics technique is used for the identification of resources in biomedical research based on DNA sequencing and fingerprinting.

Different omics techniques are:

Proteome: Whole proteins found in a given cell or tissue.

Transcriptome: Total genes that are converted into transcripts (i.e., mRNA molecules). This is a method to analyze the gene expression level by measuring the transcriptome.

Interactome: This method is based on the interaction between the given protein and biologically active metabolism products.

Phenome: This method includes all noticeable characteristics of organism levels.

Metabolome: This is a non-targeted analytical approach based on the whole metabolism products and intermediates in a cell or tissue, which is mainly the product of its gene expression (Pandey et al., 2016).

13.5.1.4 Metabolomics

Metabolomics is a comprehensive qualitative and quantitative analysis of all metabolites present in specific tissue, organism, or cells, and it has shown to be a promising tool in identifying phenotypes and quantifing the complete set of metabolites in a sample (Fiehn, 2001).

In the majority of herbal medicines, active phytoconstituents are poorly identified, and this has been considered one of the main difficulties. Hence, targeted metabolomics may be used to characterize the chemical structures and to define the precise names and amounts of bioactive compounds found in herbal material for every product (Kumar et al., 2023).

These specifics would be very helpful in regulating the uniformity of herbal medicine made in various batches.

This information would be particularly useful in the design and development of herbal medicine as well as to improve treatment and therapy (Khoomrung et al., 2017).

Metabolic analysis can be categorized into:

1. Targeted compound analysis: Specific metabolite quantification.
2. Metabolic profiling: Determination of a specific metabolic pathway or group of related compounds both quantitatively and qualitatively.
3. Metabolomic fingerprinting: Classification of samples by rapid global analysis.
4. Metabolomic analysis: Analysis of all metabolites both quantitatively and qualitatively.

13.6 EXPERIMENTAL PARTS

Experimental parts are performed in three distinct stages that are strongly interrelated. They are:

1. Preparation of sample.
2. Data acquisition through analytical methods: FT-IR spectrometry, nuclear magnetic resonance, mass spectrometry (MS), liquid chromatography (LC)-MS, gas chromatography (GC)-MS, ultra-performance liquid chromatography-mass spectrometry (UPLC-MS), GC X GC-time-of-flight (TOF), GC-TOF-MS, UPLC-quadrupole (Q)-TOF-MS, LC (HILIC)-MS, and LC-atmospheric pressure chemical ionization (APCI)-MS.
3. Data mining using suitable chemometric methods: PCA, partial least squares-discriminant analysis, and PLS regression (Zhang et al., 2012; Lee et al., 2017).

13.7 MARKER COMPOUNDS

As herbal medicines have a highly complex nature of phytoconstituents, the desired therapeutic effects are achieved only if an adequate quality of phytoconstituents is used. A review report indicated that about 12,000 secondary metabolites have been isolated; this number is estimated to be less than 10% of the total phytoconstituents in the plant kingdom (Vasu et al., 2009). The use of herbal medicines has increased remarkably, and they have become important for people's livelihoods and the economy. Herbal products are standardized based on the concentration of active phytoconstituents in a single dose. The identification of all the phytoconstituents in a herbal medicine is challenging to achieve. Therefore, a marker approach is that compound-oriented and pattern-oriented is used in assessing authenticity and ensuring quality. When the active phytoconstituent is not known, a substance called a marker, which is specific to the particular herbal drug/herbal product, must be identified with a suitable analytical method; this is crucial (see the WHO, 1998).

Marker compounds are pure, single, isolated compounds of secondary metabolites, mostly from terpene, steroid, alkaloid, flavonoid, aromatic, and heteroaromatic compounds having alcohol, carbonyl, olefinic, acid, ester, and amide functionalities, and which may or may not have therapeutic characteristics.

Markers are classified into:

1. Bioactive markers: Responsible for the activity claimed for a particular drug.
2. Chemical markers: A constituent that is neither specific to the drug nor to the activity of the herbal drug.
3. General markers: Compound distribution is wide in plants.
4. Genetic markers: Gene or DNA sequences from a known location on a chromosome (Pathak, 2017; Li et al., 2008).

Further primary and secondary metabolites, biochemical constituents, and other macromolecules are categorized under chemical markers.

Chemical markers are categorized into:

1. Analytical markers.
2. Active markers.

Markers can be prepared by chemical modification, which includes dealkylation, deacetylation, dehydration, isomerization, decomposition, degradation, and polymerization. Impurity markers are called negative markers, produced by pyrolysis. Most herbal extracts are not consistently standardized by utilizing any one marker (Garg et al., 2012). Subsequently, multi-component or fingerprint analysis approaches have been introduced to ensure the quality of herbal drugs and herbal products.

13.8 FINGERPRINT PROFILE

Different climatic conditions and other environmental factors, including light, humidity, temperature, wind, cultivation area, and harvesting time, can create variation in the chemical constituents, leading to impacts on the quality of the herbs and herbal preparations (Giri et al., 2010). Due to the multifaceted composition of the constituents, developing an appropriate analytical method for the quantification of the constituents is unrealistic.

In the last two decades, fingerprinting of herbal medicine has been considered a very powerful approach in controlling the quality of herbs and herbal preparations. Recently, the FDA and WHO both accepted the fingerprinting method as being an efficient method in controlling the consistency as well as quality of herbal drugs and herbal products internationally (Tistaert et al., 2011; Hua-Bin et al., 2012).

Fingerprinting is defined as a set of distinctive spectroscopic or chromatographic signals, whose comparison leads to the identification and recognition of certain classes of constituents in herbs and herbal preparations (Ciesla, 2012). The fundamental features of fingerprinting are "integrity" and "fuzziness" or "sameness" and "differences" (Kulkarni et al., 2014). Fingerprinting techniques have been considered a holistic approach and more suitable to assess the quality of herbs instead of analyzing individual compounds (Ni et al., 2009). It is difficult to specify the 100% similarity of signals such as bands, peaks, etc., but it can confirm the sample identity based on the Botanical Reference Material or standard compound. In addition, fingerprint analysis is applied to detect adulteration, identify closely related plant species, contaminants, and control the extraction process and the quality of the finished product.

Fingerprinting of herbal medicines is a crossover effort that includes analytical science, separation science, and bioinformatics.

Significant issues in the development of fingerprint analysis are (Liang et al., 2009):

1. How can more stable and effective information be found?
2. How can similarities and differences to the chemometric method be assessed?

Two approaches, i) compound-oriented and ii) pattern-oriented, are undertaken to resolve the issues in the fingerprint analysis technique (Zeng et al., 2008). A marker compound (a known active ingredient) is used in the compound-oriented approach, while, in assessing similarity and correlation, the fingerprint pattern is used in the pattern-oriented approach. In the pattern-oriented approach, both qualitative and quantitative detection techniques are adopted to ensure the quality of the herbs (Yongyu et al., 2011). Moreover, all herbal product manufacturers follow WHO quality control guidelines to ensure the quality of the herbal preparations. Sometimes, a single fingerprint analysis may not be adequate to give information about the herbs. In such cases, multi-component or fingerprinting analysis with different test conditions with the same or different separation principles can be applied for standardization (WHO, 1998; Meena et al., 2010). Different techniques (spectral fingerprint, chromatographic fingerprint, and hyphenated technique) (Yongyu et al., 2011; Zhou et al., 2009; Li et al., 2004a; Li et al., 2004b) are shown in Figure 13.3.

13.8.1 DNA Fingerprinting

Recently, DNA fingerprinting has also been referred to as genetic fingerprinting and has become an important analytical technique in DNA profiling and DNA typing. This technique is utilized in the identification and standardization of herbs and herbal preparations, whether adulterated or substituted with other species that are indistinguishable from the original based on nucleotide sequences as an identifying tool. Basically, a DNA fingerprint genome is like a barcode pattern obtained by chromosomal DNA amplification to differentiate and distinguish samples from each

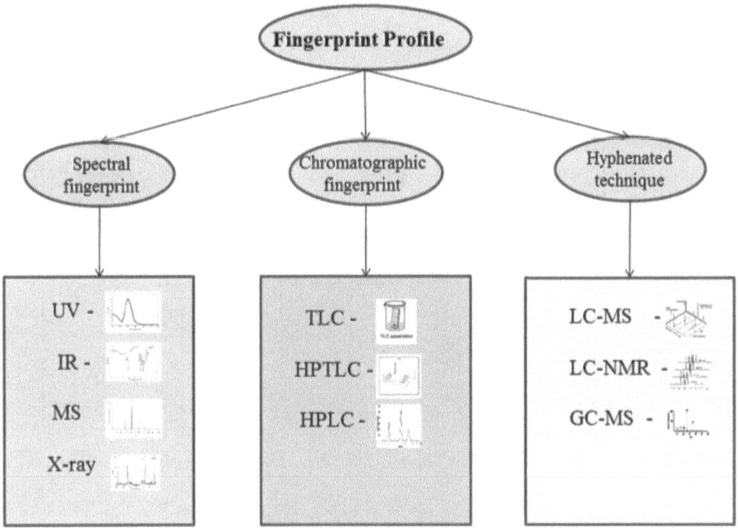

FIGURE 13.3 Classification of fingerprint analysis techniques.

other (Kulkarni et al., 2014; Yongyu et al., 2011). This genetic information will not be affected by physiological conditions, harvesting time, environmental factors, age, processing, and storage. Additionally, isolated DNA can remain stable for 3–5 years at –20°C (Yip et al., 2007). DNA fingerprinting can be performed by:

1. Hybridization-based methods: Restriction fragment length polymorphism (RFLP), variable number tandem repeats (VNTR), probe hybridization with micro and minisatellite, random genomic clone, and cDNA clone.
2. Polymerase chain reaction (PCR)-based methods: Randomly amplified polymorphic DNA (RAPD), arbitrarily primed polymerase chain reaction (AP-PCR), amplified fragment length polymorphism (AFLP), and inter simple sequence repeats (SSR).
3. Sequencing-based methods: Single nucleotide polymorphism (SNP), short tandem repeats (STR), simple sequence repeats (SSR), sequence characterized amplified region (SCAR), and cleaved amplified polymorphic sequence (CAPS) (Kumar et al., 2014; Ali et al., 2014; Selvakumari et al., 2017).

The above techniques are summarized in Figure 13.4, and the steps for the analysis of DNA fingerprinting are depicted in Figure 13.5.

13.8.2 Fingerprint Data-Handling Techniques

In order to prepare the analysis report, the required information is extracted from the obtained fingerprint data. The obtained data are pretreated, followed by analysis with either the unsupervised or supervised method.

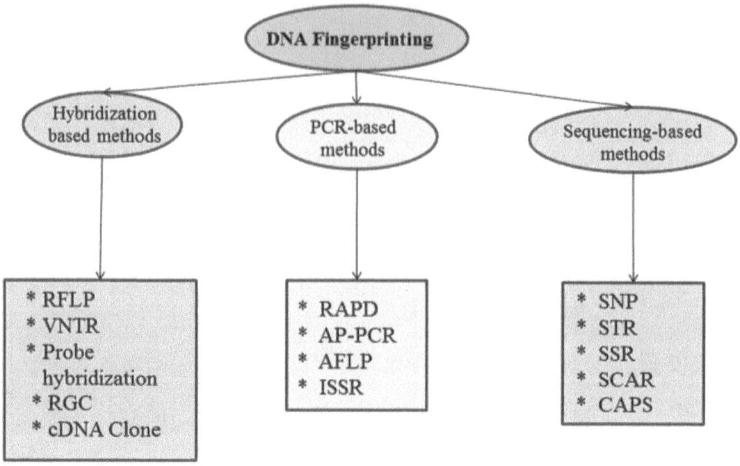

FIGURE 13.4 Techniques involved in DNA fingerprint analysis.

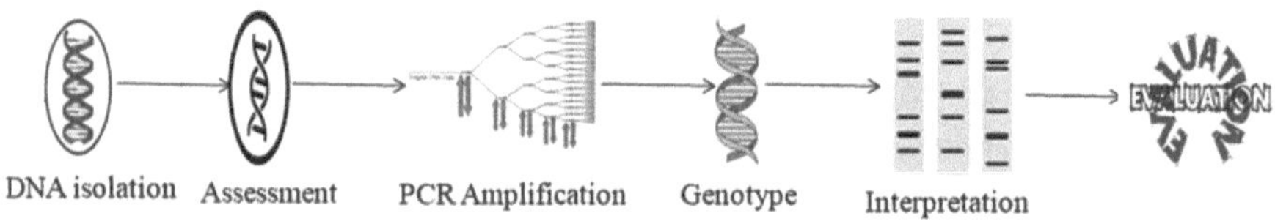

FIGURE 13.5 Steps involved in DNA fingerprinting analysis.

Unsupervised data analysis includes:

1. Similarity analysis.
2. Exploratory analysis and curve resolution.
3. Pattern recognition or classification.
4. Multivariate calibration (Tistaert et al., 2011).

13.9 HERBAL STANDARDIZATION BASED ON PHARMACOKINETICS

The efficacy of herbal resourced compounds (HRC) depends on the multiple constituents absorbed in the body and their pharmacokinetic properties. Recently, knowledge of these multiple constituents of herbs has been used for treating different diseases. However, different processing of herbs can modify the nature of crude herbal materials—they can either increase, decrease, or even disappear after processing (Sun et al., 2019). Alteration in the chemical constituents can affect the pharmacokinetic properties, maximum plasma concentration (C_{max}), time to reach C_{max} (Tmax), and halftime ($t_{1/2}$) of the herbal material (Yang et al., 2015). The pharmacological efficacy of these herbs depends on the pharmacokinetics, bioavailability of the active phytoconstituents, and its metabolites (Alolga et al., 2015). Other physiological factors such as age, external stimuli, and disease conditions can affect the pharmacokinetic properties; subsequently, this will create variations in its therapeutic effect (Yang et al., 2015). Enzymes P-glycoprotein (P-gp) and cytochrome P450 enzyme (CYP450) are common metabolizing enzymes for most of the HRC (Soleymani et al., 2017). However, differences between the single compound and herbal extract can be observed due to synergistic or competitive absorption of various phytoconstituents and their chemical complexity (Yuan and Lin, 2000).

Several herbals were demonstrated to show good biological effects *in vitro* but don't show reproducible results *in vivo* (Bonifacio et al., 2014). Thus, effective clinical research is necessary to investigate the interaction between the herbs and influence factors, in addition to confirming the mechanisms of interaction (Sun et al., 2019). Hence, more research is focused on developing sensitive bioanalytical methods for herbal samples (Li et al., 2008; Cheng et al., 2013).

Understanding phytoconstituent pharmacokinetic properties can provide basic information in designing the dosage regimens for clinical use.

Recently, the pharmacokinetic concept has been applied to herbal medicine and has the following advantages:

1. Investigating the multiple components and identifying the phytoconstituents can explain its effect.
2. Can explain the active components' combination mechanism in extraction methods.
3. Can reveal the active components' pharmacokinetic parameters *in vivo* (Zhang et al., 2017).

This pharmacokinetic and pharmacodynamic (PKPD) concept has been suggested in various studies for its dynamic process in treating different diseases. In addition, this concept has been used to investigate herb efficacy, herb–herb synergistic effects, and herbs with drug effects (Zhang et al., 2016; Li et al., 2013).

However, researchers face some issues and challenges with the herbal medicine PKPD concept that can't be overcome easily. To overcome these issues, tools such as metabolic technology and network pharmacology are being used to establish an effective PKPD correlation for the clinical use of herbal medicine (Zhang et al., 2017; Zhu et al., 2013). Thus, with a better understanding of the pharmacokinetic properties of herbal medicine and the influencing factors, we can gain a better management of herbs when treating diseases.

13.10 REGULATIONS FOR STANDARDIZATION OF CRUDE DRUGS

In 1986, the concept of herbal medicines was first included in the WHO International Conference on Drug Regulatory Authorities (ICDRA). In 1991, a herbal medicines guideline for assessment was drafted by WHO; subsequently, it was approved by the 6th ICDRA. These guidelines included the basic criteria related to the quality, safety, and efficacy of herbal medicines. The guidelines for herbal medicine were developed in two parts for assessment by WHO, based on the recommendations on the draft guideline given by the 6th ICDRA.

Internationally, various pharmacopoeias have provided guidelines for both herbal products and herbs in order to maintain quality in their respective nations. Some of the pharmacopoeias that deal with herbs and herbal products are Ayurvedic Pharmacopoeia of India, British Herbal Pharmacopoeia, British Herbal Compendium, Chinese Herbal Pharmacopoeia, United States Herbal Pharmacopoeia, and Japanese Standards for Herbal Medicine (Booksh and Kowalski 1994).

13.10.1 WHO Guidelines

The guidelines set by the WHO for herbal drug products (WHO, 2007) are summarized as:

1. Reference to the identity of the drug through botanical evaluation.
2. Reference to the physicochemical character of the drug.
3. Reference to the pharmacological parameters.
4. Toxicity details.
5. Microbial contamination.
6. Radioactive contamination.

13.11 CONCLUSION

The composition of the many phytoconstituents that are produced during a plant's growth is considered to be very beneficial. Herbal drug standardization allows comprehensive and complete control over desired phytoconstituents, including analytical methods for the identification of active phytoconstituents. The finished herbal product is quality controlled according to the established specifications, which can lead to the introduction of a standardized and therapeutically effective product to the market. The World Health Organization ensures the quality of herbal products by applying suitable quality control parameters during the manufacturing process. In addition, combining novel, intelligent methods in fingerprint analysis, both qualitative and quantitative, can ensure the quality of herbal products. Recently, toxicity and pharmacokinetic parameter assessments have also been considered as effective tools in standardizing herbal products. The ultimate goal of herbal standardization is to ensure the safety and efficacy of herbal products.

REFERENCES

Ali, M. A., Gyulai, G., Hidvegi, N., Kerti, B., Al Hemaid, F. M., Pandey, A. K., & Lee, J. (2014). The changing epitome of species identification–DNA barcoding. *Saudi Journal of Biological Sciences*, *21*(3), 204–231.

Alolga, R. N., Fan, Y., Zhang, G., Li, J., Zhao, Y. J., Lelu Kakila, J., & Qi, L. W. (2015). Pharmacokinetics of a multicomponent herbal preparation in healthy Chinese and Africanvolunteers. *Scientific Reports*, *5*(1), 12961.

Aminov, R. I. (2010) A brief history of the antibiotic era: Lessons learned and challenges for the future. *Front Microbiol. 1*, 134.

Bajpai, Y., Gaber, S. H., Muthukumar, M., Rajan, S., Bajpai, A., & Trivedi, M. (2023). Proteomic landscape presents cues for vegetative to reproductive transition in mango. *Journal of Applied Horticulture, 2023, 25*(1), 3–9. https://doi.org/10.37855/jah.2023.v25i01.01; ISSN: 0972-1045, https://horticultureresearch.net/title.php?a=1001

Bijauliya, R. K., Alok, S., Chanchal, D. K., & Kumar, M. (2017). A comprehensive review on standardization of herbal drugs. *International Journal of Pharmaceutical Sciences and Research, 8*(9), 3663–3677.

Bonifacio, B. V., da Silva, P. B., Ramos, M. A. D. S., Negri, K. M. S., Bauab, T. M., & Chorilli, M. (2014). Nanotechnology-based drug delivery systems and herbal medicines: A review. *International Journal of Nanomedicine, 9*, 1–15.

Booksh, K. S., & Kowalski, B. R. (1994). Theory of analytical chemistry. *Analytical Chemistry, 66*(15), 782A–791A.

Caballero, M. V., & Candiracci, M. (2018). Zebrafish as screening model for detecting toxicity and drugs efficacy. *Journal of Unexplored Medical Data, 3*(4), 1–14.

Calixto, J. B. (2000). Efficacy, safety, quality control, marketing and regulatory guidelines for herbal medicines (phytotherapeutic agents). *Brazilian Journal of Medical and Biological Research, 33*, 179–189.

Chakraborty, C., Sharma, A. R., Sharma, G., Lee, S. S. (2016). Zebrafish: A complete animal model to enumerate the nanoparticle toxicity. *J Nanobiotechnol. 14*, 65.

Cheng, X. L., Qi, L. W., Wang, Q., Liu, X. G., Boubertakh, B., Wan, J. Y., & Li, P. (2013). Highly efficient sample preparation and quantification of constituents from traditional Chinese herbal medicines using matrix solid-phase dispersion extraction and UPLC-MS/MS. *Analyst, 138*(8), 2279–2288.

Ciesla, L. (2012). Biological fingerprinting of herbal samples by means of liquid chromatography. *Chromatography Research International 1*, 532418.

Dhiman, A., Sharma, K., Sharma, A., & Sindhu, P. (2016). A review on the status of quality control and standardization of herbal drugs in India. *Drug Development and Therapeutics, 7*(2), 107–107.

Farah, M. H., Edwards, R., Lindquist, M., Leon, C., & Shaw, D. (2000). International monitoring of adverse health effects associated with herbal medicines. *Pharmacoepidemiology and Drug Safety, 9*(2), 105–112.

Fiehn, O. (2001). Combining genomics, metabolome analysis, and biochemical modelling to understand metabolic networks. *Comparative and functional genomics, 2*(3), 155–168.

Gangwar, P., Shukla, S., & Trivedi, M. (2021). Screening and identification of salt-tolerant lines in novel chickpea (Cicer arietinum L.) germplasms. *Research Journal of Biotechnology, 16*(5), 74–81. https://worldresearchersassociations.com/biotechcurrissue/11.pdf

Garg, V., Dhar, V. J., Sharma, A., Dutt, R. (2012). Facts about standardization of herbal medicine: a review. *J Chin Integr Med 10*(10), 1077–1083.

Gebrelibanos, Hiben M., Kamelia, L., de Haan, L., Spenkelink, B., Wesseling, S., Vervoort, J., et al. (2019). Hazard assessment of *Maerua subcordata* (Gilg) DeWolf. for selected endpoints using a battery of in vitro tests. *J Ethnopharmacol. 241*, 111978.

Gemperline, P. (2006). *Practical guide to chemometrics*. Boca Raton, FL: CRC Press.

Giri, L., Andola, H. C., Purohit, V. K., Rawat, M. S. M., Rawal, R. S., & Bhatt, I. D. (2010). Chromatographic and spectral fingerprinting standardization of traditional medicines: An overview as modern tools. *Research Journal of Phytochemistry, 4*(4), 234–241.

Gupta, A., Kumar, S., Bajpai, Y., Chaturvedi, K., Johri, P., Tiwari, R. K., Vivekanand, V., & Trivedi, M. (2024). Pharmaceutically active micropollutants: Origin, hazards and removal. *Frontiers in Microbiology, 15*, 1–11. https://doi.org/10.3389/fmicb.2024.1339469

Hua-Bin, Z., Ai-Qin, D., Xin-Ling, Z., Pei-Hai, W., Wei-Jie, L., Guo-Sheng, Y., & Aboul-Enein, H. Y. (2012). Quality control methodology and their application in analysis on HPLC fingerprint spectra of herbal medicines. *Chromatography Research International, 2012*(1), 851792.

Jain, T., Srivastava, K., Singh, R., Ansari, M. J., Singh, S., & Dutta, P. K. (2022). Role of nanotechnology 18 in the field of phytopharmaceuticals for the delivery of herbal drugs. *Phytopharmaceuticals and Biotechnology of Herbal Plants*, 317–330. https://doi.org/10.1201/b22917-18

Jaiswal, S., Chavhan, S. A., Shinde, S. A., & Wawge, N. K. (2018). New tools for herbal drug standardization. *Asian Journal of Research in Pharmaceutical Science, 8*(3), 161–169.

Jayasinghe, C. D., & Jayawardena, U. A. (2019). Toxicity assessment of herbal medicine using zebrafish embryos: A systematic review. *Evidence-Based Complementary and Alternative Medicine, 2019*(1), 7272808.

Jesubatham, P. D., VM, B. G., Viswanathan, S., & Srividya, S. (2018). Non-toxic and non teratogenic extract of Thuja orientalis L. inhibited angiogenesis in zebra fish and suppressed the growth of human lung cancer cell line. *Biomedicine & Pharmacotherapy, 106*, 699–706.

Johri, A. V., Johri, P., Hoyle, N., Nadareishvili, L., Pipia, L., & Nizharadze, D. (2023). *Case report : Successful treatment of recurrent E . coli infection with bacteriophage therapy for patient suffering from chronic bacterial prostatitis. September*, 1–8. https://doi.org/10.3389/fphar.2023.1243824

Johri, P., Riall, K., Becher, H., Excoffier, L., Charlesworth, B., & Jensen, J. D. (2021). The impact of purifying and background selection on the inference of population history: Problems and prospects. *Molecular Biology and Evolution, 38*(7), 2986–3003. https://doi.org/10.1093/molbev/msab050

Katiyar, C., Gupta, A., Kanjilal, S., & Katiyar, S. (2012). Drug discovery from plant sources: An integrated approach. *AYU (An international quarterly journal of research in Ayurveda), 33*(1), 10–19.

Khoomrung, S., Wanichthanarak, K., Nookaew, I., Thamsermsang, O., Seubnooch, P., Laohapand, T., & Akarasereenont, P. (2017). Metabolomics and integrative omics for the development of Thai traditional medicine. *Frontiers in Pharmacology, 8*, 474.

Khor, M. (2014). The growing use of herbal medicines: Issues relating to adverse reactions and challenges in monitoring safety. *Front Pharmacol. 4*(177), 1–10.

Kozioł, E., Deniz, F. S. S., Orhan, I. E., Marcourt, L., Budzyńska, B., Wolfender, J. L., & Skalicka-Woźniak, K. (2019). High-performance counter-current chromatography isolation and initial neuroactivity characterization of furanocoumarin derivatives from Peucedanum alsaticum L (Apiaceae). *Phytomedicine, 54*, 259–264.

Kulkarni, K. M., Patil, L. S., Khanvilkar, V. V., & Kadam, V. J. (2014). Fingerprinting techniques in herbal standardization. *Indo American Journal of Pharmaceutical Research, 4*(2), 1049–1062.

Kumar, S., Singh, A., Yadav, A., Bajpai, A., Singh, N. K., Rajan, S., & Trivedi, M. (2023). Identification and validation of novel genomic SSR markers for molecular characterization of guava (Psidium guajava L.). *South African Journal of Botany, 155*, 79–89. https://doi.org/10.1016/j.sajb.2023.02.005

Kumar, P. S., Ketkar, P., Nayak, S., & Roy, S. (2014). Application of DNA fingerprinting tools for authentication of ayurvedic herbal medicines-A review. *Journal of Scientific and Innovative Research, 3*, 606612.

Kunle, O. F., Egharevba, H. O., & Ahmadu, P. O. (2012). Standardization of herbal medicines: A review. *International Journal of Biodiversity and Conservation, 4*(3), 101–112.

Lee, K. M., Jeon, J. Y., Lee, B. J., Lee, H., & Choi, H. K. (2017). Application of metabolomics to quality control of natural product derived medicines. *Biomolecules & Therapeutics, 25*(6), 559.

Li, B. Y., Hu, Y., Liang, Y. Z., Huang, L. F., Xu, C. J., & Xie, P. S. (2004a). Spectral correlative chromatography and its application to analysis of chromatographic fingerprints of herbal medicines. *Journal of Separation Science*, *27*(7–8), 581–588.

Li, B. Y., Hu, Y., Liang, Y. Z., Xie, P. S., & Du, Y. P. (2004b). Quality evaluation of fingerprints of herbal medicine with chromatographic data. *Analytica Chimica Acta*, *514*(1), 69–77.

Li, H., Zhang, Y., Yu, Y. P., Wang, P., Li, F. F., & Meng, X. L. (2013). Pharmacokinetic/pharmacodynamic modeling of antipyretic and reducing plasma concentration of NO effects of Rheum palmatum in rat. *Zhongguo Zhong yao za zhi= Zhongguo Zhongyao Zazhi= China Journal of Chinese Materia Medica*, *38*(8), 1231–1236.

Li, P., Qi, L. W., Liu, E. H., Zhou, J. L., & Wen, X. D. (2008). Analysis of Chinese herbal medicines with holistic approaches and integrated evaluation models. *TrAC Trends in Analytical Chemistry*, *27*(1), 66–77.

Li, S., Han, Q., Qiao, C., Song, J., Lung Cheng, C., & Xu, H. (2008). Chemical markers for the quality control of herbal medicines: An overview. *Chinese Medicine*, *3*, 1–16.

Liang, X. M., Jin, Y., Wang, Y. P., Jin, G. W., Fu, Q., & Xiao, Y. S. (2009). Qualitative and quantitative analysis in quality control of traditional Chinese medicines. *Journal of Chromatography A*, *1216*(11), 2033–2044.

Long, H., Johri, P., Gout, J., Ni, J., Hao, Y., Licknack, T., Wang, Y., Pan, J., Jiménez-marín, B., & Lynch, M. (n.d.). Paramecium genetics, genomics, and evolution. *Annual Review of Genetics*, *57*, 391–410.

Meena, A. K., Rao, M. M., Panda, P., Kiran, Y. A., & Singh Uttam, S. B. (2010). Standardisation of ayurvedic polyherbal formulation, Pancasama Churna. *International Journal of Pharmacognosy and Phytochemical Research*, 2(1), 11–4.

Modarresi Chahardehi, A., Arsad, H., & Lim, V. (2020). Zebrafish as a successful animal model for screening toxicity of medicinal plants. *Plants*, *9*(10), 1345.

Mosihuzzaman, M., & Choudhary, M. I. (2008). Protocols on safety, efficacy, standardization, and documentation of herbal medicine (IUPAC Technical Report). *Pure and Applied Chemistry*, *80*(10), 2195–2230.

Ni, Y., Lai, Y., Brandes, S., & Kokot, S. (2009). Multi-wavelength HPLC fingerprints from complex substances: An exploratory chemometrics study of the Cassia seed

Pandey, R., Tiwari, R. K., & Shukla, S. S. (2016). Omics: A newer technique in herbal drug standardization & quantification. *Journal of Young Pharmacists*, *8*(2), 76.

Pathak, A. (2017). Q-markers or chemical markers: A new insight towards quality control of herbal medicines. *Organic & Medicinal Chemistry International Journal*, *3*(2), 62–63.

Patra, K. C., Pareta, S. K., Harwansh, R. K., & Kumar, K. J. (2010). Traditional approaches towards standardization of herbal medicines: A review. *Journal of Pharmaceutical Science and Technology*, *2*(11), 372–379.

Pravin, H. N., Joseph, K., Aruna, J., Kadam, V. (2012). Future trends in standardization of herbal drugs. *J Appl Pharma Sci* 2(6), 38-44.

Shukla, M., Tuwair, K. A.-B., Trivedi, M., & Tiwari, R. K. (2018). Status of research, regulations and challenges for genetically modified crops in India. *GM Crops & Food*, *9*, 1–16. https://doi.org/10.1080/21645698.2018.1529518

Ricci, P. F. (2006). *Environmental and health risk assessment and management: Principles and practices*. Dordrecht: Springer.

Rizzo, L. Y., Golombek, S. K., Mertens, M. E., Pan, Y., Laaf, D., Broda, J., & Lammers, T. (2013). In vivo nanotoxicity testing using the zebrafish embryo assay. *Journal of Materials Chemistry B*, *1*(32), 3918–3925.

Roggo, Y., Chalus, P., Maurer, L., Lema-Martinez, C., Edmond, A., & Jent, N. (2007). A review of near infrared spectroscopy and chemometrics in pharmaceutical technologies. *Journal of Pharmaceutical and Biomedical Analysis*, *44*(3), 683–700.

Sagar, B. P., Zafar, R., Panwar, R., Kumar, V., & Mangla, A. (2005). Herbal drugs standardization. *The Indian Pharmacist, 4*(35), 19–22.

Sanchez, F. C., Rutan, S. C., Garcia, M. G., & Massart, D. L. (1997). Resolution of multicomponent overlapped peaks by the orthogonal projection approach, evolving factor analysis and window factor analysis. *Chemometrics and Intelligent Laboratory Systems, 36*(2), 153–164.

Sanchez, F. C., Toft, J., Van den Bogaert, B., Massart, D.L. (1996). Orthogonal project approach applied to peak purity assessment. *Anal Chem. 68*, 79–85.

Sant, K. E., & Timme-Laragy, A. R. (2018). Zebrafish as a model for toxicological perturbation of yolk and nutrition in the early embryo. *Current Environmental Health Reports, 5*, 125–133.

Selvakumari, E., Jenifer, J., Priyadharshini, S., & Vinodhini, R. (2017). Application of DNA fingerprinting for plant identification. *JAIR, 5*(10), 149–151.

Singh, I., Juneja, P., Kaur, B., & Kumar, P. (2013). Pharmaceutical applications of chemometric techniques. *International Scholarly Research Notices, 2013*(1), 795178.

Singh, S., Datta, R., Johri, P., & Trivedi, M. (Eds.). (2022). *Phytopharmaceuticals and biotechnology of herbal plants.* Boca Raton, FL: CRC Press.

Smilde, A. K., Bro, R., & Paul, G. (2004). *Multi-Way analysis with applications in the chemical sciences.* NewYork, NY: John Wiley and Sons.

Soleymani, S., Bahramsoltani, R., Rahimi, R., & Abdollahi, M. (2017). Clinical risks of St John's Wort (Hypericum perforatum) co-administration. *Expert Opinion on Drug Metabolism & Toxicology, 13*(10), 1047–1062.

Sun, S., Wang, Y., Wu, A., Ding, Z., & Liu, X. (2019). Influence factors of the pharmacokinetics of herbal resourced compounds in clinical practice. *Evidence-Based Complementary and Alternative Medicine, 2019*(1), 1983780.

Tistaert, C., Dejaegher, B., & Vander Heyden, Y. (2011). Chromatographic separation techniques and data handling methods for herbal fingerprints: A review. *Analytica Chimica Acta, 690*(2), 148–161.

Trivedi, M., Singh, S., Johri, P., & Sanchez, P. L. (Eds.). (2024). *Nutraceuticals inspiring the contemporary therapy for lifestyle diseases.* Boca Raton, FL: CRC Press.

Vasu, K., Goud, J. V., Suryam, A., & Charya, M. S. (2009). Biomolecular and phytochemical analyses of three aquatic angiosperms. *African Journal of Microbiology Research, 3*(8), 418–421.

WHO. (1998). WHO guidelines for selecting marker substances of herbal origin for quality control of herbal medicines. https://www.who.int/traditional-complementary-integrativemedicine/publications/trs1003_annex1.pdf?ua=1 (Accessed on 20.03.2022) WHO Quality control methods for medicinal plant Materialshttps://apps.who.int/iris/handle/10665/41986

WHO. (2007). WHO guidelines for assessing quality of herbal medicines with reference to contaminants and residues. Geneva: World Health Organization. https://www.who.int/publications/i/item/9789241594448

Yang, T., Liu, S., Wang, C. H., Tao, Y. Y., Zhou, H., & Liu, C. H. (2015). Comparative pharmacokinetic and tissue distribution profiles of four major bioactive components in normal and hepatic fibrosis rats after oral administration of Fuzheng Huayu recipe. *Journal of Pharmaceutical and Biomedical Analysis, 114*, 152–158.

Yip, P. Y., Chau, C. F., Mak, C. Y., & Kwan, H. S. (2007). DNA methods for identification of Chinese medicinal materials. *Chinese Medicine, 2*, 1–19.

Yongyu, Z., Shujun, S., Jianye, D., Wenyu, W., Huijuan, C., Jianbing, W., & Xiaojun, G. (2011). Quality control method for herbal medicine-chemical fingerprint analysis. *Quality Control of Herbal Medicines and Related Areas, 2011*, 171–194.

Yuan, R., & Lin, Y. (2000). Traditional Chinese medicine: An approach to scientific proof and clinical validation. *Pharmacology & Therapeutics*, *86*(2), 191–198.

Yumnamcha, T., Nongthomba, U., & Devi, M. D. (2014). Phytochemical screening and evaluation of genotoxicity and acute toxicity of aqueous extract of Croton tiglium L. *International Journal of Scientific and Research Publications*, *4*(1), 2250–3153.

Yumnamcha, T., Roy, D., Devi, M. D., & Nongthomba, U. (2015). Evaluation of developmental toxicity and apoptotic induction of the aqueous extract of Millettia pachycarpa using zebrafish as model organism. *Toxicological & Environmental Chemistry*, *97*(10), 1363–1381.

Zeng, Z., Chau, F. T., Chan, H. Y., Cheung, C. Y., Lau, T. Y., Wei, S., & Liang, Y. (2008). Recent advances in the compound-oriented and pattern-oriented approaches to the quality control of herbal medicines. *Chinese Medicine*, *3*, 1–7.

Zhang, A., Sun, H., Wang, P., Han, Y., & Wang, X. (2012). Modern analytical techniques in metabolomics analysis. *Analyst*, *137*(2), 293–300.

Zhang, K., Yan, G., Zhang, A., Sun, H., & Wang, X. (2017). Recent advances in pharmacokinetics approach for herbal medicine. *RSC Advances*, *7*(46), 28876–28888.

Zhang, L., Yan, J., Liu, X., Ye, Z., Yang, X., Meyboom, R., & Duez, P. (2012). Pharmacovigilance practice and risk control of traditional Chinese medicine drugs in China: Current status and future perspective. *Journal of Ethnopharmacology*, *140*(3), 519–525.

Zhang, Z., Qin, L., Peng, L., Zhang, Q., Wang, Q., Lu, Z., & Gao, X. (2016). Pharmacokinetic-pharmacodynamic modeling to study the antipyretic effect of Qingkailing injection on pyrexia model rats. *Molecules*, *21*(3), 317.

Zhou, J. L., Qi, L. W., & Li, P. (2009). Herbal medicine analysis by liquid chromatography/time-of-flight mass spectrometry. *Journal of -Chromatography A*, *1213*(44), 7582–7594.

Zhou, J., Ouedraogo, M., Qu, F., & Duez, P. (2013). Potential genotoxicity of traditional Chinese medicinal plants and phytochemicals: An overview. *Phytotherapy Research*, *27*(12), 1745–1755.

Zhu, H. X., Pan, L. M., Zhang, Q. C., Tang, Y. P., & Guo, L. W. (2013). Study on PK/PD model for traditional Chinese medicine biopharmaceutics based on principle of "correspondence of prescriptions and syndromes". *Zhongguo Zhong yao za zhi= Zhongguo Zhongyao Zazhi = China Journal of Chinese Materia Medica*, *38*(12), 2033–2038.

14 Herbal and Natural Remedies for the Treatment of Diabetes

*Umakant Sahu, Vishal Jain, Kamal Shah, Arun K. Singh Parihar, Sekhar Verma, Shraddha Manish Gupta, Durgesh Nandini Chauhan, and Nagendra Singh Chauhan**

14.1 INTRODUCTION

Diabetes has emerged as one of the fastest-growing and most prevalent chronic metabolic and endocrine disorders throughout the globe, and is characterised by hyperglycemia and glucose intolerance. The World Health Organization (WHO) and the International Diabetes Federation estimate that more than 537 million adults globally, between the ages of 20 and 79, constituting 10.5% of all adults in this age range, have diabetes. The projections for the future are alarming, with an estimated 643 million people expected to have diabetes globally by 2030. The prevalence rate is anticipated to exceed 12.8%, and the number is projected to further increase to 783 million by the year 2045. The incidence of diabetes has been steadily increasing in Southeast Asian (SEA) nations for at least two decades. Current estimates surpass previous forecasts, indicating a significant and sustained rise in the prevalence of diabetes in this region. The study suggests that by 2045, the incidence of diabetes in the world, Southeast Asia, and India will rise to 12.5%, 11.5%, and 10.9%, respectively. These figures highlight the anticipated further escalation of diabetes cases in these regions (Kumar et al., 2024). Type 2 diabetes mellitus (T2DM) is the diagnosis for over 90% of diabetic people worldwide. When the beta cells in your pancreas that create insulin are unintentionally attacked and destroyed by your immune system, the damage is irreversible. Insulin resistance is the initial stage of type 2 diabetes. When the body is unable to utilise insulin as effectively as it should, the pancreas must create more insulin until it is unable to meet the body's needs. High blood sugar results from a subsequent decrease in insulin production (Gothai et al., 2016). Type 2 diabetes is acknowledged as a critical public health issue that has a significant impact on human life and medical costs. Urbanisation and rapid economic development

* Corresponding author

 DOI: 10.1201/9781003392910-14

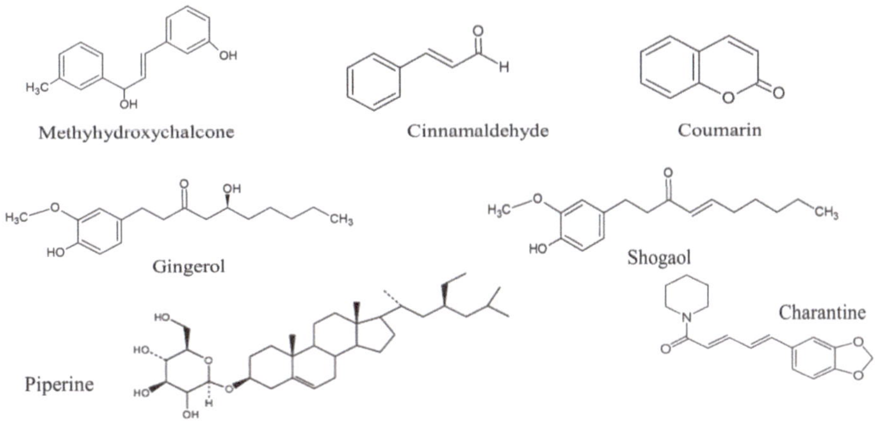

FIGURE 14.1 Phytoconstituents having antidiabetic activity.

have increased the burden of diabetes in many parts of the world. These changes have been linked to the increased prevalence of sedentary lifestyles and poor dietary habits, which increase fasting plasma glucose levels and body mass index (BMI) (Abdul et al., 2020). The most common effect of long-term, poorly controlled diabetes is diabetic neuropathy, which manifests as pain and paraesthesia (Kitada et al., 2010; Petrie et al., 2018; Forbes et al., 2013). There have been numerous new treatments for diabetes complications evaluated in clinical trials, with generally underwhelming outcomes. Indeed, the exact mechanisms underlying diabetic complications are still not fully understood. Alternatives that are feasible and have no side effects for the management of T2DM and its consequences include natural products and their active ingredients (Figures 14.1 and 14.2). Numerous active medicinal plants and their naturally occurring bioactive molecules have already been reported to have therapeutic effects in diabetic patients (Table 14.1). Since antiquity, a number of medicinal plants have been used to treat and prevent diabetes and related conditions (Mukherje et al., 2020; Singh et al., 2022).

14.2 PATHOPHYSIOLOGY OF DIABETES

Diabetes, often known as diabetes mellitus (DM), is a complex metabolic disease that is characterised by hyperglycemia. This physiological anomaly signifies consistently elevated blood glucose levels, stemming from irregularities in insulin secretion, insulin action, or both (Gupta al, 2022). Both glucagon and insulin hormones are released by the pancreas. The islets of Langerhans contain beta (ß) and alpha (α) cells, which release insulin and glucagon, respectively. Insulin transfers glucose into the muscles, liver, and adipose tissue, lowering blood glucose levels through glycogenesis. The alpha (α) cells are crucial for blood glucose regulation because they produce glucagon, which raises blood glucose levels by speeding up the process of glycogenolysis (Trivedi et al., 2024). The incapacity of insulin-sensitive organs

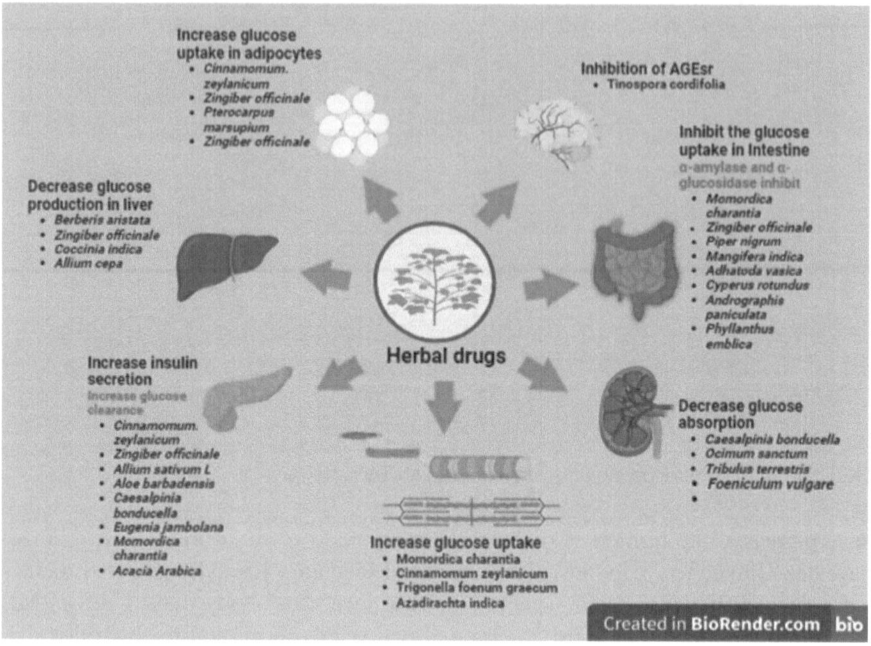

FIGURE 14.2 Mechanism of herbal drugs as antidiabetic agents.

to react to insulin and the abnormalities in pancreatic β-cell insulin production are key cause of diabetes (Chattopadhyay et al., 2022). The major etiological factors for diabetic complications involve lipid and protein abnormalities (Moini et al., 2019). Diabetes leads to anomalies in the metabolism of very low-density lipoprotein (VLDL), low-density lipoprotein (LDL), and high-density lipoprotein (HDL) due to free radical-induced lipoprotein oxidation. Lipid peroxidation is caused by increased oxidative stress in diabetes (Giacco et al., 2010; Wijenayaka et al., 2022). The integration of herbal medicine into modern practices faces challenges due to insufficient scientific and clinical data on efficacy and safety. Further research, including clinical trials, standardisation bioassays, pharmacological and toxicological evaluations, and identification of active components, is crucial for establishing the potential of herbal drugs in diabetes management (Singh et al., 2019; Ozougwu et al., 2013; Galicia-Garcia et al., 2022).

14.3 LIST OF POTENTIAL HERBAL DRUGS AND THEIR ROLE IN THE MANAGEMENT OF DIABETES

14.3.1 Cinnamon (*Cinnamomum zeylanicum* Blume)

Cinnamon, belonging to the Lauraceae family, has been consumed since ancient times and is derived mainly from two species: *Cinnamomum zeylanicum and Cinnamomum cassia*. It contains various essential oils and compounds like

TABLE 14.1
List of Some Potential Herbal Drugs in Diabetes Management

S.No.	Biological Source	Family	Part Used	Chemical Constituents	References
1.	*Allium sativum* (garlic)	*Alliaceae*	Bulb	Allicin, allicin, alliin, and S-allyl-cysteine	(Shang et al., 2019)
2.	*Caesalpinia bonducella* (sagargota)	*Caesalpiniaceae*	Seed, leaves	α-amyrin, β-amyrin, lupeol acetate, and lupeol	(Widhiantara et al., 2018)
3.	*Ocimum sanctum* (tulsi)	*Lamiaceae*	Seed, leaves, whole plant	Eugenol, P-coumarine, rosemarinic acid	(Hussain et al., 2017)
4.	*Aegle marmelos L.* (bael)	*Rutaceae*	Fruit, leaves	Aurapten, imperatorin, psoralen, marmesinin, aeglin	(Anandhu et al., 2021)
5.	*Zingiber officinale* (ginger)	*Zingiberaceae*	Tuberous rhizome	6-gingerol, 6-shogaol, 10-gingerol, 12-gingerol	(Arcusa et al., 2022)
6.	*Trigonella foenum-graecum L.* (fenugreek)	*Fabaceae*	Seed, leaves	Trigonelline, naringenin, kaempferol, fenugrin, foenugracin	(Nimbekar et al., 2022)
7.	*Piper nigrum* (Marich)	*Piperaceae*	Fruit	Piperine	(Wang et al., 2023)
8.	*Bauhinia variegate* (kachnara)	*Fabaceae*	Flower, leaves, stem-bark	Roseoside	(Tripathi et al., 2019)
9.	*Tinospora cordifolia*	*Menispermaceae*	Stem, leaf, satva	Berberine, tinosporin, inocordifolin, cordifolisides	Kumar et al., 2024)
10.	*Withania somnifera* (Ashwagandha)	*Solanaceae*	Root	Withanolides, withaferin A	(Macharia et al., 2023)
11.	*Azadirtachta indica* (neem)	*Meliaceae*	Seed, leaf, bark	Azadirachtin, meliacin, gedunin, salanin, nimbin	(Modi et al., 2023)
12.	*Momordica charantia* (karela)	*Cucurbitaceae*	Fruit, leaves	Charantia, momordicolide, monordicophenoide A	(Richter et al., 2023)
13.	*Phyllanthus amarus* (bhuiawala)	*Euphorbiaceae*	Leaves	Isobubbialine and epibubbialine, phyllanthin	(Verma et al., 2014)

(Continued)

TABLE 14.1 (CONTINUED)
List of Some Potential Herbal Drugs in Diabetes Management

S.No.	Biological Source	Family	Part Used	Chemical Constituents	References
14.	*Swertia chirayita* (chiretta)	*Gentianaceae*	Leaves	Amarogentin, swertiamarin, swerchirin, sweroside, gentianine	(Kumar et al., 2016)
15.	*Eugenia jambolana* (jamuna)	*Myrtaceae*	Kernel, leaves, septum	Anthocyanins, glucoside, ellagic acid, isoquercetin, kaempferol, and myrecetin	(Mandala et al., 2023)

cinnamaldehyde, cinnamic acid, coumarin, and eugenol (Sneka et al., 2022). Studies suggest that cinnamon acts as a natural insulin sensitizer, benefiting both in vitro cell line research and in vivo animal studies. A significant compound in cinnamon, methyl hydroxychalcone polymer (MHCP), mimics insulin, enhancing insulin signaling and sensitivity (Wijenayaka et al., 2022). Its water-soluble extract boosts insulin receptor autophosphorylation while reducing tyrosine phosphatase activity, thereby improving insulin sensitivity. Cinnamon also helps in managing hyperglycemia and inflammation through several mechanisms. It delays gastric emptying, inhibits α-glycosidase enzyme activity, reduces glucose absorption, and promotes glycogen synthesis (Vanschoonbeek et al., 2006).

14.3.2 Ginger (*Zingiber officinale*)

Zingiber officinale, commonly known as ginger and belonging to the Zingiberaceae family, is a widely consumed spice globally. It has a history of use in herbal therapy for various conditions like pain, indigestion, vomiting, and cold-induced disorders. Alongside its antioxidant, antibacterial, hepatoprotective, neuroprotective, gastroprotective, anticancer, anti-inflammatory, and analgesic properties, ginger has shown significant benefits in managing glycaemic levels in individuals with diabetes mellitus (Jayathilake et al., 2022). Key compounds found in ginger include a-pinene, borneol, 1,8-cineole, citronellol, geranial, geraniol, and zingerone, along with gingerols, shogaols, gingerdiols, ingenol, and paradols (Mukjerjee et al., 2022). Trikatuk, a Thai traditional medicine containing ginger, has been utilised for diabetes control, with its extract demonstrating alpha-glucosidase inhibitory activity through compounds like 6-shogaol and 6-gingerol (Nutmakul et al., 2023).

14.3.3 Bitter Gourd (*Momodica charantia*)

Bitter gourd, a plant of the Cucurbitaceae family, contains prevalent flavonoids and phenolic chemicals such as gallic acid, ellagic acid, myricetin, protocatechuic acid, gentisic acid, (+)-catechin, vanillic acid, syringic acid, (π)-epicatechin, p-coumaric acid, benzoic acid, sinapinic acid, o-coumaric acid, chlorogenic acid, t-cinnamic acid, and t-ferulic acid. The bitter gourd is one of the top herbs (19.2%) commonly used in herbal formulations with an emphasis on the effects on glycaemic control (HbA1c values). Bitter gourd has shown a strong correlation with better glycaemic control due to increased insulin sensitivity of charantin, especially when used with 500 mg of metformin per day. Potential benefits were indicated by the synergistic impact reported in mice administered 300 mg/kg of bitter gourd along with 15 mg/kg of metformin. Higher doses of metformin and herbal medicine, however, were not associated with better glycaemic control in this trial (Krawczyk et al., 2022). An ethanol extract from the powdered dry form of bitter gourd was investigated as an antidiabetic drug in rats that had been given alloxan-induced diabetes. The extract, having potency to control glycaemia, resulted in a drop in serum glucose levels in the diabetic rats (Alam et al., 2022).

14.3.4 Marich (*Piper nigrum*)

Piper nigrum (Family Piperaceae) shows various properties, such as being antihypertensive, antioxidant, antiplatelet, anticancer, anticonvulsant, antithyroid, analgesic, anti-inflammatory, antidiarrheal, antispasmodic, antidepressant, immunomodulatory, antibacterial, antifungal, and hepatoprotective (Damanhouri et al., 2014).

The Ayurvedic formulation Trikatu Churna, which contains *Piper nigrum*, demonstrated alpha-glucosidase inhibitory action similar to acarbose, especially from the extract mixture. *Piper nigrum*, especially with piperine, displayed potent alpha-glucosidase inhibition, confirmed by molecular docking studies. *Piper nigrum* and *Piper retrofractum* contribute to alpha-glucosidase inhibition due to high piperine contents. This study suggests Trikatuk's potential in managing postprandial hyperglycemia (Nutmakul et al., 2023).

14.3.5 Neem (*Azadirachta indica*)

The *Azadirachta indica* or neem, native to the Indian subcontinent, is renowned as a "miracle tree." It contains diverse properties, such as being anti-allergenic, antidermonic, anti-feedant, antifungal, anti-inflammatory, antipyorrheic, and antiscabic, as well as being a cardiotonic, diuretic, insecticide, larvicide, nematicide, and spermicide (Rahman et al., 2023). In an oral glucose tolerance test (OGTT) on Wistar albino rats, the alcoholic extract of neem root bark exhibited anti-hyperglycemic effects, notably at an 800 mg/kg dosage (Sarkar et al., 2021). A comparative study on streptozotocin (STZ)-induced diabetic rats revealed that the aqueous extract of *A. indica* leaves significantly increased hemoglobin concentration, packed cell volume, red blood cell count, and platelet count. Histopathological studies showed regeneration of organs compared to diabetic-untreated animals (Ezeigwe et al., 2020). The bioactive compounds of neem, including nimbolide, azadirachtin, and gedunin, exhibit antidiabetic activity by inhibiting glycogenolytic activity (Sahu et al., 2021; Christian et al., 2020; Satyanarayana et al., 2015). *A. indica* leaf extract enhanced insulin signaling molecules, glycogen, and glucose oxidation in skeletal muscle in a study on high-fat diet-induced diabetic rats, suggesting a potential function in the management of type 2 diabetes mellitus (Siddiqui et al., 2023, Trivedi et. al., 2024). The hydroalcoholic extraction of neem exhibited anti-hyperglycaemic activity in STZ-induced diabetic rats, particularly at 800 mg/kg dosage. Neem extract at different doses significantly decreased postprandial blood sugar levels, fasting blood sugar levels, glycosylated hemoglobin, insulin resistance, endothelial dysfunction, oxidative stress, systemic inflammation, IL-6, TNF-α, platelet aggregation, and lipid profile in subjects with type 2 diabetes mellitus in a randomised controlled clinical study (Pingali et al., 2020).

14.3.6 Sehjan (*Moringa oleifera*)

Moringa oleifera, a plant renowned for its health benefits, exhibits antidiabetic properties, particularly in combating diabetes mellitus. Moringa leaves are rich in minerals, proteins, vitamins, polyphenols, flavonoids, glucosinolates, isothiocyanates, and alkaloids, and demonstrate bioactive compounds with anticancer, anti-inflammatory, antioxidant, and hypolipidemic characteristics (Mthiyane et al., 2022). Effective doses in animal models range from 100 to 300 mg/kg, while individuals with prediabetes or diabetes may benefit from around 20 g of *Moringa oleifera* leaf powder for at least two weeks. Limited clinical investigations exist, but current findings provide a foundation for future research on utilising Moringa as a functional meal for diabetic complications (Villarruel-López et al., 2018). To assess toxicity and genotoxicity, powdered leaf doses were administered, revealing a hypoglycemic effect (<250 mg/dL) with increased body weight (>30 g) (Fatoumata et al., 2020). In STZ-induced diabetic rats, aqueous-ethanol extracts, particularly the 95% ethanol extract and butanol fraction, showed immediate drops in blood glucose levels. Significant drops in blood glucose levels were observed in other trials that used ethyl acetate extract at 200 mg/kg and powdered moringa leaf at 50 mg/day. *M. oleifera* leaf extract decreased glucose, insulin, and inflammatory cytokine levels in type 2 diabetic rats caused by STZ and a high-fat diet, suggesting its potential as a therapeutic (Anwer et al., 2021).

14.3.7 Bael (*Aegle marmelos*)

The tree *Aegle marmelos*, also known as bael (or bel), is a moderately big, slender, and aromatic member of the Rutaceae family. The Indian Ayurvedic Pharmacopoeia lists *A. marmelos* as a dietary supplement with the ability to treat diabetes and related conditions. Alpha-amylase and alpha-glycosidase were shown to be significantly inhibited by *A. marmelos* in the enzyme inhibition assay, with IC50 values of 46.21 and 42.07 mg/mL, respectively (Ahmad et al., 2021). Additionally, *A. marmelos* leaf juice supplementation improved all biochemical parameters in type 2 diabetes mellitus, suggesting its efficacy as an adjunct to oral hypoglycemic medications. This is likely because the leaf contains the active ingredients aegelin 2, scopoletin, and ß -sitosterol (Nigam et al., 2019). A. *marmelos* fruit extract was tested for its antidiabetic effects in Wistar rats that had type I diabetes brought on by STZ. Potential therapeutic effects of *A. marmelos* extract included enhanced beta cell density and the repair of damaged pancreatic architecture in treated groups. Assessments of the antioxidant enzymes catalase, superoxide dismutase, and lipid peroxidation levels in the extract provide more evidence that it has the potential to be used as a diabetes treatment (Al-shami et al., 2022). The *A. marmelos* leaf and callus extracts showed a noteworthy decrease in rabbit blood sugar levels. According to the study, *A. marmelos* may have the ability to act as a hypoglycemic agent by bringing rats' blood sugar and body weight back to normal, just like insulin does (Arumugam et al., 2008). In STZ-induced diabetic rats, the aqueous extract of *A. marmelos* fruits increased insulin secretion and produced better results than glibenclamide. *A. marmelos* can be used therapeutically because it successfully lowers blood glucose levels and enhances glucose tolerance in both diabetic and non-diabetic rats (Monika et al., 2023; Birudu et al., 2021). Another study investigated the impact of *A. marmelos* plant extract on STZ-induced rats with diabetes mellitus type 2 affecting the kidney and liver. The findings revealed decreased glucose levels, improved pancreas histopathology, and reduced inflammatory cellular lining in the liver, suggesting *A. marmelos* possesses renoprotective, hepatoprotective, and hypoglycemic therapeutic properties (Anandhu et al., 2021).

14.3.8 Guduchi (*Tinospora cordifolia*)

An important role in Ayurvedic medicine is played by *Tinospora cordifolia*, a herbaceous or climbing deciduous shrub that is part of the Menispermaceae family and is found in India, China, Myanmar, Bangladesh, and Sri Lanka. *T. cordifolia* is well-known for treating a wide range of ailments, including fevers, jaundice, diabetes, dysentery, urinary infections, and skin disorders. Chemical, pharmacological, pre-clinical, and clinical research on the plant has revealed new therapeutic potential benefits (Ahsan et al., 2023). The Ayurvedic Pharmacopoeia of India recognises its antidiabetic uses, supported by clinical research on isolated phytoconstituents like tinosporin, berberine, and jatroporrhizine, demonstrating both preventive and curative antidiabetic characteristics (Sharma et al., 2015). In particular, *T. cordifolia*'s diterpenoid tinosporaside demonstrates anti-hyperglycemic activity by promoting GLUT4 translocation to the plasma membrane, improving skeletal muscle cells' absorption of glucose, and raising AMP-activated protein kinase and protein kinase B/AKT phosphorylation (Mishra et al., 2023).

14.3.9 Tulsi or Holy Basil (*Ocimum sanctum*)

Tulsi, scientifically known as *Ocimum sanctum*, is a sacred herb native to India, belonging to the Lamiaceae family. Recognised as the "queen of herbs" and the "elixir of life" in Ayurveda, tulsi has been traditionally used for various ailments such as cold, headache, cough, and more (Islam, 2022). There are approximately 64 species in the Ocimum genus, including important ones like Rama thulasi and clove basil (Karthıka et al., 2023). Tulsi exhibits tremendous biological activity and has been used in Ayurveda and Siddha systems (Hasan et al., 2023) for conditions like diabetes, arthritis, bronchitis, and skin diseases due to its antimicrobial substances (Srivastava et al., 2023). Preclinical research indicates that Tulsi and its phytochemicals, such as eugenol, rosmarinic acid, and luteolin, prevent cancers caused by chemicals. Some components in Ocimum species, especially eugenol, show anti-hyperglycemic effects by inhibiting α-glucosidase (Upadhyay, 2017; Reddy et al., 2023). The synthesis of colloidal silver nanoparticles from tulsi leaves demonstrates anti-hyperglycemic properties. Tulsi extracts stimulate insulin secretion and have shown potential in preventing diabetic-induced complications (Antora et al., 2017; Malapermal et al., 2017). Additionally, the aqueous extract of *Ocimum basilicum* exhibits hypoglycemic activity through inhibition of sucrase, maltase, and pancreatic α-amylase. Studies on tulsi root methanolic extract reveal its effectiveness in elevating serum insulin levels. Basil leaf powder consumption by type 2 diabetes patients reduces blood glucose levels and improves endothelial protection, comparable to the effects of metformin (Hannan et al., 2006; El-Beshbishy et al., 2012).

Experimental animal models demonstrate that tulsi extracts lower blood glucose levels, improve pancreatic cell health, and enhance antioxidant enzyme function. The antioxidants present in tulsi leaves inhibit carbohydrate absorption in the intestine, providing potential benefits for glycaemic control (Ahmad et al., 2012).

14.3.10 Bhuiawala (*Phyllanthus amarus*)

Phyllanthus amarus, a herb belonging to the Euphorbiaceae family, is also known as Bhuiamala. With a history of over 2000 years in Ayurvedic medicine, this herb exhibits diverse traditional applications (Adedapo et al., 2014). Its methanolic extract demonstrates potent antioxidant activity and reduces blood sugar in diabetic rats (Mohite et al., 2020). Many medical benefits of *P. amarus* include the treatment of hepatitis B, jaundice, cirrhosis of the liver, intestinal infections, diabetes, persistent fever, and appetite loss (Sadique et al., 2014). It also possesses anti-inflammatory, antimutagenic, anticarcinogenic, and antidiarrheal properties. The plant contains beneficial compounds such as tannins, ellagitannins (geranin, amarin), alkaloids, flavonoids, lignans (phyllanthin, hypophyllanthin, nirurin), and other bioactive constituents (Patel et al., 2011). In studies, diabetic mice were given an oral dose of 400 mg/kg body weight of *P. amarus* ethanolic leaf extract for 45 days, and there were notable improvements in body weight and a decrease in blood glucose (Singh et al., 2019). *P. amarus* shows promise in combating oxidative stress associated with diabetes, acting as an alternative approach due to its antioxidant properties (Adeneye, 2012). The methanol extract of *P. niruri* aerial parts exhibits antidiabetic potential by reducing fasting blood sugar and suppressing postprandial glucose rise in rats. Overall, *P. amarus* holds therapeutic potential in managing diabetes and related complications, supported by its long-standing use in traditional medicine and scientific studies (Moshi et al., 2001).

14.4 CONCLUSION AND FUTURE PERSPECTIVES

The exploration of herbal and natural remedies for diabetes presents a promising avenue for the holistic management of this chronic condition. Traditional herbal medicines have shown considerable potential in lowering blood sugar levels, improving insulin sensitivity, and mitigating diabetic complications. Moreover, their widespread availability, affordability, and relatively low risk of adverse effects make them an attractive option for many individuals seeking complementary or alternative approaches to conventional diabetes treatment. These herbal drugs require comprehensive investigation and integration into mainstream healthcare practices. Collaborative efforts among researchers, healthcare providers, regulatory agencies, and traditional medicine practitioners are essential to bridge the gap between traditional knowledge and modern evidence-based medicine. Furthermore, advancements in technology, such as metabolomics, genomics, and bioinformatics, hold significant potential for identifying the active compounds in medicinal plants, elucidating their mechanisms of action, and optimising their therapeutic potential. Ultimately, by fostering interdisciplinary collaboration, promoting scientific rigor, and embracing cultural diversity, we can harness the healing power of nature to enhance diabetes management and improve the quality of life for millions of individuals worldwide.

REFERENCES

Islam, M. T. (2022). Anti-diabetic potential of Ocimum sanctum Linn. *International Journal of Advanced Chemistry Research*, 6, 38–41.

Abdul Basith Khan, M., Hashim, M. J., King, J. K., Govender, R. D., Mustafa, H., & Al Kaabi, J. (2020). Epidemiology of type 2 diabetes—global burden of disease and forecasted trends. *Journal of Epidemiology and Global Health*, 10(1), 107–111.

Adedapo, A., Ofuegbe, S., & Oguntibeju, O. (2014). The antidiabetic activities of the aqueous leaf extract of Phyllanthus amarus in some laboratory animals. In O. Oguntibeju (Ed.), *Antioxidant-antidiabetic agents and human health* (pp. 115–137). IntechOpen.

Adeneye, A. A. (2012). The leaf and seed aqueous extract of Phyllanthus amarus improves insulin resistance diabetes in experimental animal studies. *Journal of Ethnopharmacology*, 144(3), 705–711.

Ahmad, M. Z., Ali, M., & Mir, S. R. (2012). Anti-diabetic activity of Ocimum sanctum L. roots and isolation of new phytoconstituents using two-dimensional nuclear magnetic resonance spectroscopy. *Journal of Pharmacogn Phytother*, 4, 75–85.

Ahmad, W., Amir, M., Ahmad, A., Ali, A., Ali, A., Wahab, S., & Alam, P. (2021). Aegle marmelos leaf extract phytochemical analysis, cytotoxicity, in vitro antioxidant and antidiabetic activities. *Plants*, 10(12), 2573.

Ahsan, R., Mishra, A., Badar, B., Owais, M., & Mishra, V. (2023). Therapeutic application, phytoactives and pharmacology of Tinospora cordifolia: An evocative review. *Chinese Journal of Integrative Medicine*, 29(6), 549–555.

Alam, S., Sarker, M. M. R., Sultana, T. N., Chowdhury, M. N. R., Rashid, M. A., Chaity, N. I., & Mohamed, I. N. (2022). Antidiabetic phytochemicals from medicinal plants: Prospective candidates for new drug discovery and development. *Frontiers in Endocrinology*, 13, 800714.

Al-shami, A. S., Jha, D. K., Babukutty, B. E., Haque, M., & Khadka, J. (2022). Evaluation of protective effect of aegle marmelos fruits extracts against streptozotocin-induced type 1 diabetes mellitus in rat's model. *Evaluation*, 15(12), 95–102.

Anandhu, K. S., Jose, M., Kuriakose, S., & Jayalakshmi, P. M. (2021). Phytochemical analysis and In vitro Antidiabetic activity of aqueous extract of Lagerstroemia speciosa and Aegle marmelos. *Research Journal of Pharmacy and Technology*, 14(9), 4697–4701.

Antora, R. A., & Salleh, R. M. (2017). Antihyperglycemic effect of Ocimum plants: A short review. *Asian Pacific Journal of Tropical Biomedicine*, 7(8), 755–759.

Anwer, T., Safhi, M. M., Makeen, H. A., Alshahrani, S., Siddiqui, R., Sivakumar, S. M., & Alam, M. F. (2021). Antidiabetic potential of Moringa oleifera Lam. leaf extract in type 2 diabetic rats, and its mechanism of action. *Tropical Journal of Pharmaceutical Research*, 20(1), 95–103.

Arcusa, R., Villaño, D., Marhuenda, J., Cano, M., Cerdà, B., & Zafrilla, P. (2022). Potential role of ginger (Zingiber officinale Roscoe) in the prevention of neurodegenerative diseases. *Frontiers in Nutrition*, 9, 809621.

Arumugam, S., Kavimani, S., Kadalmani, B., Ahmed, A. B. A., Akbarsha, M. A., & Rao, M. V. (2008). Antidiabetic activity of leaf and callus extracts of Aegle marmelos in rabbit. *Science Asia*, 34(3), 317–321.

Birudu, R. B., Pamulapati, P., & Manoharan, S. K. (2021). Effects of Aegle marmelos (L.) Methanolic Leaf Extracts on Biochemical Parameters in Diabetic Rats. *Journal of Reports in Pharmaceutical Sciences*, 10(2), 209–215.

Chattopadhyay, K., Wang, H., Kaur, J., Nalbant, G., Almaqhawi, A., Kundakci, B., & Leonardi-Bee, J. (2022). Effectiveness and safety of Ayurvedic medicines in type 2 diabetes mellitus management: a systematic review and meta-analysis. *Frontiers in Pharmacology*, 13, 821810.

Christian, E. O., Onyeaghana, O. C., Ngozi, N. N., Olivia, O. O., Okwukwe, E. P., Ann, O. E., & Henrietta, O. C. (2020). Evaluation of the effects of Azadirachta indica leaf on haematology, lipid profile, body weight and organ-system functions of streptozotocin-induced diabetic male rats. *African Journal of Biochemistry Research*, 14(3), 57–71.

Damanhouri, Z. A., & Ahmad, A. (2014). A review on therapeutic potential of Piper nigrum L. Black Pepper): The King of Spices. *Medicinal & Aromatic Plants*, 3(3), 161.

El-Beshbishy, H. A., & Bahashwan, S. A. (2012). Hypoglycemic effect of basil (Ocimum basilicum) aqueous extract is mediated through inhibition of α-glucosidase and α-amylase activities: An in vitro study. *Toxicology and Industrial Health*, 28(1), 42–50.

Ezeigwe, O. C., Ezeonu, F. C., Okani, C. O., Onwusulu, D. N., & Onuegbu, M. E. (2020). Aqueous extract of Azadirachta indica leaves favorably alters the course of streptozotocin-induced diabetes in rats: A comparative prospective cohort study. *Biomedical Research and Therapy*, 7(7), 3877–3889.

Fatoumata, B. A., Mohamet, S. E. N. E., Sambou, J. K., MbackÃ, M., & El HadjiMakhtar, B. A. (2020). Antidiabetic properties of Moringa oleifera: A review of the literature. *Journal of Diabetes and Endocrinology*, 11(1), 18–29.

Forbes, J. M., & Cooper, M. E. (2013). Mechanisms of diabetic complications. *Physiological Reviews*, 93(1), 137–188.

Galicia-Garcia, U., Benito-Vicente, A., Jebari, S., Larrea-Sebal, A., Siddiqi, H., Uribe, K. B., & Martín, C. (2020). Pathophysiology of type 2 diabetes mellitus. *International Journal of Molecular Sciences*, 21(17), 6275.

Giacco, F., & Brownlee, M. (2010). Oxidative stress and diabetic complications. *Circulation Research*, 107(9), 1058–1070.

Gothai, S., Ganesan, P., Park, S. Y., Fakurazi, S., Choi, D. K., & Arulselvan, P. (2016). Natural phyto-bioactive compounds for the treatment of type 2 diabetes: Inflammation as a target. *Nutrients*, 8(8), 461.

Gupta, G., Paul, S., Singh, S., Pietramellara, G., Pathan, S. I., Danish, S., & Jha, P. N. (2022). Exploring functional diversity and community structure of diazotrophic endophytic bacteria associated with Pennisetum glaucum growing under field in a semi-arid region. *Land*, 11(7), 991.

Hannan, J. M. A., Marenah, L., Ali, L., Rokeya, B., Flatt, P. R., & Abdel-Wahab, Y. H. A. (2006). Ocimum sanctum leaf extracts stimulate insulin secretion from perfused pancreas, isolated islets and clonal pancreatic β-cells. *Journal of Endocrinology*, 189(1), 127–136.

Hasan, M. R., Alotaibi, B. S., Althafar, Z. M., Mujamammi, A. H., & Jameela, J. (2023). An update on the therapeutic anticancer potential of Ocimum sanctum L.:"Elixir of life". *Molecules*, 28(3), 1193.

Hussain, A. I., Chatha, S. A. S., Kamal, G. M., Ali, M. A., Hanif, M. A., & Lazhari, M. I. (2017). Chemical composition and biological activities of essential oil and extracts from Ocimum sanctum. *International Journal of Food Properties*, 20(7), 1569–1581.

Jayathilake, A. L., Jayasinghe, M. A., & Walpita, J. (2022). Development of ginger, turmeric oleoresins and pomegranate peel extracts incorporated pasteurized milk with pharmacologically important active compounds. *Applied Food Research*, 2(1), 100063.

Karthıka, M., & Rasmi, A. R. (2023). Pharmacological potential of fungal endophytes associated with the genus Ocimum L. *International Journal of Secondary Metabolite*, 10(1), 1–10.

Kitada, M., Zhang, Z., Mima, A., & King, G. L. (2010). Molecular mechanisms of diabetic vascular complications. *Journal of Diabetes Investigation*, 1(3), 77–89.

Krawczyk, M., Burzynska-Pedziwiatr, I., Wozniak, L. A., & Bukowiecka-Matusiak, M. (2022). Evidence from a systematic review and meta-analysis pointing to the antidiabetic effect of polyphenol-rich plant extracts from Gymnema montanum, Momordica charantia and Moringa oleifera. *Current Issues in Molecular Biology*, 44(2), 699–717.

Kumar, A., Gangwar, R., Ahmad Zargar, A., Kumar, R., & Sharma, A. (2024). Prevalence of diabetes in India: A review of IDF diabetes atlas 10th edition. *Current Diabetes Reviews*, 20(1), 105–114.

Kumar, M. V., Reddy, H., Swamy, A. N., Jayappa, K., Reddy, M. S., Kumar, D., & Kiran, V. U. (2024). A therapeutic evaluation of anticancer and pharmacological abilities on Tinospora cordifolia: A systematic review. *Uttar Pradesh Journal of Zoology*, 45(13), 1–8.

Kumar, V., & Van Staden, J. (2016). A review of Swertia chirayita (Gentianaceae) as a traditional medicinal plant. *Frontiers in Pharmacology*, 6, 308.

Macharia, J. M., Káposztás, Z., & Bence, R. L. (2023). Medicinal characteristics of Withania somnifera L. In colorectal cancer management. *Pharmaceuticals*, 16(7), 915.

Malapermal, V., Botha, I., Krishna, S. B. N., & Mbatha, J. N. (2017). Enhancing antidiabetic and antimicrobial performance of Ocimum basilicum, and Ocimum sanctum (L.) using silver nanoparticles. *Saudi Journal of Biological Sciences*, 24(6), 1294–1305.

Mandal, S. K., Das, A., Devkota, H. P., & Das, N. (2023). Syzygium cumini (L.) Skeels. In T. Belwal, I. D. Bhatt, & H. P. Devkota (Eds.), *Himalayan fruits and berries* (pp. 403–418). Academic Press.

Mishra, A., Sharma, K., Pandey, J., Dev, K., Kadan, S., Sahai, M., & Maurya, R. (2023). Tinosporaside from Tinospora cordifolia Encourages Skeletal Muscle Glucose Transport through Both PI-3-Kinase-and AMPK-Dependent Mechanisms. *Molecules*, 28(2), 483.

Modi, G., & Soni, R. K. (2023). Azadirachta indica in focus: Investigating Neem's diverse applications. *Journal of Chemical Health Risks*, 13(3), 1500–1510.

Mohite, S. S., Gajare, R., & Khose, N. B. (2020). Comparative study of treatment of gestational diabetes mellitus cited in various ayurvedic and modern research papers published in last 5 years. *Global Journal for Research Analysis*, 64–67. https://doi.org/10.36106/gjra/7902212

Moini, J. (2019). Pathophysiology of diabetes. *Epidemiology of Diabetes*, 30(1), 25–43.

Monika, S., Thirumal, M., & Kumar, P. R. (2023). Phytochemical and biological review of Aegle marmelos Linn. *Future Science OA*, 9(3), FSO849.

Moshi, M. J., Lutale, J. J., Rimoy, G. H., Abbas, Z. G., Josiah, R. M., & Swai, A. B. (2001). The effect of Phyllanthus amarus aqueous extract on blood glucose in non-insulin dependent diabetic patients. *Phytotherapy Research*, 15(7), 577–580.

Mthiyane, F. T., Dludla, P. V., Ziqubu, K., Mthembu, S. X., Muvhulawa, N., Hlengwa, N., & Mazibuko-Mbeje, S. E. (2022). A review on the antidiabetic properties of Moringa oleifera extracts: focusing on oxidative stress and inflammation as main therapeutic targets. *Frontiers in Pharmacology*, 13, 940572.

Mukherjee, P. K., Kar, A., Banerjee, S., & Katiyar, C. K. (2020). Antidiabetic natural products. In *Annual reports in medicinal chemistry* (Vol. 55, pp. 373–409). Academic Press.

Mukjerjee, S., & Karati, D. (2022). A mechanistic view on phytochemistry, pharmacognostic properties, and pharmacological activities of phytocompounds present in Zingiber officinale: A comprehensive review. *Pharmacological Research-Modern Chinese Medicine*, 5, 100173.

Nigam, V., & Nambiar, V. S. (2019). Aegle marmelos leaf juice as a complementary therapy to control type 2 diabetes–Randomised controlled trial in Gujarat, India. *Advances in Integrative Medicine*, 6(1), 11–22.

Nimbekar, T., & Sanghi, D. (2022). Wild Edible vegetables used for health benefit by rural people of gondia district in Maharashtra State, India. *Traditional Medicine*, 3(1), 1–9.

Nutmakul, T., & Chewchinda, S. (2023). Synergistic effect of Trikatuk, a traditional Thai formulation, on antioxidant and alpha-glucosidase inhibitory activities. *Heliyon*, 9(1), e13063.

Ozougwu, J. C., Obimba, K. C., Belonwu, C. D., & Unakalamba, C. B. (2013). The pathogenesis and pathophysiology of type 1 and type 2 diabetes mellitus. *Journal of Physiology and Pathophysiology*, 4(4), 46–57.

Patel, J. R., Tripathi, P., Sharma, V., Chauhan, N. S., & Dixit, V. K. (2011). Phyllanthus amarus: ethnomedicinal uses, phytochemistry and pharmacology: A review. *Journal of Ethnopharmacology*, 138(2), 286–313.

Petrie, J. R., Guzik, T. J., & Touyz, R. M. (2018). Diabetes, hypertension, and cardiovascular disease: Clinical insights and vascular mechanisms. *Canadian Journal of Cardiology*, 34(5), 575–584.

Pingali, U., Ali, M. A., Gundagani, S., & Nutalapati, C. (2020). Evaluation of the effect of an aqueous extract of Azadirachta indica (Neem) leaves and twigs on glycemic control, endothelial dysfunction and systemic inflammation in subjects with type 2 diabetes mellitus–a randomized, double-blind, placebo-controlled clinical study. *Diabetes, Metabolic Syndrome and Obesity*, 13, 4401–4412.

Rahman, M. A., Hidayat, S., Jamaludin, U., & Leksono, S. M. (2023). Utilization of neem trees (Azadirachta indica) as biopesticide from ancient India to Indonesia: An ethnographic study based on literature review. *Gravity: Jurnal Ilmiah Penelitian dan Pembelajaran Fisika*, 9(2). https://doi.org/10.30870/gravity.v9i2.20454

Reddy, Y. V. (2023). Swiss ADME predictions of phytoconstituents present in Ocimum sanctum Linn. *Journal of Pharmacognosy and Phytochemistry*, 12(5), 17–26.

Richter, E., Geetha, T., Burnett, D., Broderick, T. L., & Babu, J. R. (2023). The effects of Momordica charantia on type 2 diabetes mellitus and Alzheimer's disease. *International Journal of Molecular Sciences*, 24(5), 4643.

Sadique Husain, M. A. A., Ahmed, S., Quamri, A., & Khan, M. A. (2014). Hepatoprotective, anticancer & antiviral effects of Bhui Amla in unani medicine: an overview. *Journal of Medicinal Plants*, 2(6), 50–52.

Sahu, M., Kumar, V., & Joshi, V. (2021). Indian medicinal plants with antidiabetic potential: An overview. *Research Journal of Pharmacy and Technology*, 14(4), 2328–2335.

Sarkar, S., Singh, R. P., & Bhattacharya, G. (2021). Exploring the role of Azadirachta indica (neem) and its active compounds in the regulation of biological pathways: An update on molecular approach. *3 Biotech*, 11(4), 178.

Satyanarayana, K., Sravanthi, K., Shaker, I. A., & Ponnulakshmi, R. (2015). Molecular approach to identify antidiabetic potential of Azadirachta indica. *Journal of Ayurveda and Integrative Medicine*, 6(3), 165.

Shang, A., Cao, S. Y., Xu, X. Y., Gan, R. Y., Tang, G. Y., Corke, H., & Li, H. B. (2019). Bioactive compounds and biological functions of garlic (Allium sativum L.). *Foods*, 8(7), 246.

Sharma, R., Amin, H., & Prajapati, P. K. (2015). Antidiabetic claims of Tinospora cordifolia (Willd.) Miers: critical appraisal and role in therapy. *Asian Pacific Journal of Tropical Biomedicine*, 5(1), 68–78.

Siddiqui, S. A., Khan, S., & Wani, S. A. (2023). Controlling diabetes with the aid of medicinal herbs: A critical compilation of a decade of research. *Critical Reviews in Food Science and Nutrition*, 63(33), 12552–12566.

Singh, A., Sengar, R. S., Singh, A. K., Singh, S., Kumar, A., & Ahmad, K. (2019). Impact of genetic engineering on agriculture food crops. *Biotech Today: An International Journal of Biological Sciences*, 9(1), 17–24.

Singh, S., Datta, R., Johri, P., & Trivedi, M. (Eds.). (2022). *Phytopharmaceuticals and biotechnology of herbal plants*. Boca Raton, FL: CRC Press.

Singh, S., Mazumder, A., & Chakaraborthy, G. S. (2019). Potential herbs against diabetes mellitus—An update. *International Journal of Pharmaceutical Sciences and Research*, 10(8), 3619–362.

Sneka, S., Cecil, A., & Rajeshkumar, S. (2022). Antidiabetic activity symplocos racemosa (lodhra) and cinnamon cassia (cinnamon bark) formulation synthesised silver nanoparticles. *Journal of Pharmaceutical Negative Results*, 3114–3120. https://doi.org/10.47750/pnr.2022.13.s08.386

Srivastava, A., Subhashini, Pandey, V., Yadav, V., Singh, S., & Srivastava, R. (2023). Potential of hydroethanolic leaf extract of Ocimum sanctum in ameliorating redox status and lung injury in COPD: An in vivo and in silico study. *Scientific Reports*, 13(1), 1131.

Tripathi, A. K., Gupta, P. S., & Singh, S. K. (2019). Antidiabetic, anti-hyperlipidemic and antioxidant activities of Bauhinia variegata flower extract. *Biocatalysis and Agricultural Biotechnology*, 19, 101142.

Trivedi, M., Singh, S., Johri, P., & Sanchez, P. L. (Eds.). (2024). *Nutraceuticals inspiring the contemporary therapy for lifestyle diseases*. Boca Raton, FL: CRC Press.

Upadhyay, R. K. (2017). Tulsi: A holy plant with high medicinal and therapeutic value. *International Journal of Green Pharmacy (IJGP)*, 11(1).

Vanschoonbeek, K., Thomassen, B. J., Senden, J. M., Wodzig, W. K., & van Loon, L. J. (2006). Cinnamon supplementation does not improve glycemic control in postmenopausal type 2 diabetes patients. *The Journal of Nutrition*, 136(4), 977–980.

Verma, S., Sharma, H., & Garg, M. (2014). Phyllanthus amarus: A review. *Journal of Pharmacognosy and Phytochemistry*, 3(2), 18–22.

Villarruel-López, A., López-De La Mora, D. A., Vázquez-Paulino, O. D., Puebla-Mora, A. G., Torres-Vitela, M. R., Guerrero-Quiroz, L. A., & Nuño, K. (2018). Effect of Moringa oleifera consumption on diabetic rats. *BMC Complementary and Alternative Medicine*, 18, 1–10.

Wang, H., Li, H., Hou, Y., Zhang, P., & Tan, M. (2023). Plant polysaccharides: Sources, structures, and antidiabetic effects. *Current Opinion in Food Science*, 51, 101013.

Widhiantara, I. G., Arunngam, P., & Siswanto, F. M. (2018). Ethanolic extract of Caesalpinia bonducella f. seed ameliorates diabetes phenotype of streptozotocin-nicotinamide-induced type 2 diabetes rat. *Biomedical and Pharmacology Journal*, 11(2), 1127–1133.

Wijenayaka, G. M. U. D., Bulugahapitiya, V. P., & Jayasinghe, S. (2022). Cinnamon, a promising herbal plant for combatting diabetes and Its anti-diabetes mechanisms. *Ceylon Journal of Science*, 51(4), 335–346.

15 Molecular Farming

A Sustainable Approach for Biomolecule Production

*Parul Johri, Manish Dwivedi, Rajesh K Tiwari, and Mala Trivedi**

15.1 INTRODUCTION

Molecular farming is a biotechnological process that consists of genetic modification of agricultural products to produce proteins, chemicals, and other components for commercial and pharmaceutical use. Many developing countries cannot carry the costs of various medical treatments; hence, we need to synthesize new drugs. Molecular farming offers effective solutions for the current growing need for biomedicines (Ahmad, 2014).

When compared to alternative production systems, plants enable us to generate valuable, effective, and recombinant proteins on a large scale. They also provide a number of benefits in terms of cost, safety, and application.

Typically, the technology of molecular farming in plants has dominated the manufacture of pharmaceutical proteins. However, plants can also be used to produce industrial enzymes, dietary supplements, biopolymers, and proteins for research (such as β-glucuronidase and avidin, for example). Transgenic plants have supplanted earlier production processes that included bacteria, eukaryotes (yeasts, double-stranded fungus), transgenic animals, and animal cells. In 1986 and 1989, respectively, transgenic plants produced the first recombinant antibodies and the first recombinant (pharmaceutical) proteins extracted from plants. Avidin, or egg protein, was the first recombinant protein to be generated for industrial use in transgenic maize in 1997 (Hiatt et al., 1989).

15.2 METHODS OF PLANT TRANSFORMATION

Plant molecular farming relies on the production of transgenic plants and can be done in two general ways, which are described in the sections that follow.

* Corresponding author

DOI: 10.1201/9781003392910-15

15.2.1 Stable or Permanent Transformation

15.2.1.1 Stable Nuclear Transformation

Stable nuclear transformation means the addition or integration of new genes or foreign genes into the nuclear genome of plants, causing a change of genetic makeup and consequent expression of transgenes after joining with the host genomes. Using this method, the largest number of recombinant proteins have been produced. However, a long production cycle for a few crops and food products has restricted the wide acceptance of this method (Fischer et al., 2004).

15.2.1.2 Stable Plastid Transformation

Plastid transformation is a better way than nuclear transformation because it has various benefits, including preventing transgene escape and the absence of chloroplasts in pollen, which consequently reduces the improbability of transfer and decreases environmental concerns (Meyers et al., 2010). Scientists have already extracted a human pharmaceutical protein, at a concentration of more than 3% to 6% of total soluble proteins, from the chloroplasts of tobacco (Reddy et al., 2002; Sadhu & Reddy, 2003).

15.2.1.3 Plant Cell Suspension Culture

The plant cell suspension culture method includes the removal of the cell wall and the transfer of genes to the isolated protoplasts and suspension culture. The purification system and its downstream processing are cheaper and easier (Kim et al., 2008). Additionally, because this is a rapid technique, transgenic plant production is not required; however, cell lines can be developed in a few months. Plant suspension culture samples can be utilized to create biomedicines, such as recombinant human glucocerebrosidase to treat diabetes and vaccines against Newcastle disease in poultry. Cell suspension has not yet been chosen or employed as the best option for production in plant systems despite the fact that it is less expensive, safer, and simpler than other genetic modification techniques.

15.2.2 Temporary or Transient Transformation

Transient or temporary transformation production is the fastest system for plant molecular farming. A number of the methods used are described in the following sections.

15.2.2.1 Agrobacterium-Mediated Transformation

Without steady gene transfer, recombinant Agrobacterium suspension infiltration into tobacco leaf tissue results in the transmission of T-DNA to a very high percentage of cells, where transgenes are produced at high levels without stable gene transfer. This method is very effective for the production of clinical biomedicines (Vezina et al., 2009; Regnard et al., 2010).

15.2.2.2 Viral-Mediated Transformation

The viral-mediated transformation method depends on the ability of plant viruses, such as tobacco mosaic virus (TMV) and potato virus X, to work as vectors to transfer foreign genes into plant genomes.

15.2.2.3 Magnification Systems

Transformation systems based on viral vectors and Agrobacterium-mediated methods suffer from a few constraints for the co-expression of two or more polypeptides required for the production of hetero-oligomeric proteins (Giritch et al., 2006).

Therefore, a new expression system known as MagnICON technology was developed by Icon Genetics. This method involves the removal of coat proteins (responsible for systemic movement) of non-competitive virus strains and delivery of the derived viral vectors to plants using Agrobacterium-mediated transformation.

15.3 THE BENEFITS OF USING TRANSGENIC PLANTS AS BIOREACTORS

The benefits of using transgenic plants as bioreactors are:

- The health and viability of the products that are generated (bacteria and human diseases cannot exist in plants).
- The ability to undergo post-translational modifications (taking into account the properties of eukaryotic cells).
- The possibility of using breeding and sexual crosses to produce active recombinant multi-chain proteins.
- Reducing the costs of production.
- Reducing the costs of storage and transportation of recombinant proteins.
- Removing the purification step (when the plant tissues containing recombinant protein are edible). (Ahmad, 2014)

15.4 PROTEINS AND BIOMEDICINES OBTAINED FROM PLANTS

Bacterial and viral recombinant antigens, which are engineered to preserve the biological activity of proteins, can be produced by plants. These antigens can go through post-translational changes. The development of plant based vaccines is still a challenging job, however, the future of such vaccines is reliable and secure. The following are biomedicines made by plants:

- Antigens needed to make palatable vaccines

Antigens are expressed from various pathogens in plants and are often employed to initiate an immune response leading to immunization against illnesses or foreign particles in human proteins (Aviezer et al., 2009). Thus far, plant-based vaccinations have generated protection against several pathogenic agents such as rotavirus, HIV, hepatitis B, and rabies virus.

- Monoclonal antibodies (Mab)

Using transgenic plants as bioreactors is the most effective method, as it strengthen the activity and reduces the cost of production of antibodies. In the process of creating recombinant expression systems, the design of therapeutic antibodies necessitates the inclusion of antibody functional mechanisms.

Antibodies' primary purpose is to attach to antigens, although this process does not have a protective effect. The antibodies produced in plants incorporate immunoglobulin G (IgG) and immunoglobulin A (IgA), IgA and IgG shimmer molecules, IgG and IgA secreted molecules, single-chain variable fragments, fragment antigen-binding, and second variable of heavy and light chains (Thomas, 2002; Julian et al., 2003).

- Pharmaceutical proteins

Biomedicines such as erythropoietin, interferon, hirudin, aprotinin, leu-enkephalin, and somatotropin of human growth hormone have recently been expressed in plants.

- Industrial proteins or non-pharmaceutical proteins derived from plants are classified as enzymes and include cellulase, β-glucocerebrosidase, trypsin, aprotinin, and avidin

Molecular farming of destructive enzymes of the cell wall, such as cellulase, hemicellulase, xylanase, and particularly ligninase, provides a great way for the biofuel industry to produce cellulosic ethanol (Sticklen, 2008; Chatterjee et al., 2010).

15.5 MOLECULAR FARMING: AN OPPORTUNITY FOR PRODUCING PLANTS WITH HIGH TECHNOLOGY

Production of new high-technology plants has become possible because of molecular farming and metabolic engineering. Molecular farming has led to significantly lower costs than traditional farming. Franklin and Mayfield's most recent production experiment involves the unicellular alga *Chlamydomonas reinhardtii*. This is the only plant where transformation was carried out in all of its DNA-containing parts (nucleus, plastid, and mitochondria) (Rochaix, 2002).

15.6 PURIFICATION AND DOWNSTREAM PROCESSING OF ISOLATED RECOMBINANT PROTEINS

While purification comprises safety precautions, liquid-liquid extraction, membrane filtration, chromatography, etc., recovery processes often involve the breakdown of plant tissues, protein extraction, solid-liquid separation, and protein concentration. Utilizing the secretory systems of cells might also be advantageous since it eliminates the need to disintegrate plant cells during the recovery process, which prevents the release of phenolic chemicals. However, recombinant proteins may become unstable in culture media. Continuous protein labels are another method of aiding in the recovery of proteins. After purification, protein labels need to be taken off in order for the purified protein's structure to return to its original form.

15.7 COSTS OF SUBSEQUENT PROCESSING

The costs of subsequent processing of the recombinant proteins derived from plants have been estimated to be about 80% of total production costs (Buonaguro et al., 2010). That is why so much attention has been paid to efficient strategies for reducing costs to the lowest possible level.

15.8 REVOLUTIONIZING MOLECULAR FARMING: A PROMISING OUTLOOK ON COMMERCIAL PRODUCTION OF MEDICINES AND PHARMACEUTICAL RECOMBINANT PROTEINS

In the second phase of clinical trials, consideration is given to the product's developmental phases and eventual commercialization. The goal of many small biotechnology businesses has been to market the antibodies made by plants. The annual income from manufacturing IgA in crops was projected in 2008 to reach $25 billion, and it has been calculated that the annual requirement for secretory IgA will increase by 13% (Sticklen, 2008).

15.9 CONCLUSION

Molecular farming has so far created a wide range of useful and efficient pharmaceutical proteins and antibodies, which has significantly aided in the treatment of patients, particularly in underdeveloped nations where the expenses associated with the creation and maintenance of such medications are prohibitive. Though there are some issues that make it difficult to fully implement, such as public acceptance, transgene escape, biosecurity, and clinical and commercial analysis of products, it is hoped that with the efforts of researchers and scientists, molecular farming will soon make great advancements.

REFERENCES

Ahmad K. Molecular farming: Strategies, expression systems and bio-safety considerations. *Plant Breed.* 2014; 50:1–10.

Aviezer D, Brill-Almon E, Shaaltiel Y, et al. A plant-derived recombinant human glucocerebrosidase enzyme--a preclinical and phase I investigation. *PLoS One.* 2009; 4(3):e4792.

Buonaguro FM, Butler-Ransohoff JE. PharmaPlant: The new frontier in vaccines. *Forward. Expert Rev Vaccines.* 2010; 9(8):805–7.

Chatterjee A, Das N, Raha S, et al. Production of xylanase in transgenic tobacco for industrial use in bioenergy and biofuel applications. *Invitro Cell Dev Biol Plant.* 2010; 46:198–209.

Fischer R, Emans N, Twyman R, Schillberg S. Molecular farming in plants: Technology platforms. In: Goodman RB, editor. *Encyclopedia of Plant and Crop Science.* New York: Marcel Dekker. 2004; pp. 753–6.

Giritch A, Marillonnet S, Engler C, et al. Rapid high-yield expression of full-size IgG antibodies in plants coinfected with noncompeting viral vectors. *Proc Natl Acad Sci U S A.* 2006; 103(40):14701–6.

Hiatt A, Cafferkey R, Bowdish K. Production of antibodies in transgenic plants. *Nature.* 1989; 342(6245):76–8.

Julian K, Ma P, Christou P. The production of recombinant pharmaceutical proteins in plants. *Nature.* 2003; 4:794–805.

Kim TG, Baek MY, Lee EK, Kwon TH, Yang MS. Expression of human growth hormone in transgenic rice cell suspension culture. *Plant Cell Rep.* 2008; 27(5):885–91.

Reddy V, Sadhu L, Selvapandiyan A, et al. Analysis of chloroplast transformed tobacco plants with cry1Ia5 under rice psb A transcriptional elements reveal high level expression of Bacillus thuringenesis toxin without imposing yield penalty and stable inheritance of transplastome. *Mol Breed.* 2002; 9:259–69.

Regnard GL, Halley-Stott RP, Tanzer FL, Hitzeroth, II, Rybicki EP. High level protein expression in plants through the use of a novel autonomously replicating geminivirus shuttle vector. *Plant Biotechnol J.* 2010; 8(1):38–46.

Rochaix JD. Chlamydomonas, a model system for studying the assembly and dynamics of photosynthetic complexes. *FEBS Lett.* 2002; 529(1):34–8.

Sadhu L, Reddy V. Chloroplast expression of His-tagged GUS-fusions: A general strategy to overproduce and purify foreign proteins using transplastomic plants as bioreactors. *Mol Breed.* 2003; 11:49–58.

Sticklen MB. Plant genetic engineering for biofuel production: towards affordable cellulosic ethanol. *Nat Rev Genet.* 2008; 9(6):433–43.

Thomas B. Production of therapeutic proteins in plants. *Agricult Biotechnol.* 2002. https://doi.org/10.3733/ucanr.8078

Vezina LP, Faye L, Lerouge P, et al. Transient co-expression for fast and high-yield production of antibodies with human-like N-glycans in plants. *Plant Biotechnol J.* 2009; 7(5):442–55.

Index